MICROBIAL GROWTH ON C_1 COMPOUNDS

Microbial Growth on C_1 Compounds

Proceedings of the 5th International Symposium

edited by

H.W. VAN VERSEVELD
Vrije Universiteit Amsterdam, The Netherlands

J.A. DUINE
Technical University Delft, The Netherlands

1987 **MARTINUS NIJHOFF PUBLISHERS**
a member of the KLUWER ACADEMIC PUBLISHERS GROUP
DORDRECHT / BOSTON / LANCASTER

Distributors

for the United States and Canada: Kluwer Academic Publishers, P.O. Box 358, Accord Station, Hingham, MA 02018-0358, USA
for the UK and Ireland: Kluwer Academic Publishers, MTP Press Limited, Falcon House, Queen Square, Lancaster LA1 1RN, UK
for all other countries: Kluwer Academic Publishers Group, Distribution Center, P.O. Box 322, 3300 AH Dordrecht, The Netherlands

Library of Congress Cataloging in Publication Data

Microbial growth on C_1 compounds.

Proceedings of the 5th International Symposium on Microbial Growth on C_1 Compounds, held at the Biological Center of the University of Groningen, Haren, The Netherlands, 11-16 August 1986.
On t.p. the 1 is subscript following C in the title; also in the note and added entry.
Includes indexes.
1. Microbial growth--Congresses. 2. Bacteria, Methylotrophic--Congresses. 3. Yeast--Congresses. 4. Carbon compounds--Metabolism--Congresses. 5. Methane--Metabolism--Congresses. I. Verseveld, H. W. (Henk W. van) II. Duine, J. A. III. International Symposium on Microbial Growth on C_1 Compounds (5th : 1986 : University of Groningen)
QR86.M523 1987 576'.131 86-31079
ISBN 90-247-3459-2 (U.S.)

ISBN 90-247-3459-2

PRINTED IN THE NETHERLANDS

CONTENTS

PHYSIOLOGY, BIOCHEMISTRY AND GENETICS OF METHYLO-TROPHIC YEASTS

ENERGETICS OF C_1-METABOLISM

GENETICS AND MOLECULAR BIOLOGY OF C_1-UTILIZERS

APPLIED ASPECTS OF ONE-CARBON METABOLISM

5th C1 SYMPOSIUM

LIST OF PARTICIPANTS

G. AB, D. Ahring, N. Al Taho, N. Al-Awadhi, D. Ankel-Fuchs, C. Anthony, N. Arfman, J.A. Asenjo, H. Ashworth, M.M. Attwood, K.A. Auton, W. Babel, P. van Beelen, D. Best, M. Blaut, L. de Boer, D. Bonam, M. Bott, B. Bowien, A.G. Brooke, P.G. Bruineberg, S.M. Burton, L.V. Bystrykh, D.L.N. Cardy, G. Cecchini, W. Clement, G.A. Codd, A. Cornish, J. Cregg, L.G.M. Croes, H. Dalton, V.L. Davidson, K.J. Davis, G.E. de Vries, A.M. Dean, D. DerVartanian, G. Diekert, J.P. van Dijken, L. Dijkhuizen, G.L.Dilworth, N. Dixon, A.C. Douma, C.S. Dow, H.L. Drake, E. Drent, M. Duchars, J.A. Duine, Th. Egli, B. Eikmanns, E.J. Elliott, S.M. Evans, K.F. Fannin, J.G. Ferry, I. Forrester, B.G. Fox, J. Frank, C. Friedrich, B. Friedrich, S. Froud, G. Fuchs, S. Fukui. F. Gasser, D. Geiger, O. Ghisalba, J.L. Gibson, M. Giuseppin, I.N. Gogotov, P. Goodwin, G. Gottschalk, J. Green, P.N. Green, T.A. Grinberg, T.A. Hansen, W. Harder, N. Harms, H. Heijthuisen, I.J. Higgins, H. Hippe, B. Holloway, D. Hood, J. Hoogendrijk, M. de Hoop. T.R. Hopkins, M. Husain, Y. Igerashi, M.V. Ivanov, M. Janvier, K. Jarrell, S.G. Jezequel, N. Kato, D.B. Kell, D.P. Kelly, J.T. Keltjens, S. Kengen, P.P. Kesseler, I.J. van der Klei, W. de Koning, J.G. Kuenen, V. Lalov, P.J. Large, D.J. Leak, M. E. Lidstrom, N. Lindley, J.F. Linton, L.G. Ljungdahl, A.J. Lloyd, A. Long, P. Loubiere, S. Machlin, J. Macy, Y.R. Malashenko, P. Mazodier, C. McQaid, W. Meijer, L. Meile, O. Meyer, V. Muller, J.C. Murrell, Y. Nakamura, A. Netrusov, M. Neumann, H.G.D. Niekus, K.M. Noll, A.N. Nozhevnikova, D.N. Nunn, C. Oakley, D. Page, P. Pellerin, D. Prozzi, J.R. Quayle,J.N. Reeve, N.A. Renshaw, I.Y. Sacharova, Y. Sakai, G.F.C. Salmond, J.K.M. Sanders, B. Schink, P. Schönheit, A.C. Schwarts, R.V. Serauskas, A.A. Sibirny, S. Siemon, A.T. Slabospitskaya, A.P. Sokolov, I. Sokolov, G. Southgate, K.O. Stetter, D. Stirling, S. Stone, A.H. Stouthamer, D.W. Stroman, E. Stupperich, T. Suijlen, B. Svensson, H. Tabak, F.R. Tabita, K. Takeda, Y. Tani, R.K. Thauer, G. Thill, J.P. Touzel, Yu.A. Trotsenko, Y.D. Tsygankov, S. Ueda, R.L. Uffen, C.J. Unkefer, D. Ursi, H.W. van Verseveld, C. van der Drift, M. Veenhuis, H. Veldkamp, R. Vilu, G.D. Vogels, M. Vrdoljak, P. Warner, G.H. Wegner, M.B. Whitman, F. Widdel, W. v.d. Wijngaard, J.F. Wilkinson, A. Williams, R.S. Wolfe, A.P. Wood, N. Woods, J.R. Woodward, T. Yamanaka, L.J. Zatman, A.B.J. Zehnder, C. ggGliesche.

PREFACE

The 5th International Symposium on Microbial Growth on C_1 Compounds was held at the Biological Center of the University of Groningen, Haren, The Netherlands, 11-16 August 1986. The meeting attracted well over 200 participants from 15 countries. This volume contains the formal presentations made at that time, which, because of the breadth of topics covered, were divided into seven sections of related papers.

This meeting, under the chairmanship of Wim Harder, was both scientifically and socially very successful. This success cannot only be credited to the main presentations, but also to the well cared-for 121 poster presentations, whereof the abstracts have been published separately. The series of Symposia will be continued in 1989, in the Federal Republic of Germany.

We wish to acknowledge the invaluable help of Joke Daniëls, Roberta Ströer-Schneider, Karin Uyldert, Hansje Bartelson and Josine van Verseveld-Ströer, who retyped the manuscripts resulting in a uniform presentation of these proceedings.

The Foundation " C_1 Symposium The Netherlands" wishes to acknowledge the following organizations for their financial contributions to the Symposium: Gas Research Institute (major sponsor), Royal Dutch Academy of Sciences, N.V. Nederlandse Gasunie, G.D.Searle & Co. Ltd, Netherlands Convention Bureau, Phillips Petroleum Company, Dutch State Mines, Faculteit der Wiskunde en Natuurwetenschappen van de Rijksuniversiteit Groningen, Methanor v.o.f., Nederlandse Aardolie Maatschappij b.v., Gist brocades NV/ Shell, Stichting Antonie van Leeuwenhoek, Exxon Research and Engineering Company, AKZO nv, Imperial Chemical Industries PLC, Ciba-Geigy AG, Nederlandse Unilever Bedrijven B.V., Océ-Andeno B.V., Groninger Universiteits Fonds

September 1986

Henk W. van Verseveld
Vrije Universiteit
Amsterdam
The Netherlands

J.A. Duine
TU Delft
The Netherlands

AN EIGHTIETH ANNIVERSARY OF THE STUDY OF MICROBIAL C_1 METABOLISM

J.R. QUAYLE

I feel greatly honoured to have been asked to give the Opening Address to the Fifth International C_1 Symposium and it gives me great personal pleasure to welcome you all to one of the great European centres of microbiological study here in Haren. To very many of you, myself included, this is by no means the first visit here and this adds to the feeling of warmth at being in the company of so many friends and colleagues.

I am not going to give you a formal lecture on microbial C_1 metabolism, you are going to listen to 35 of those in the next four days, plus a closing address, nor do I have the presumption to review the programme in advance, rather am I going to look back and pause at some of the milestones that mark the way to Haren in 1986.

Like so many speakers I must first point out that the title of my lecture is somewhat misleading. I had chosen it because of the discovery in 1906 of *Bacillus methanicus* by Söhngen, working in Beijerinck's laboratory in Delft in Holland. I considered this to be the first definitive isolation and characterisation of a methylotroph. However, I was reminded by Dr van Dijken that Söhngen was not the first to discover and characterize a methylotroph, it was Loew [1] at the University of Munich in 1892 who isolated what was clearly *Pseudomonas* AMI. He carried out an enrichment in a medium containing mineral salts and 0.5% of the bisulphite addition compound of formaldehyde, standing it in air for 1-2 weeks. A pink colour developed which was due to growth of short, thick rods capable of growth on methyl alcohol, methylamine, formic acid or formaldehyde as well as a variety of polycarbon substrates. He called it *Bacillus methylicus* and proposed that glyoxylate was an intermediary metabolite of formate:

$$2HCO_2H \rightarrow CHO.CO_2H + H_2O$$

$$CHO.CO_2H \rightarrow HCHO + CO_2$$

Hence the real anniversary for the birth of microbial C_1 metabolism is in six years time when we ought to celebrate the centenary in Munich.

Nevertheless, it is still true that 1986 is the eightieth anniversary of the discovery of *Pseudomonas methanica* in Holland and this organism became of key importance in the later development of the subject of methylotrophy. I might spend a few minutes on this work. In his first paper of many, Söhngen [2] pointed out that methane is made in unending amounts on the Earth and yet only trace amounts are to be found in the atmosphere. Since methane is chemically stable he suggested that some organism was responsible for its removal, possibly making use of the large release of energy consequent on its oxidation. The first candidate for such scavenging organisms were plants and since most methane appeared to be generated in stagnant water he chose aquatic plants. He sealed up a range of aquatic plants in flasks of water, half-filled with a methane-oxygen mixture, and looked for gas absorption over a 14-day period. Many plants showed considerable methane/oxygen uptake but the starting time was variable even with different specimens of the same species. Prewashing the plants slowed the process down and he noted that the gas absorption really started when a pellicle formed on the surface of the liquid and the liquid turned turbid. He found that he could then dispense with the plants, recognised that he was dealing with microorganisms, and isolated therefrom in pure culture *Bacillus methanicus*, to be subsequently lost and re-isolated 50 years later by Dworkin & Foster in 1956 and renamed *Pseudomonas methanica*.

Considering the very wide natural distribution of C_1 compounds at all redox levels from CH_4 to CO_2 it is curious to reflect how little attention was paid to the microbiology of C_1 utilizers for the next 50 years, apart from much detailed study of autotrophs using CO_2. Sporadic reports appeared e.g. the stalked bacterium *Hyphomicrobium vulgare* was mentioned as long ago as 1898 by Stutzer & Hartleb in Breslau before reappearing in Kingma-Boltjes' laboratory in Holland in 1936, several C_1 utilizers were mentioned by the indefatigable den Dooren de Jong in 1926, and so on. It was in 1956 when I joined Krebs' MRC Unit in Oxford that he suggested to me that it might be worthwhile seeing how some organisms were capable of growth on reduced C_1 compounds. Was it indeed just a special case of autotrophy as often suggested, e.g. by Bhat & Barker in 1948 [3] in which the energy of oxidation of the C_1 substrate was coupled to the incorporation of CO_2, just as a chemosynthetic autotroph? I looked around for a suitable organism to start with, obtained a culture of *Hyphomicrobium vulgare* from Mevius in Hamburg but found its growth too slow and awkward for easy experimentation. I then obtained a culture of *Pseudomonas oxalaticus*, isolated by Khambata and Bhat in 1953 from earthworms' intestines and experimented with its growth on formate. It was easy to show that it was a *bona fide* autotroph and I reported to Krebs that there was little novel to be learned. However, there was a nagging feeling that autotrophic growth on a highly reduced carbon substrate such as methanol seemed energetically very wasteful. So I re-suspended a culture of formate-grown *Pseudomonas oxalaticus* in 10 litres of methanol growth medium and left sterile air sparging through it for a couple of days. I hoped to persuade the *P. oxalticus* to adapt to growth on methanol. To my astonishment the culture turned a beautiful coral pink as an airborne contaminant took over and we named it *Pseudomonas* AM1, A for airborne, M for methanol and 1 for first isolate. It is interesting to note, on reflection, that I had unwittingly followed in Loew's footsteps and carried out an air-borne enrichment. It was clear that *Pseudomonas* AM1 was no autotroph and so began the long series of investigations of the serine pathway, one variant of which, the isocitrate lyase$^-$ variant, has still defied solution.

I then had a narrow escape in 1965 when Pat Johnson and I started work on methane assimilation. We chose *Pseudomonas methanica* from Foster's laboratory in Texas as our test organism. Our isotope studies established that carbon from methane was rapidly incorporated by the organism via sugar phosphates. At the time this suggested operation of an autotrophic pathway. Accordingly we tested very carefully for the presence of RuBP carboxylase in the organism, found none, and then pressed on with our studies, confident that there must be a new pathway operating. Almost at the same time Foster & Davis [4] isolated a new methane utilizer which they called *Methylococcus capsulatus*, and they sent us a culture in order for us to look at its carbon metabolism. We did so and found it to be apparently identical with that in *P. methanica*.. By that time, with the work on the RuMP cycle well in hand and the possibility of the Calvin cycle in *P. methanica* having been eliminated we did not test *M. capsulatus* for the presence of RuBP carboxylase, it had high levels of hexulose monophosphate synthase. In 1977 Taylor at Warwick[5] showed that *M. capsulatus* did indeed have low levels of RuBP carboxylase while *P. methanica* had none. Had we examined *M. capsulatus* first, instead of *P. methanica*, and found RuBP carboxylase I am sure that we would have concluded that growth on methane was autotrophic in nature and we would have missed, at that time, the RuMP cycle. I mention this story particularly to research students in the audience in order to assure them that there is a strong element of chance in a scientific investigation and they should always be prepared for it.

By the late 1960's the scene was beginning to change quite quickly. Although many big American oil companies had long had an interest in hydrocarbon microbiology and investigators such as Jackson Foster and Zobell in the States had published widely, the Europeans had not paid much attention to it. This changed quickly, partly with the recognition that microbial attack on lubricating oils had very serious industrial consequences and partly with the discovery of the North Sea oil and gas fields which led oil and chemical companies to realise that they had a new and cheap chemical and biological feedstock on their doorstep. In 1967 the Institute of Petroleum organised a Symposium in London on Hydrocarbon Microbiology. Both Ribbons and I had included methane microbiology in our talks and after the symposium I was approached by two chemists from ICI, Billingham, P.P. King and D. Watchorn , who felt tthat ICI might be interested in the possibility of very large scale microbial conversion of methane to bacterial protein for use as an animal foodstuff and invited me up to Billingham. Out of these beginnings the ICI Pruteen project was born, leading to the development of the now famous, world's largest fermenter on Teeside in the astonishingly short time of 10 years. Once the industrial giants like ICI, BP and Shell had become interested in methane and methanol microbiology, large research funds began to flow, and the subject rapidly gathered pace in Europe and the USA, with research groups forming in Sittingbourne, Canterbury, London, Hull, Edinburgh, Reading, Grangemouth, Göttingen, Madison,

Dallas,Puschino, Kiev, Tokyo, Kyoto and last but by no means least, Haren. Then in 1970 another of the great milestones in C_1 metabolism occurred when Roger Whittenbury, John Wilkinson and their colleagues transformed the field by their detailed study of elective culture of methanotrophs which resulted in the isolation and characterisation of over 100 new strains of methane utilizers. Overnight the field was transformed from one with only about 4 classic organisms available (*Pseudomonas methanica, Methanomonas methanooxidans, Pseudomonas methanitrificans* and *Methylococcus capsulatus*) to a wealth of new species.

By 1973, there was sufficient interest for John Wilkinson, Roger Whittenbury, and I to raise about £1,000 from BP, ICI and Shell and we organised in Edinburg what must be rightfully considered as the first authentic C_1 Symposium. It had 50 participants and lasted 21/2 days. You may well feel that 13 years and five symposia later, the programmes and participants have not changed all that much. That surely says something about the interest of the subject and the dedication of its *aficionados*. Be that as it may, following the Edinburgh symposium in 1973, the official C_1 symposia started off with number 1 in Tokyo 1974, number 2 in Puschino 1977, number 3 in Sheffield 1980, mumber 4 in Minneapolis 1983 and now mumber 5 in Haren 1986. They have steadily grown in size, rightfully embracing autotrophy, methanogens, methylotrophic yeasts, genetics of C_1 utilizers, and biotechnological applications. By 1980 and the Sheffield Symposium the field had seen the elucidation of the icl$^+$-serine pathway mainly by Hersh & Bellion, the elucidation of the RuMP cycle mainly by the Sheffield group, the chemical structure of the co-factor of methanol dehydrogenase had been solved independently by Duine & Frank's group in Delft and Forrest and Kennard's group in Caltech and Cambridge, the assimilation pathway for methanol-grown yeasts had been elucidated, the ICI single cell protein process had just gone on stream. New directions were already visible, large scale interest in the genetic systems of methylotrophs was starting up in the laboratories of Bruce Hollloway, Bill Brammar and Mary O'Connor. Also it was quite clear that great things were stirring in the methanogenic field with new C_1 co-factors coming out of the laboratories of Thauer & Fuchs, Vogels and Ralph Wolfe and the first indications of an exciting new pathway of net C_1 assimilation in *Methanobacterium thermoautotrophicum*. 1982 also saw another milestone with publication of the first definitive text book of the whole subject produced by Chris Anthony [6], showing up those of us who had taken cowardly refuge in writing multiauthored reviews and hiding away from publishers' agents when they visited the laboratories. Anthony's book joins Stanier, Doudoroff and Adelberg's 'Allgemeine Mikrobiologie' as books that no thinking microbiologist should be without

And so to Minneapolis in1983 with the limelight on the new co-factors in methanogenesis from the American, Dutch and German groups and the new step by step linear biosynthetic sequence in *M. thermoautotrophicum* and the flood of new genetic systems in methylotrophs. What the 1986 meeting holds in store for us, the next 4 days will reveal. One wonders too at what meeting will some group tell so many of us where we failed with the icl serine$^-$ pathway over the last 20 years.

We often hear the questions asked, will or should the C_1 symposia continue in the future and for how long? The answer is simple, while ever there are about 100 people who are interested enough to want to come together somewhere in the world and talk about C_1 science and while ever there are grant-giving bodies who are prepared to help with the funding then they should continue. There will always be unexpected, exciting things to talk about - who could have guessed the involvement of C_1 microbiology in those unbelievable ecosystems clustered round the deep sea vents? Biotechnology still awaits a really profitable, high volume, low cost process - can the C_1 molecular geneticists give an organism to the process engineers that will provide a fortune for us all? What lessons can the biologists learn from the American, Dutch and French petrochemists [7-9] who can pass methanol through a zeolite and get alkanes out of the other end?

And so, Chairman, let the C_1 symposia continue, keep them out of the hands of bureaucratic committees of learned societies or acronymic international bodies, keep them small and hold them somewhere where the science is good, the scenery and culture pleasing, the food and drink plentiful, and the international company congenial. Mr. Chairman, we have all those things here in Haren and we are all grateful to you and your staff for your labours which have made it possible for us to be here this week.

REFERENCES

[1] Loew I. (1892) Z.Bakteriol.12,462.
[2] Söhngen N.L.(1906).Z.Bakteriol.,Parasitenk(Abt.II)15,513
[3] Bhat J.V. and Barker H.A.(1948) J.Bacteriol.55,359.
[4] Foster J.W. and Davis R.H.(1966) J.Bacteriol.72,646.
[5] Taylor S.C.(1977) FEMS Microbiol.Lett.2,305.
[6] Anthony C.(1982) The Biochemistry of Methylotrophs.Academic Press.London.
[7] Miesel S.L.,McCullough J.P.,Lechthaler C.H. and Weisz P.B.(1976) Chemtech.6,86.
[8] Lee C.S. and Wu M.M.(1985) J.Chem.Soc.,Chem.Commun.,250.
[9] Olah G.A.,Prakash G.K.S.,Ellis R.W.and Olah J.A.(1986) J.Chem.Soc.,Chem.Commun.,9.

PHYSIOLOGY AND BIOCHEMISTRY OF AUTOTROPHIC BACTERIA

PHYSIOLOGY AND BIOCHEMISTRY OF AUTOTROPHIC BACTERIA

G.A. CODD[a],J.G. KUENEN[b]

a, Department of Biological Sciences, Dundee University , Dundee DD1 4HN, UK.

b, Laboratory of Microbiology and Enzymology, Delft University of Technology, 2628, BC Delft, The Netherlands.

1. INTRODUCTION

The three years since the previous International Symposium on Microbial Growth on C_1 Compounds have seen numerous advances in our knowledge of bacterial autotrophy. Significant advances have been made in the recognition and characterization of "new" groups of autotrophs and in the understanding of the pathways of CO_2 assimilation, including the enzymes involved. Several routes of CO_2 assimilation in prokaryotes are recognised. The reductive tricarboxylic acid cycle in the green sulphur bacterium *Chlorobium* is well established [1], and a variant of this cycle, in which phosphoenolpyruvate carboxylase is replaced by pyruvate carboxylase, has been found in the aerobic obligate autotroph, *Hydrogenobacter thermophilus* [2]. The total synthesis of acetyl-CoA from CO_2 and H_2 is achieved by a linear pathway in the anaerobic, autotrophic bacterial methanogens, acetogens and most sulphate-reducers [3]. Advances on these organisms, and with autotrophic sulphate-reducing bacteria and sulphur-reducing archaeabacteria which assimilate CO_2 via a reductive citric acid cycle, are fully discussed elsewhere [3, Fuchs this symposium].

The Calvin cycle does not operate in the above organisms, but is responsible for CO_2 assimilation in most other known groups of autotrophic prokaryotes, besides algae and plants [4, 5]. Interest in the Calvin cycle in microbes continues since its regulation at the physiological and molecular levels varies between bacterial groups. Furthermore, the facility with which the key enzymes of the cycle can be studied and manipulated at the molecular and genetic levels, may be useful in attempts to alleviate constraints on higher plant productivity. Here we briefly review the diversity and physiological spectrum of CO_2-fixing bacteria and consider the occurrence and operation of the Calvin cycle in prokaryotes. Details of the regulation of the Calvin cycle at the molecular and genetic levels are considered elsewhere in these proceedings [Dow, Bowien, Friedrich, Tabita]. We provide an overview of molecular and genetic advances on the key enzymes, inorganic carbon accumulation and the compartmentation of CO_2 fixation in Calvin cycle bacteria. Finally, we briefly consider patterns of physiological regulation of the Calvin cycle in obligate and facultative autotrophs and some ecophysiological aspects. For a detailed discussion of physiological aspects the reader is referred to two excellent reviews by Dijkhuizen and Harder [6,7].

2. DIVERSITY AND THE PHYSIOLOGICAL SPECTRUM AMONG THE CO_2-FIXING BACTERIA

Among the aerobic and anaerobic litho-(autho)trophs as well as among the photo-(auto)trophs we see a spectrum of physiological types with respect to their energy source and/or their carbon source. Many organisms use CO_2 as the exclusive or major carbon souce under all growth conditions. Organisms in this category are also obligately dependent on light or their lithotrophic energy source. In a second category with a facultative metabolism, the energy source and/or carbon source can be replaced by organic compounds. Among the lithotrophs we find a third category in which CO_2 cannot be used as primary carbon source, but only in heterotrophic (anaplerotic) carbon metabolism. Among the phototrophic microbes the only organisms in this category are the halophilic bactaeria which use bacteriorhodopsin for energy generation. The spectrum has been long known for many lithotrophs and phototrophs and as knowledge increases, more and more microbial families or genera turn out to harbour these different metabolic types. For example, the complete spectrum is known for the aerobic sulphur oxidizers [8], nitrifying bacteria [9] iron oxidizers, and also for aerobic H_2 bacteria [10, 11]. Recently facultatively autotrophic *Beggiatoa* species have been described [12] as well as obligately and facultatively autotrophic methane producers [13], hydrogen/sulphur autotrophs [14, Stetter, this symposium] and autotrophic sulphate reducers [15,16]. Many of the anaerobic phototrophs are facultative autotrophs, for example the Rhodospirillacae. However, all known Chromatiaceae and Chlorobiaceae have had a very limited organic substrate range, while the known *Chlorobium* spp. cannot grow without light [17]. The importance of organic compounds in the metabolism of the facultative organisms may vary significantly. Many of the facultative lithotrophs grow much faster on organic compounds, i.e. heterotrophically or mixotrophically, than lithoautotrophically. By contrast the facultative photoautotrophs are all primarily phototrophs, which grow faster in the light, autotrophically or mixotrophically. In the latter group, heterotrophic potential merely seems to serve as a survival mechanism [17, 18].

3. OCCURRENCE OF THE CALVIN CYCLE IN AUTOTROPHIC BACTERIA

There is abundant evidence that the Calvin cycle is used by the colourless sulphur-oxidizing bacteria and carboxydotrophic bacteria among the chemolithoautotrophs [5, 6]. The nutritional capabilities of filamentous sulphide oxidizers such as *Beggiatoa* are beginning to be clarified. Freshwater *Beggiatoa* strains have not been grown autotrophically [19, W.R. Strohl pers. commun.] However, autotrophic growth by marine strains of *Beggiatoa* and the presence of the carboxylating enzyme of the Calvin cycle, ribulose 1,5-bisphosphate carboxylase/oxygenase (RuBisCo) has been shown in this organism [12, Nelson pers. commun.] It seems, therefore, likely that these are Calvin cycle organisms. Chemolithoautotrophic growth of the iron-oxidizer *Gallionella* has also been obtained recently. The presence of adequate levels of phosphoribulokinase (PRK) and RuBisCo in *G. ferruginea* suggests that this organism can fix CO_2 via the Calvin cycle [20, D. Hanert pers. commun].

Besides the purple sulphur bacteria, purple non-sulphur bacteria and cyanobacteria among the phototrophs the Calvin cycle appears to operate in the chlorophyll b-containing prokaryote *Prochloron* [21, 22]. The unequivocal demonstration of autotrophy in this organism has been prevented by the inability to obtain growth without its animal host. However, the recent discovery of a free-living filamentous chlorophyll-b containing oxygenic prokaryote which grows well in laboratory culture [23] provides exciting prospects for the study of CO_2 fixation and photosynthesis in these organisms of evolutionary importance. Although low PRK and RuBisCo activities were initially found in the green non-sulphur bacterium *Chloroflexus* [see 5], recent attempts to measure these enzymes in autotrophic cultures were negative and a novel CO_2 fixation pathway with acetyl-CoA as an intermediate has been indicated [24].

4. CALVIN CYCLE ENZYMES IN AUTOTROPHIC BACTERIA

Research on bacterial Calvin cycle enzymes has mainly concentrated on the 2 enzymes which are unique to, and essential for CO_2 fixation by this route: RuBisCo and PRK. In contrast to RuBisCo, PRK has received little attention until recently, although this enzyme is a major control point *in vivo*.

4.1 Phosphoribulokinase

PRK catalyzes the phosphorylation of ribulose 5-phosphate to produce the CO_2-acceptor, ribulose 1,5-biphosphate (RuBP). Allosteric regulation of PRK is chemolithoautotrophic and purple photosynthetic bacteria (activation by NADH; inhibition by AMP and/or PEP) is well established and consistent with the production of NADH during chemolithotrophy and anoxygenic photosynthesis [see 5]. In contrast, algal and plant chloroplast PRK's lack a requirement for NAD(P)H but their activities are increased in the dark by reduced sulphydryl compounds and by ferredoxin/thioredoxin photoactivation. Cyanobacterial PRK's are of particular interest in view of the postulated origin of chloroplasts from ancestral endosymbiotic cyanobacteria. The PRK's from the cyanobacteria *Chlorogloeopsis fritschii* and *Nostoc muscorum* (*Anabaena* 7119) resemble chloroplast PRK's in regulatory characteristics [25,26]. This apparent regulatory dichotomy can be viewed against the structural properties of the few PRK's so far purified (Table 1).

All of the prokaryotic PRK's are of high M_r and contain one class of subunit in a hexameric or octomeric enzyme. The eukaryotic PRK's are all low M_r dimers (Table 1). It will be of interest to know whether the *C. fritschii* PRK structure is typical for cyanobacteria and whether a molecular evolution of this enzyme in microbial autotrophs can be discerned. Heterologous DNA hybridization using probes to PRK genes will be of interest and a synthetic DNA probe to part of the *Alcaligenes eutrophus* H16 PRK [37] provides good opportunities for this approach. This probe hybridises with chromosomal and megaplasmid pHG1 DNA from *A. eutrophus* H16 indicating the dual location of PRK genes in this organism. Powls et al [38] have shown that light-activated *Scenedesmus* PRK is derived from a single higher M_r protein which shows latent activites of PRK and NADPH-glyceraldehyde 3-phosphate dehydrogenase, thus raising the possibility of a common genetic origin for these enzymes in this alga.

Table 1. Molecular properties of phosphoribulokinase

Source	Native enzyme M_r	Subunit (Su) M_r	Proposed quaternary structure	Ref
A. PROKARYOTES				
Chromatium D	240,000	-	-	27
Rhodopseudomonas capsulata	220,000	36,000	6 Su	28
Rhodopseudomonas acidophila	248,000	32,000	8 Su	29
Alcaligenes eutrophus	256,000	33,000	8 Su	30
Chlorogloeopsis fritschii	230,000	40,000	6 Su	25
B. EUKARYOTES				
Chlamydomonas reinhardtii	-	42,000	-	31
Scenedesmus obliquus	84,000[a]	42,000	2 Su	31
Bryopsis maxima	90,000	41,000	2 Su	33
Tobacco	90,000	46,000	2 Su	34
Spinach	90,000	44,000	2 Su	35
Wheat	83,000	42,000	2 Su	36

[a] a latent 470,000 M_r hexadecamer also occurs in this alga [32].

4.2 Ribulose 1,5-bisphosphate carboxylase/oxygenase

RuBisCO's have been purified and characterized to varying extents from at least 40 prokaryotes [5] and recombinant DNA technology has been adopted for their study. The topic of RuBisCO is very large; entire symposia are devoted to it periodically [e.g. 39], and comprehensive coverage of RuBisCO research is not possible here. As in eukaryotes, most bacterial RuBisCO's are high M_r (ca. 500,000) proteins containing 8 large (L; 55,000-64,000 Daltons) plus 8 small (S; 10,000-16,500) subunits [5]. In all cases, RuBisCO is a bifunctional enzyme, catalyzing the carboxylation and oxygenation of RuBP, the latter being an apparently wasteful and currently unavoidable process. O_2 is a linear competitive inhibitor of carboxylation and vice versa [5,39, 40]. All RuBisCO's can exist in an inactive and active form, activation proceeding via an ordered reversible binding of a CO_2 at lysine residue 201 on the L subunit, followed by a divalent metal cation. These findings, plus the location of the single catalytic site for carboxylation and oxygenation at lysine 175 of the L subunits of apparently all enzymes [5, 39, 40] suggests a uniform mechanism of RuBisCO function throughout the Calvin cycle autotrophs. This is consistent with the high degree of amino acid sequency homology among L subunits of diverse prokaryotes and eukaryotes [see 41] except *Rhodospirillum rubrum*. The *R. rubrum* enzyme, the only known 2L RuBisCO, may have taken a different evolutionary course.

S subunits are clearly not needed for activation and catalysis of the *R. rubrum* enzyme, or of 6L RuBisCO's from *Rhodopsendomonas sphaeroides* [42] and *Rps. capsulata* [43]. However, an essentiality for S subunits exists among 8L8S RuBisCO's. 8L "catalytic cores" from 3 cyanobacterial enzymes and *Chromatium vinosum* RuBisCO show negligible catalysis, although this is restored fully by reconstitution with homologous S subunits and partially by heterologous S subunit hybridization [44-46]. These findings, and differences in the activation and catalytic characteristics between 6L and 8L8S enzymes from *Rps. sphaeroides* [47] indicate a functional role of S subunits in maintaining the conformation of the active site on the L subunits. Strategies to improve plant RuBisCO should involve attempts to modify both subunits.

Several bacteria contain 2 forms of RuBisCO. *Rps. sphaeroides, Rps. capsulata* and*Rps. blastica* contain 8L8S enzymes besides the 6L form [42, 43, 48]. Differences in kinetic properties [42, 43], enzyme synthesis regulation [49, 50] and in L subunit structures in the 2 forms [51] strongly suggest that the L subunits of the different RuBisCO's in a single organism are different gene products. Multiple forms of RuBisCO have been reported in nitrifying bacteria and *Rhizobium* [see 52, 41] and the location and number of the respective genes in these organisms will be of interest.

The localization, cloning and heterologous expression of RuBisCO genes has been achieved from several purple bacteria, cyanobacteria and hydrogen bacteria [e.g. 30, 41, 53-57]. We have constructed a gene bank for *Chlorogloeopsis fritschii*, cloned the RuBisCO genes into Charon 4A and obtained expression of the enzyme in *Escherichia coli* with high activity [56]. Principles to emerge from the cloning and expression of the microbial 8L8S RubisCO's are that the L and S subunit genes are linked, they are under the control of a single promoter and are co-transcribed. Heterologoous expression of 8L8S bacterial RuBisCO's appears to be possible without special assembly factors, in contrast to the chloroplast enzyme. Foundations have been laid for the application of bacterial RuBisCO genetics to the study of enzyme assembly, subunit interaction and enzyme modification.

The possibility of achieving a 50% increase in net plant photosynthesis by selective abolition of the oxygenase reaction [58] is an attractive, though uncertain prospect. The oxygenase reaction is a feature of all extant RuBisCO's and clearly arose early during the evolution of the enzyme [5, 39]. The oxygenase reaction may be an inevitable and unavoidable feature of RuBisCO [40]. However, differential effects of metal ions and temperature on the carboxylase and oxygenase activities of the *R. rubrum* [59, 60] and *Euglena* enzymes [61] and variations in carboxylase/oxygenase (C/O) specificities between C_3 plants, C_4 plants and phototrophic microbes [62] indicate that manipulation of the enzyme in favour of carboxylation may be possible . Initial attempts to change the C/O specificity of *Alcaligenes eutrophus* RuBisCO by mutant selection were unsuccessful [63]. However, autotrophic bacteria, including extremophiles, provide a diverse resource and further attempts by traditional selection and site-directed mutagenesis would appear worthwhile.

5. CO_2-CONCENTRATION MECHANISMS AND CARBONIC ANHYDRASE IN CALVIN CYCLE BACTERIA

The oxygenase reaction of RuBisCO would not have imposed a constraint on the earliest Calvin cycle prokaryotes which developed in anaerobic environments. Indeed, the C/O specificities of RuBisCO's of extant purple bacteria are considerably lower than those of higher plants [62]. The increase in O_2 tensions due to the evolution of oxygenic cyanobacteria would have provided a selection pressure to increase the C/O specificity of RuBisCO itself. This has occurred to some extent in C_3 plants [62]. However, other strategies to reduce photorespiration caused by the oxygenase reaction have been used. The development of the C_4 pathway by increasing CO_2 supply to RuBisCO occurred in some plants, but this does not account for the low photorespiration rates in cyanobacteria and algae [5]. The C/O specificies of the few cyanobactaerial RuBisCO's examined for this ratio are lower than in C_3 and C_4 plants [60, 62]. It is well established that cyanobactaeria and microalgae have developed "CO_2-concentrating mechanisms" which increase photosynthetic efficiency and minimise photorespiration at CO_2 concentrations which would otherwise be limiting [5, 64].

An active inorganic transport mechanism has been identified in several cyanobacteria [5, 64]. Recently a membrane potential-driven active transport of CO_2 has been found in the chemolithoautotroph *Thiobacillus neapolitanus* (Y.A. Holthuizen, F.F. M. Dissel-Emiliana, J.G. Kuenen, W.N. Konings, in prepn.). The cyanobacterial photosystem I-driven system involves special proteins in the cytoplasmic membrane. Omata and Ogawa [65, 66] have identified a 42,000 Dalton polypeptide which is synthesized in the plasmalemma of *Anacystis nidulans* when cells are transferred from high to low CO_2 growth conditions and evidence suggests that this is involved in active inorganic carbon transport. Several microalgae and cyanobacteria can use CO_2 and bicarbonate for growth and Aizawa and Miyachi [64] have summarized the evidence that carbonic anhydrase (CA) also functions in inorganic carbon concentration and the suppression of cyanobacterial and algal photorespiration. CA activities in microalgae tend to be higher in low CO_2- than high CO_2-grown cultures. Among cyanobacteria, CA has been found in extracts of *Microcoleus lyngbyaceus,* 3 *Anabaena variabilis* strains, *Anacystis nidulans, Coccochloris peniocystis* and an *Oscillatoria* sp [67-69]. CA activities were higher in extracts of low CO_2-, than high CO_2- grown cultures and were detected in broken cell homogenates but not in whole cells. Soluble (cytoplasmic) CA's have been inferred in *A. variabilis* ATCC 29413 [68]. *A. nidulans, C. peniocystis* and *Ocillatoria* sp. [69]. The *A. variabilis* strain M-2 and M-3 CA's are both cytoplasmic and associated with cell membranes [68].

Air-grown *Chlorogloeopsis fritschii* cultures show high CA activity and the possibility was considered that this enzyme is associated with carboxysomes [70] for which a CO_2-concentrating function has been considered [71, 72]. Although more than 95% of the *C.fritschii* CA was partculate *in vitro*, the enzyme was not associated with the carboxysomes or thylakoids. Recently, we have shown that the *C. fritschii* CA is associated with the cell surface. The specific activity of this enzyme is increased between 2.6 and

6.8 times by transferring cultures from growth on 5% CO_2 in air, to air, and inhibition by the CA inhibitors ethoxyzolamide, acetozolamide and sulphanilamide occurs.

This is the first report of a cell surface CA in the cyanobacteria and a role for this enzyme in the uptake of bicarbonate by this organism has been indicated. Inhibition of the external CA results in a 60-80% decrease in bicarbonate-dependent photosynthesis by low CO_2-grown *C. fritschii* [A.M. Hawthornthwaite, K. Okabe and G.A. Codd, in prepn.]. External CA as in *C. fritschii* and several green algae may be particularly importan t in neutral and alkaline pH waters [64]. The presence of CA in other Calvin cycle bacteria jhas been little studied. The enzyme occurs in *Rhodospirillum rubrum* [73], *Nitrosomonas europea* [74] and *Thiobacillus thiooxidans* [75]. CA activities were high in phototrophic cultures and undetectable in chemoheterotrophic cells of *R. rubrum* [73]. The increased activity in low CO_2-, versus high CO_2-grown *N. europea* [74] further indicates a role for CA in the growth of these organisms on inorganic carbon.

6. CARBOXYSOMES

Polyhedral bodies occur in some but not all members of the colourless sulphur-oxidizing, nitrite-oxidizing and ammonia-oxidizing bacteria and in all cyanobacteria examined. These bodies have been isolated from members of the above groups and shown to contain RuBisCO, a fact recognized by their renaming as carboxysomes [71,76, 8]. Initial demonstrations of the presence of RuBisCO in carboxysomes depended upon isolation of the organelles and enzyme localization *in vitro*. This approach has been complemented recently by immunoelectronmicroscopy of cyanobacterial cell sections using gold-labelled RuBisCO antibodies [77]. Polyhedral bodies, though of diverse shape, size and number, also exist in *Prochloron* strains and cyanelles, photosynthetic endosymbionts of free-living cyanobacterial origin [71]. *In vitro* RuBisCO localization and cell section immunogold labelling has shown that the polyhedral bodies from these phototrophs are also carboxysomes (Table 2).

PRK is not present in the carboxysomes of *C. fritschii* [77, 78] *Prochloron,* cyanelles (Table 2) or *Thiobacillus neapolitanus* [79, 80]. The absence of PRK from carboxysomes indicates that RuBP would have to enter the organelles if carboxysomal RuBisCO participates in CO_2 fixation *in vivo* RuBP-dependent CO_2 fixation into 3-phosphoglyceric acid has been obtained using intact *T. neapolitanus* carboxysomes *in vitro* [79, 80] although whether this occurs *in vivo* is unknown. An active role for carboxysomes in CO_2 fixation is one of several possibilities which also include RuBisCO protective and storage roles [71, 72]. Evidence for and against all of these possibilities exists and identification of the several crypic peptides in these organelles, in addition to their most abundant protein RuBisCO, is currently limiting advances in the elucidation of carboxysome function.

Table 2. Localization of Calvin cycle enzymes in *Prochloron* and the cyanelles of *Cyanophor paradoxa*

Organism	Enzyme	Fraction	Sp. Act[a]	% Distribution	EM Gold labelling of cell section[c]
Prochloron	RuBisCO	Cytoplasm	0.30	79	+
		PB[b]	0.08	21	+
	PRK	Cytoplasm	0.29	100	+
		PB	0.00	0	-
Cyanophora	RuBisCO	Cytoplasm	0.44	47	+
cyanelle		PB	0.50	53	+
	PRK	Cytoplasm	0.57	93	+
		PB	0.04	7	-

[a]. specific activity moles substrate transformed min^{-1} mg $protein^{1}$. b, polyhydral body fraction. c, methods as in 77 and E. Mangeney, A.M. Hawthhornthwaite, G.A. Codd, S.P. Gibbs in prepn.).

7. PHYSIOLOGICAL CONTROL OF CO_2 FIXATION

Autotrophic CO_2 assimilation is a very energy expensive process and in order to control this process economically the facultative autotrophs need an accurate system to gear the rate of CO_2 fixation to their requirements. Control of the Calvin cycle takes place primarily at the level of the two key enzymes, RuBisCO and PRK. In particular, the latter is subject to allosteric control by the redox status and the energy charge of the cell. Metabolic intermediates which may play an important role in the (de)repression of the two key enzymes are likely to be among the first products of the Calvin cycle, for example, phosphoglycerate and phophoenolpyruvate, or related metabolites [5, 6].

In the obligate autotrophs, control takes place primarily by modulation of existing pathways. However, a general pattern is that obligate autotrophs will respond to limitation by CO_2 by an increase in the concentration of RuBisCO. For example, in *T. neapolitanus*, a 3-5 fold increase occurs which is accompanied by a similar increase in the number of carboxysomes [72]. This is also true for *T. ferrooxidans* W. Hazeu, P. Bos and J.G. Kuenen unpublished].

In both the phototrophs and lithotrophs CO_2 limitation, and in general a high O_2/CO_2 ratio, may lead to increased oxygenase activity of RuBisCO, which may result in glycollate production/excretion [5]. In these organisms one finds no, or hardly any response of RuBisCO levels to the presence of organic compounds in the growth medium [8].

The response of facultative autotrophs to changes in their nutritional environment is generally much more elaborate and flexible than that found among the obligate autotrophs. While the principles of control may be very similar in most facultative autotrophs, the details of reponse and enzyme adjustment are very dependent on the metabolic strategy of the organisms. They respond to CO_2 limitation in a way similar to the obligate autotrophs [81, 82]. The facultative autotrophs often show diauxic responses when presented with excess concentrations of mixtures of organic compounds and the lithotrophic energy source, resulting in complete repression of RuBisCO, But this is not seen under growth limitation by these mixtures. For example, in *Thiobacillus versutus* (formerly A2), grown on a mixture of acetate and thiosulphate, one observes a very accurate tuning of CO_2-fixing capacity to need [8]. This applies to most other facultatives [6]. However, with certain organic carbon sources; fructose in *Alcaligenes eutrophus* [82], and oxalic acid in *Pseudomonas oxalaticus* [6]. CO_2 fixation capacity may not be repressed, or may be only partly repressed, by the organic compound. Also in phototrophs (*Rhodopseudomonas* spp.) a similar pattern can be seen [81]. An interesting situation exists in *Chromatium* and *Chlorobium* species, in which acetate can almost entirely replace CO_2 as carbon source, but CO_2 fixation capacity is not repressed. In fact, when presented with excess sulphide and acetate, *Chlorobium* consumes sulphide preferentially, implying that it must fix CO_2 in the presence of excess acetate [83].

8. ECOPHYSIOLOGICAL ASPECTS

With the exception of the oxygenic cyanobacteria, all other phototrophic and chemolithotrophic bacteria are limited to environments where there is a reduced energy and/or electron donor available in addition to light for the phototrophs [84, 17]. Of course, many of the chemolithotrophs require the simultaneous presence of oxygen as terminal electron acceptor. Such habitats are found at the interface of aerobic and anaerobic environments. These interfaces may be relatively stable when there is a continuous supply of both the inorganic source of energy reducing power in the dark, e.g. around submarine volcanic springs, but in most known cases they are highly dynamic, due to diurnal or tidal rhythms. Examples are tidal mudflats with algal mats, or interfaces of stratified water bodies.

Studies by Loogman and Mur, and Riegman and Mur have shown that cyanobacteria and green algae are extremely well adapted to changes in the light-dark cycle. These organisms adjust the rate of glycogen synthesis during the light period in such a way that in the dark a growth rate can be maintained at the expense of reserve material to equal the rate of growth in the light [85].

Facultative chemolithotrophs appear to be well equipped to survive in environments where mixtures of organic and inorganic carbon and/or energy sources are available, or in the case of phototrophs, where organic compounds and light are simultaneously present. The specialized, obligate autotrophs may become dominant in environments where fluxes of the reduced inorganic energy source, or light in the case of phototrophs, are high relative to the flux of organic substrates. This has been shown to be the case among the sulphur-oxidizing bacteria [8]. Few studies have been made of the metabolic flexibility of the facultative organisms, i.e. how fast they can switch from heterotrophic to autotrophic growth. *T. versutus* [8] can grow well under alternating supply of organic and inorganic substrates, but with increasing length of the heterotrophic periods its autotrophic potential is repressed progressively, resulting in long lags before autotrophic growth can be resumed. This implies that such organisms are primarily competitive under environmental conditions during which mixtures of organic and inorganic substrates are available. A somewhat similar case has been investigated by Wijbenga and Van Gemerden [cf. 17] who found that *Rhodopseudomonas capsulata* behaved in a manner analogous to that of *T. versutus* during alternating supply of sulphide and acetate and lost in competition with *Chromatium* under these conditions. Further information on the ecophysiological aspects of autotrophic growth can be found in refs.17,85, 86.

We are pleased to acknowledge the contributions and comments of Drs. D. Vakeria and K. Okabe and many colleagues who provided unpublished manuscripts.

References

[1] Evans, M.C.W., Buchanan, B.B., Aron, D.I. (1960) Proc. Natl. Acad. Sci. USA 55: 241-244.
[2] Shiba, H., Kawasumi, T., Igarashi, Y., Kodama, T., Minoda,Y. (1985) Arch. Microbiol. 141: 198-203.
[3] Fuchs, G. (1986) FEMS Microbiol. Reviews 39: 181-213.
[4] Bassham, J.A. (1979) In: Encycl. Plant Physiol. New Ser. 6: 9-30.
[5] Codd, G.A. (1984) In: Aspects of Microbial Metabolism and Ecology, ed. G.A. Codd, London, Academic Press, 129-173.
[6] Dijkhuizen, L., Harder, W. (1984) Antonie van Leeuwenhoek 50: 473-487.
[7] Dijkhuizen, L, Harder, W. (1985) In: Comprehensive Biotechnology, vol. 1 The Principles ofBiotechnology ed. H. Dalton, Pergamon Press, Oxford, 409-423.
[8] Kuenen, J.G., Beudeker, R.F. (1982) Phil. Trans. Roy. Soc. Lond. B 298: 373-379.
[9] Watson, S.W., Bock, E., Valois, F.W., Waterbury, J.B., Schlosser, U. (1986) Arch. Microbiol. 144: 1-7.
[10] Kawasami, T., Igarashi, Y., Kodama, T., Minoda, Y. (1984) Int. J. Syst. Bacteriol. 34: 5-10.
[11] Bonjour, F., Aragno, M. (1986) FEMS Microbiol. Letters 35: 11-15.
[12] Nelson, D.G., Jannasch, H.W. (1983) Arch. Microbiol. 136: 262-269.
[13] Blotevogel, K.H., Fischer, U., Mocha, M., Jansen, S. (1985) Arch. Microbiol. 142: 211-217.
[14] Fisher, F., Zillig, W., Stetter, K.O., Schreiber, G. (1983) Nature 301: 511-513.
[15] Klemps,R., Cypionka, H., Widdel, F., Pfennig, N. (1986) Arch. Microbiol. 143: 203-208.
[16] Cypionka, H., Pfennig, N. (1986) Arch. Microbiol. 143: 396-399.
[17] Kuenen, J.G., Robertson, L.A., Van Gemerden, H. (1985) In: Advances in Microbiol Ecology vol. 8. Ed. K.C. Marshall, New York, Plenum Press, 1-59.
[18] Smith, A.J. (1983) Ann. Microbiol. Inst. Pasteur 134B: 93-113.
[19] Strohl, W. R., Larkin, J.M. (1978) Appl. Environ. Microbiol. 36: 755-770.
[20] Hanert, D. (1982) Abstr. XIII. Int. Congr. Microbiol. Boston, 61.
[21] Akazawa, T., Newcomb, E.H., Osmond, C.B. (1978) Mar. Biol. 47: 325-330.
[22] Berhow, M.A., McFadden,B.A. (1983) Planta 158: 281-287.
[23] Burger-Wiersma, T., Veenhuis, M, Korthals, H.J., van de Wiel, C.C.M., Mur, L.R. (1986) Nature 320: 262-264.
[24] Holo, H., Sirevag, R. (1936) Arch. Microbiol. in the press.
[25] Marsden, W.J.N., Codd, G.A. (1984) J. Gen. Microbiol. 130: 999-1006.
[26] Crawford, N.A., Sutton, C.W., Yee, B.C., Johnson, T.C., Carlson, D.C., Buchanan, B.B. (1984) Arch. Microbiol. 139: 124-129.
[27] Hart, B.A., Gibson, J.(1971) Arch. Biochem. Biophys. 144: 308-321.
[28] Tabita, F.R. (1980) J. Bacteriol. 143:1275-1280.
[29] Rippel, S., Bowien, B. (1984) Arch. Microbiol. 139: 207-212.
[30] Siebert, S., Bowien, B. (1984). Biochim. Biophys. Acta 787: 208-214.
[31] Salvucci, M. E., Ogren,W.L. (1985) Planta 165: 340-347.
[32] Lazaro, J.J., Sutton, C.W., Nicholson, S., Powls, R. (1986) Eur. J. Biochem. 156: 423-429.
[33] Satoh, H., Okada, M., Nakayama, K., Murata, T. (1985) Plant Cell Physiol. 26: 931-940.

[34] Kagawa, T. (1982) In Methods in Chloroplast and Molecular Biology, Eds, M. Edelman. R. B. Hallick, N.H. Chua, Amsterdam, Elsevier, 695-705.
[35] Porter, M.A., Milanez, S., Stringer, C. D., Hartman, F.C. (1985) Arch. Biochem. Biophys. 245: 14-23.
[36] Surek, B., Heilbronn, A., Austen, A., Latzko, E. (1985) Planta 165: 507-512.
[37] Klintworth, R., Husemann, M., Salnikow, J., Bowien, B. (1985) J. Bacteriol. 164: 954-956.
[38] Nicholson, S., Easterbym J.S., Powls, R. (1986) FEBS Letters 202: 19-22.
[39] Ellis, R.J., Gray, J.C. (1986) Phil. Trans. Roy. Soc. Lond. B 313: 303-467.
[40] Lorimer, G. H. (1981) Annu. Rev.Plant Physiol. 32: 349-383.
[41] Andersen, K., Wilke-Douglas, M., Caton, J. (1986) Biochem. Soc. Trans. 14: 29-31.
[42] Gibson, J.L., Tabita, F.R. (1977) J. Biol. Chem. 252: 943-949.
[43] Gibson, J.L., Tabita, F.R. (1977) J. Bacteriol. 140: 1023-1027.
[44] Andrews, T.J., Ballment, B.(1983) Biol. Chem. 258: 7514-7518.
[45] Asami, S., Takabe, T., Akazawa, Codd, G.A. (1983) Arch. Biochem. Biophys. 225: 713-721.
[46] Incharoensakdi, A., Takabe, T., Akazawa, T. (1985) Arch. Biochem. Biophys. 237: 445-453.
[47] Gibson, J.L., Tabita, F.R. (1979) J. Bacteriol. 140: 1023-1027.
[48] Sani, A. (1985) Ph. D. Thesis, University of Warwick, Coventry, England.
[49] Weaver, K.E., Tabita, F.R. (1985) J. Bacteriol. 164: 147-154.
[50] Jouanneau, Y., Tabita, F.R. (1986) J. Bacteriol. 165: 620-624.
[51] Gibson, J.L., Tabita, F. R. (1985) J. Bacteriol. 164: 1188-1193.
[52] Ebert, A. (1982) Ph. D. Thesis, University of Hamburg, W. Germany.
[53] Nierzwicki-Bauer, S.A., Curtis, S.E., Haselkorn, R. (1984) Proc. Natl. Acad. Sci. USA 81: 5961-5965.
[54] Somerville, C.R., Somerville, S.C. (1984) Mol. Gen. Genet.193: 214-219.
[55] Viale, A.M., Kobayashi, H., Takabe, T., Akazawa, T.(1985) FEBS Letters 192: 283-288.
[56] Vakeria, D., Codd, G.A., Hawthornthwaite, A.M., Stewart, W.D.P. (1986) Arch. Microbiol. in the press.
[57] Tabita, F.R., Small, C.L. (1985) Proc. Natl. Acad. Sci. USA 82: 6100-6103.
[58] Somerville, C., Fitchen, J., Somerville, S., McIntosh, L., Nargang, F. (1983) In: Advances in Gene Technology: Molecular Genetics of Plants and Animals San Diego, Academic Press, 295-307.
[59] Robinson, R.D., Martin, M,N., Tabita, F.R. (1979) Biochemistry 18: 4453-4458.
[60] Jordan, D.B., Ogren, W.L. (1984) Planta 161: 308-313.
[61] Wildner, G.F., Henkel, J. (1978) FEBS Letters 91: 99-103.
[62] Jordan, D.B., Ogren, W.L. (1983) Arch. Biochem. Biophys. 227: 425-433.
[63] Andersen, K. (1979) Biochem. Biophys. Acta 585: 1-11.
[64] Aizawa, K., Miyachi, S. (1986) FEMS Microbiol. Reviews 39: 215-233.
[65] Omata, T., Ogawa, T. (1985) Plant Cell Physiol. 26: 1075-1081.
[66] Omata, T., Ogawa, T. (1986) Plant Physiol. 80: 525-530.
[67] Graham, D., Smillie, R.M. (1976) Aust. J. Plant Physiol. 3: 113-119.
[68] Yagawa, Y., Shiraiwa, Y., Miyachi, S. (1984) Plant Cell Physiol. 25: 775-783.
[69] Ingle, R.K., Colman, B. (1975) Can. J. Bot. 53: 2385-2387.
[70] Lanaras,T., Hawthornthwaite, A,M., Codd, G.A. (1985) FEMS Microbiol. Letters 26: 285-288.
[71] Codd, G.A., Marsden, W.J.N. (1984) Biol. Reviews 59: 389-422.
[72] Beudeker, R.F., Codd, G.A., Kuenen, J.G. (1981) Arch. Microbiol. 129: 361-367.
[73] Gill, S.R., Fedorka-Gray, P.J., Tweteu, R.K., Sleeper, B.P. (1984) Arch. Microbiol. 138: 113-118.
[74] Jahnke, L.S., Lyman, C., Hooper, A. B. (1984) Arch. Microbiol. 140: 291-293.
[75] Romanova, A.K., Zykalova, K.A., Kostrikina, N.A., Emnova, E.E., Biryuzova,V.I. 1982. Mikrobiologiya 51: 296-301.
[76] Shively, J.M. (1974) Annu. Rev. Microbiol. 28: 167-187.
[77] Hawthornthwaite, A. M. , Lanaras, T., Codd, G.A. (1985) J. Gen. Microbiol. 131: 2497-2500.
[78] Marsden, W.J.N., Lanaras, T., Codd, G.A. (1984) J. Gen. Microbiol. 130: 2089-2093.
[79] Cannon, G.C., Shively, J. M. (1983) Arch. Microbiol. 134: 52-59.
[80] Holthuizen, Y.A., van Breemen, J.F.L., Kuenen, J.G., Konings, W.N. (1986) Arch. Microbiol. 144: 398-404.
[81] Tabita, F.R., Martin, M.N., Beudeker, R.F., Quivey R.G., Sarles, L.S., Weaver, K.E. (1984) In: Microbial Growth on C_2 compounds. Eds. R.L. Crawford, R.S. Hanson, Washington D.C., Am. Soc. Microbiol. 3-8.

[82] Friedrich, C.G., Friedrich, B., Bowien, B. (1981) J. Gen. Microbiol. 122: 69-78.
[83] Gemerden, H. Van (1983) Ann. Microbiol. Inst. Pasteur 134B: 73-92.
[84] Jorgensen, B.B., Revsbech, N.P., Cohen, Y. (1983) Limnol. Oceanog. 28: 1075-1093.
[85] Mur, L.R. (1983) Ann. Microbiol. Inst. Pasteur 134B: 61-72.
[86] Gemerden, H. van, de Wit, R. (1986) In: Microbes in Extreme Environments. Ed. R.A. Herbert, G.A. Codd, San Diego, Academic Press, 111-127.

METABOLIC AND MOLECULAR REGULATION OF THE CO_2-ASSIMILATING ENZYME SYSTEM IN AEROBIC CHEMOAUTOTROPHS

B. BOWIEN, M. GUSEMANN, R. KLINTWORTH AND U. WINDHÖVEL
Institut für Mikrobiologie, Georg-August-Universität Göttingen,Federal Republic of Germany

1. INTRODUCTION

Primary production of organic matter by autotrophic organisms is based on fixation and subsequent reduction of CO_2 to yield metabolites for the synthesis of cellular constituents. There are three different biosynthetic pathways known by now that enable eukaryotes and prokaryotes to grow autotrophically. Among these pathways, the reductive pentose phosphate cycle [Calvin cycle; 1] is most widespread because it occurs not only in photosynthetic higher plants and algae but also in most groups of photo- and chemosynthetic bacteria. The remaining two routes, the reductive tricarboxylic cycle [2, 3] and the newly discovered acetyl-CoA pathway [4] seem to be confined to certain groups of anaerobic bacteria.

The Calvin cycle is an autocatalytic sequence of thirteen reactions catalyzed by enzymes most of which are also common to the central carbohydrate metabolism of heterotrophic organisms. However, two enzymes, phosphoribulokinase (PRK) and ribulose-1,5-bisphosphate carboxylase/oxygenase (Rubisco), are unique to the cycle, and their presence is generally taken to indicate the function of the cycle in an autotroph. They catalyze consecutive reactions in this reaction sequence. The CO_2-accepting intermediate ribulose 1,5-bisphosphate that is synthesized by the action of PRK, is converted by the catalysis of Rubisco to yield the first product of CO_2 fixation, 3-phosphoglycerate. Rubisco is characterized by its bifunctionality. Besides the carboxylation of ribulose 1,5-bisphosphate it also catalyzes the competing reaction of the oxygenolytic cleavage of this substrate, resulting in the formation of 2-phosphoglycolate which is oxidized to CO_2 and 3-phosphoglycerate via the pathway of glycolate metabolism [5]. The photorespiration of plants is a manifestation of the oxygenase activity of Rubisco [6]. Other key activities of the Calvin cycle, fructose-1,6-bisphosphatase (FBPase) and sedoheptulose-1,7-bisphosphatase (SBPase), are involved in the rearrangement reactions leading to the regeneration of pentose phosphates. Although very limited information is available on the latter enzymes from autotrophic bacteria, FBPase/SBPase activities seem to be properties of a single protein (7-10).

The structural properties of PRK from organisms so far studied are quite different. The enzymes from the chemoautotrophic hydrogen-oxidizing bacterium *Alcaligenes eutrophus* and the phototrophic nonsulfur purple bacterium *Rhodopseudomonas acidophila* are 250-kDa proteins consisting of eight

33-kDa subunits [11,12], whereas the cyanobacterium *Chlorogloelopsis fritschii* contains a hexameric enzyme of 230-kDa (40-kDa subunits; 13) and the plant PRK is a 90-kDa dimer (44-kDa subunits; 14). In contrast, the subunit structure of Rubisco from the various autotrophs studied with respect to this enzyme is relatively uniform. The plant-type enzyme which also occurs in algae, cyanobacteria as well as photo- and chemoautotrophic bacteria has a molecular mass of approx. 550 kDa with eight large (L) catalytic subunits (50-55 kDa) and eight small (S) subunits (12-16 kDa) of unknown but essential function comprising the native protein [15].

In order to establish and maintain a balanced metabolic situation during CO_2 assimilation, autotrophs have to regulate the carbon flux through the Calvin cycle on the level of both enzyme activity and enzyme synthesis. Principal targets of these control mechanisms are the key enzymes of the cycle. We focus here on discussing various modes of regulation affecting the operation of the Calvin cycle in facultative and obligate chemoautotrophs. In addition, some new data of our studies on the genetic organization and regulation of the CO_2-fixing enzyme system (Cfx enzymes), i.e. PRK and Rubisco, of *A. eutrophes* H16 are presented.

2. REGULATION OF ENZYME ACTIVITY

Since the synthesis of cell material from CO_2 requires a high input of metabolic energy and reducing power, central intermediates of the energy and carbon metabolism in both facultative and obligate chemoautotrophs are effectors of target enzymes in the Calvin cycle. PRK from hydrogen bacteria [11, 16] and from the nitrifier *Nitrobacter winogradskyi* [17] is strongly activated by NADH, however, the enzyme from the sulfur oxidizer *Thiobacillus neapolitanus* was reported not to be affected by this reductant of the cycle [18]. On the other hand, AMP and phosphoenolpyruvate (PEP) are general inhibitors of PRK [18-22]. This regulatory pattern is consistent with the physiological needs of a CO_2-assimilating cell. The availability of reducing power and energy is the primary signal for PRK-directed control of CO_2 fixation via the Calvin cycle. PEP as an indirect end-product of the cycle can be considered a feedback inhibitor of an early reaction of the sequence catalyzed by PRK, thus supporting the balance between synthesis of cycle intermediates and their utilization in other biosynthetic pathways.

Activity control of FBPase/SBPase by metabolites would be expected to complement the regulation of PRK. In fact, FBPase from *T. neapolitanus* is subject to inhibition by AMP and PEP [23]. The enzymes from *A. eutrophus* and another hydrogen bacterium, *Nocardia opaca*, however, are relatively insensitive to both metabolites and exhibit inhibition by ATP [8; 9]. The metabolic significance of the ATP inhibition is difficult to understand since the Calvin cycle should operate at maximal rates when the energy supply is in excess and since the PRK is active under these conditions. More metabolic importance may be attached to the inhibition of FBPases/SBPases by pentose phosphates. The enzyme from *A. eutrophus is* inhibited by ribulose 5-phosphate, whereas ribulose 1,5-bisphosphate affects the *N. opaca* enzyme [8, 9]. In metabolic terms, this regulation can be interpreted as feedback inhibition similar to the rationale used in the case of the PEP inhibition of PRK. Support for the concept of a metabolic requirement of an autotroph for a special Calvin cycle FBPase/SBPase comes from the observation that autotrophically grown cells of *N. opaca* contain two isoenzymes [9] of which the form A isoenzyme has the regulatory properties mentioned above. Form B occurs only in heterotrophic cells of the organism and its regulation (very strong inhibition by AMP and activation by PEP) is typical for the gluconeogenetic role of this enzyme. Moreover, *in vivo* the latter enzyme seems to act only as FBPase.

The regulation of PRK and FBPase/SBPase activities by metabolites in chemoautotrophs resembles the control of these enzymes in anoxygenic photosynthetic bacteria such as the nonsulfur and the sulfur purple bacteria [24, 25]. In the nonsulfur purple bacterium *Rhodospirillum rubrum*, these two regulatory enzymes form a complex [26], underscoring their roles in the control of the Calvin cycle. In cyanobacteria and chloroplasts, PRK and FBPase/SBPase together with other enzymes, are regulated by light, involving a reductive activation upon illumination [25,27]. Interestingly, the PRK of *A. eutrophus* is also regulated by an inactivation-reactivation mechanism [28]. Imposing mixotrophic conditions upon autotrophically growing cells of the organism by the addition of pyruvate to the culture caused a rapid partial inactivation of the enzyme. Reactivation occurred after exhaustion of the pyruvate. It was shown that a direct metabolic product pf pyruvate, PEP, and ATP are likely to play a role in the inactivation process by a mechanism that remains to be elucidated [29]. Covalent modification of the enzyme is an obvious possibility. The physiological significance of the PRK inactivation is suggested by the fact that it is paralleled by a decrease in the CO_2 fixation capacity of the cells. PRK activity in *A. eutrophus* is thus modulated by PEP via two modes of regulation, inhibition and inactivation. The interesting question arises as to the possible occurrence of inactivation-reaction control of PRK in other autotrophic bacteria, especially facultative autotrophs.

It is well established that plant Rubisco can exist in both an active and an inactive form.The activation is a reversible process involving the formation of a ternary complex of the enzyme with CO_2 and Mg^{2+}. This activating CO_2 molecule is different from the substrate CO_2 which participates in catalysis. A carbamate complex is formed on the ε-amino group of a lysylresidue in the L subunit during activation [15]. The activation is also characteristic of bacterial Rubiscos and has been demonstrated for the enzyme from *A. eutrophus* [30]. Conformational changes of the *A. eutrophus* enzyme were observed to occur upon changing its activation state [30, 31]. Regulation of Rubisco activity by metabolites may be of minor importance. 6-Phosphogluconate has been described as an inhibitor of the enzyme from a number of autotrophs including chemoautotrophs [32,33]. The physiological significance of this inhibition is uncertain. It may assist in curtailing CO_2 assimilation when "heterotrophic" substrates such as carbohydrates are available to the cells.

3. REGULATION OF ENZYME SYNTHESIS

Unlike obligate chemoautotrophs, the facultative chemoautotrophs are able to very effectively adjust the synthesis of Calvin cycle enzymes, especially PRK and Rubisco, according to the nutritional situation. However, even obligate autotrophs regulate enzyme synthesis to a certain extent [34]. As a general rule, the synthesis of these enzymes in facultative autotrophs is repressed during heterotrophic growth on organic substrates. This regulation serves to avoid the energetically expensive autotrophic CO_2 fixation under more favorable nutritional conditions. The repression of PRK and/or Rubisco is often complete as in *A. eutarophus* [35], the sulfur oxidizer *Thiobacillus versutus* [36] or the formate-oxidizing autotroph *Pseudomonas oxalaticus* [37]. However, depending on the nature of the "heterotrophic" substrate, the repression may only be partial. This applies also for mixotrophic growth of the organisms in chemostat cultures [36-38]. Derepression/repression rather than induction/repression mechanisms are therefore likely to control the synthesis of PRK and Rubisco in some but not all facultative chemoautotrophs. Many hydrogen bacteria require the presence of hydrogen and CO_2 as energy and carbon sources to induce to synthesis of the two key enzymes of the Calvin cycle [39]. Regulation of Rubisco synthesis has been observed in some obligate autotrophs. When *T. neapolitanus* is grown under carbon starvation by limiting the CO_2 supply, Rubisco synthesis is strongly derepressed beyond the normal autotrophic rate [40].

A. eutrophus has been studied more thoroughly with the aim to recognize (the) metabolic signal(s) directing the synthesis of the Cfx enzymes in this organism. Strong repression prevails in cells grown on pyruvate, succinate or acetate whereas PRK and Rubisco synthesis in fructose- or gluconate-grown cells is partially derepressed [35, 41]. These findings suggest that the breakdown of the various substrates may result in (a) critically different intracellular concentration(s) of (a) metabolite(s) which act(s) as signal(s) for the regulatory system. PEP was proposed to be a negative effector and a reduced metabolite (possibly NADH) to be a positive effector of Rubisco expression [42]. Consistent with this view is the strong derepression of Rubisco synthesis (3-to-5-fold of normal autotrophic level) upon carbon starvation of autotrophically growing cells by CO_2 limitation [43]. Studies performed with *P. oxalaticus* also led to the conclusion that the pool size of a C_3 metabolite determines the degree of Rubisco repression [37]. If PEP really acts as signal (co-repressor) in this end-product repression, it would participate in the control of CO_2 assimilation in *A. eutrophus* on the enzyme activity as well as the expression level. PEP would thus be a key metabolite in the regulation of autotrophic CO_2 fixation.

Another aspect in the regulation of CO_2 assimilation in facultative autotrophs is the synthesis under autotrophic growth conditions of special isoenzymes functioning in the reductive and/or regenerative parts of the Calvin cycle. A first example was the detection of FBPase/SBPase [9] and aldolase isoenzymes (unpublished results) in *N. opaca*. It should be worth-while to search how widespread this phenomenon of isoenzyme formation is among autotrophic bacteria.

4. GENETIC REGULATION AND ORGANIZATION OF THE CO_2-FIXING ENZYME SYSTEM IN *ALCALIGENES EUTROPHUS*

The expression of PRK and Rubisco, the Cfx enzymes, in *A. eutrophus* H16 is coordinately regulated [35, 41]. An understanding of the control mechanisms involved at the molecular level requires a genetic analysis of the system that has been initiated in our laboratory.

A. eutrophus H16 contains the self-transmissible 450-kilobase (kb) megaplasmid pHG1 which is essential for lithoautotrophic growth of the strain since it encodes the two hydrogenases of the organism [44]. Plasmid-cured strains retained, however, the ability for organoautotrophic growth on formate, indicating that the chromosome of the organism carries structural and regulatory *cfx* genes or has both functions.

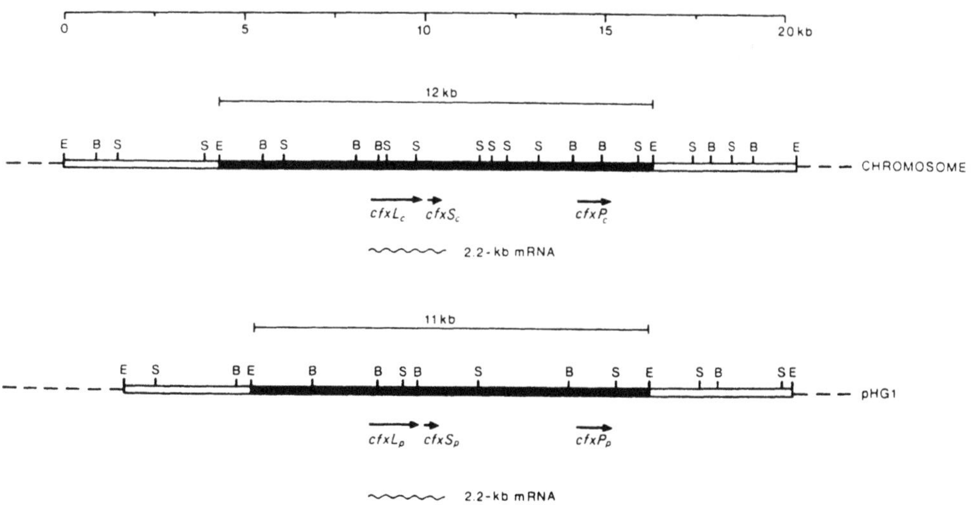

Figure 1. Restriction maps of the chromosomal and megaplasmid (pHG1) regions of *A. eutrophus* H16 carrying *cfx* genes. Only the restriction sites for the endonucleases *Eco*RI (E), *Bam*HI (B) and *SalI* (S) are shown. The arrows indicate the positions, sizes and transcriptional directions of the genes. Subscripts c and p refer to chromosomal and plasmid genes, respectively.

Evidence for the existence of megaplasmid-coded *cfx* genes in *A. eutrophus* H16 was obtained from the localization of chromosomal and plasmid structural genes for the subunit of PRK [*cfx*P; 46]. In additional Southern blot hybridizations, the Rubisco gene from *R. rubrum* [47] was used as heterologous probe for the detection of the L subunit gene (*cfx*L). A synthetic oligodeoxynucleotide, the sequence of which was derived from the amino-terminal amino acid sequence of the S subunit, served as homologous probe for the subunit gene (*cfx*S). The Rubisco genes were also localized in the chromosome and the plasmid of the organism. The chromosomal set of *cfx* genes as detected on a 12-kb, the plasmid set on an 11-kb *Eco*RI restriction fragment (Fig, 1).

The duplicate *cfx* regions in the genome of *A. eutrophus* H16 were found to be highly homologous and very similarly organized. In both regions, the Rubisco genes form an operon as judged from Northern blot hybridizations that revealed a transcript size of 2.2 kb (Fig. 1). The *cfxL*-*cfx*S operon showed the same direction of transcription as the *cfx*P_gene located about 3.5 kb downstream of the *cfx*S gene. *cfx*P apparently belongs to a different transcriptional unit. Cloning of suitable subfragments in the expression vector pUC9 allowed the expression of active pHG1-coded PRK and Rubisco in *Escherichia coli*. This indicated that the plasmid *cfx*_ genes of *A. eutrophus* H16 are indeed functional.
We also addressed the question if the expression of the four identified *cfx* operons is under the control of common regulatory loci. Transposon Tn5-induced regulatory mutants of *A. eutrophus* H39 (derivative of wild-type H16), deficient in autotrophic CO_2 fixation, were characterized. All mutants obtained carried chromosomal Tn5 insertions. So far, two classes of mutants could be recognized (Tab.1) (i) mutants with an insertion in the 12-kb *Eco*RI fragment outside the structural *cfx*_genes were totally deficient in the formation of the Cfx enzymes; (ii) mutants with an insertion i n an 13-kb *Eco*RI fragment were deficient only in the heterotrophic derepression of Cfx enzyme synthesis but nevertheless unable to grow autotrophically on formate. It is conclluded that the four cfx operons form a cfx regulon the expression of which is controlled by at least two chromosomal regulatory genes. The products of these genes act presumably as activators of transcription.

Table 1. Properties of transposon Tn5-induced regulatory mutants of *A. eutrophus* deficient in autotrophic CO_2 assimilation.

Strain	Chromosomal EcoRI fragment carrying Tn5 insertion	Specific activity (U/mg protein)			
		Fructose[a]		Fructose + Formate	
		PRK	Rubisco	PRK	Rubisco
HF39	-	0.058	0.037	0.132	0.065
HF39/13	12 kb[b]	<0.001	<0.001	<0.001	<0.001
HF39/3	13kb	<0.001	<0.001	0.073	0.154

[a]Growth substrate; [b]Fragment carrying the chromosomal *cfx* genes

In summary, the cfx regulon of *A. eutrophus* H16 represents a novel system in that its component transcriptional units are duplicated with one set on the chromosome and the other on the plasmid of the organism. A similar reiteration of Rubisco genes was also shown to occur in the related strain *A. eutrophus* ATCC 17707 [48, 49]. Further studies on this system will certainly contribute to the elucidation of the complex genetic regulation of autotrophic CO^2 fixation in a facultative autotroph.

References

[1] Bassham JA, Benson AA, Kay LD, Harris AZ, Wilson AT, Calvin M. (1954) J. Am. Chem. Soc. 76:1760-1770.
[2] Evans MCW, Buchanan BB, Arnon DI. (1966) Proc. Natl.. Acad. Sci. U.S.A. 55: 928-934.
[3] Fuchs G, Stupperich E, Eden C. (1980) Arch. Microbiol. 128: 64-71.
[4] Wood HG, Ragsdale SW, Pezacka E. (1986) Trends Biochem. Sci. 11: 14-18.
[5] Lorimer GH. (1981) Annu. Rev. Plant. Physiol. 32: 349-383.
[6] Ogren WL. (1984) Annu. Rev. Plant. Physiol. 35: 415-442.
[7] Springate CF, Stachow CS. (1972) Arch Biochem. Biophys. 152: 1-12.
[8] Abdelal AT, Schlegel HG. (1974) J. Bacteriol. 120: 304-310.
[9] Amachi T, Bowien B. (1979) J. Gen. Microbiol. 113: 347-356.
[10] Gerbling KP, Steup M, Latzko E. (1986) Plant Physiol. 80: 716-720.
[11] Siebert K, Bowien B. (1984) Biochim. Biophys. Acta 787: 208-214
[12] Rippel S, Bowien B. (1984) Arch. Microbiol. 139: 207-212.
[13] Marsden WJN, Codd GA. (1984) J. Gen. Microbiol. 130: 999-1006.
[14] Porten MA, Milanez S, Stringer CD, Hartman FC. (1986) Arch. Biochem. Biophys. 245: 14-23.
[15] Miziorko HM, Lorimer GH. (1983) Annu. Rev. Biochem. 52: 507-535.
[16] MacElroy RD, Johnson MK. (1969) Arch. Biochem. Biophys. 131: 272-275.
[17] Kiesow LA, Kindsley BF, Bless JW. (1977) J. Bacteriol. 130: 20-25.
[18] MacElroy RD, Mack HM, Johnson EJ. (1972) J. Bacteriol. 112: 532-538.
[19] Gale NL, Beck JV. (1966) Biochem. Biophys. Res. Commun. 24: 792-796.
[20] MacElroy RD, Johnson EJ, Johnson MK. (1968) Biochem. Biophys. Res. Commun. 30: 678-682.
[21] Ballard RW. MacElroy; RD. (1971) Biochem. Biophys. Res. Commun. 44: 614-618.
[22] Abdelal ATH, Schlegel HGA. (1974) Biochem. J. 139: 481-489.
[23] Johnson E (1980) J. Bacteriol. 143: 1275-1280.
[25] Crawford NA, Sutton CW, Yee BC, Johnson TC, Carlson DC, Buchanan BB. (1984) Arch.Microbiol. 139: 124-129.
[26] Joint IR, Morris I, Fuller RC. (1972) J. Biol. Chem. 247: 4833-4838.
[27] Flügge UI, Stitt M, Freisl M, Heldt HW. (1982) Plant Physiol. 69: 263-267.
[28] Leadbeater L, Bowien B. (1984) J. Bacteriol. 157: 95-99
[29] Bowien B, Leadbeater L. (1984) *In* Microbial Growth on C_1 compounds (Crawford DL, Hanson RS, eds.) pp. 9-13. American Society for Microbiology, washington DC.
[30] Bowien B, Gottschalk E-M. (1982) J. Biol. Chem. 257: 11845-11847.
[31] Meisenberger O, Pilz I, Bowien B, Pal GP, Saenger W. (1984) J. Biol. Chem. 259: 4463-4465.
[32] Tabita FR, McFadden BA. (1972) Biochem. Biophys. Res. Commun. 48: 1152-1159.
[33] McCarthy JT, Charles AM. (1975) Arch. Microbiol. 105: 51-59.
[34] Smith AJ. Hoare DS. (1977) Bacteriol. Rev. 41: 419-448.
[35] Friedrich CG, Friedrich B, Bowien B. (1981) J. Gen. Microbiol. 122: 69-78
[36] Smith AL, Kelly DP, Wood AP. (1980) J. Gen. Microbiol. 121: 127-138.
[37] Dijkhuizen L, Harder W. (1984) J. Gen. Microbiol. 130: 447-457.
[38] Kärst U, Griedrich CG. (1984) J. Gen. Microbiol. 130: 1987-1994.
[39] Bowien, B, Schlegel HG. (1981) Annu. Rev. Microbiol. 35: 405-452.
[40] Beudeker RF, Cannon CG Kuenen JG, Shively JM. (1980) Arch. Microbiol. 124: 185-189.
[41] Leadbeater L, Siebert K, Schobert P, Bowien B. (1982) FEMS Microbiol. Lett. 14: 263-266.
[42] Im D-S, Friedrich CG. (1983) J. Bacteriol. 154: 803-808.
[43] Friedrich CG. (1982) J. Bacteriol. 149: 203-210.
[44] Friedrich B, Hogrefe C, Schlegel HG. (1981) J. Bacteriol. 147: 198-205.
[45] Bowien B, Friedrich B, Friedrich CG. (1984) Arch. Microbiol. 1f39: 305a-310.
[46] Klintworth R, Husemann M, Salnikow J, Bowien B. (1985) J. Bacteriol. 164: 954-956.
[47] Somerville CR, Somerville SC. (1984) Mol. Gen. Genet. 193: 214-219.
[48] Andersen K, Wilke-Douglas M. (1984) J. Bacteriol. 159: 973-978.
[49] Andersen K, Wilke-Douglas M, Caton J. (1986) Biochem. Soc. Trans. 14: 29-31.

CO_2 FIXATION IN *RHODOPSEUDOMONAS BLASTICA*

C.S. DOW
Department of Biological Sciences, University of Coventry, Coventry CV4 7AL, England

The Calvin reductive phosphate cycle constitutes the primary pathway for the generation of intermediates of central metabolism from CO_2 in plants, algae and virtually all photosynthetic and autotrophic microbes (figure 1a) [1, 2]. Of the two key enzymes, phosphoribulokinase and D-ribulose 1,5 bisphosphate carboxylase-oxygenase (RuBisCO), the latter is estimated to be the most abundant protein in the biosphere with the greatest number of associated publications [3]. Such singular attention, primarily by plant biochemists, is attributable to the following: (a) RuBisCO is not only responsible for the fixation of CO_2 via the enediol of ribulose bisphosphate (RuBP) to yield two molecules of D-3-phosphoglycerate but it is also the key enzyme of the competing photorespiratory pathway catalyzing the oxygenation of RuBP to yield one molecule each of 3-phosphoglycerate and 2-phosphogluconate (figure 1b). Consequently photosynthesis and photorespiration are opposing metabolic cycles - the former effects a net fixation of carbon from CO_2 whilst the latter effects a net loss of carbon from central metabolism via glycolate, which is either metabolised or excreted, with an attendant reduction in growth yield [4].

Therefore, an understanding of the physiology, function and potential for modification of RuBisCo has some considerable practical significance. (b) RuBisCO synthesis (in plants) involves the interaction of light with two distinct genetic systems the combined products of which yield active enzyme [4]. Not only is the Calvin cycle energetically expensive, the flow of carbon through it is predicated by two antagonistic reactions inherent in one multimeric enzyme, consequently this must be subject to very strict control at both the transcriptional and physiological levels [5].

It is the application of genetic and protein engineering to the elucidation of structure, function and regulation that has engendered detailed analysis of bacterial, as opposed to plant, RuBisCO [6. 7. 8. 9. 10]. Particular attention is being focussed on the anoxygenic phototrophic bacteria, particularly the Rhodospirillaceae, which offer considerable potential for the elucidation of RuBisCO molecular biology since they present the following features - a range of multimeric enzymes [11], from the form I, similar to the enzyme produced by phototrophic eukaryotes and most prokaryotes, which is a heteromultimer of two nonidentical subunits - eight large (L) and eight small (S) subunits (L_8S_8), to the homodimeric enzyme (L_2) of *Rhodospirillum rubrum* and the dual enzymes (form I - L_8S_8 and form II - L_6) characteristic of *Rhodobacter sphaeroides* [12] and *R. capsulata* [13]; they are metabolically extremely versatile [14] being capable of photoheterotrophic, photolithotrophic and chemoheterotrophic growth with the attendant variations in RuBisCO synthesis and expression; several species have well established

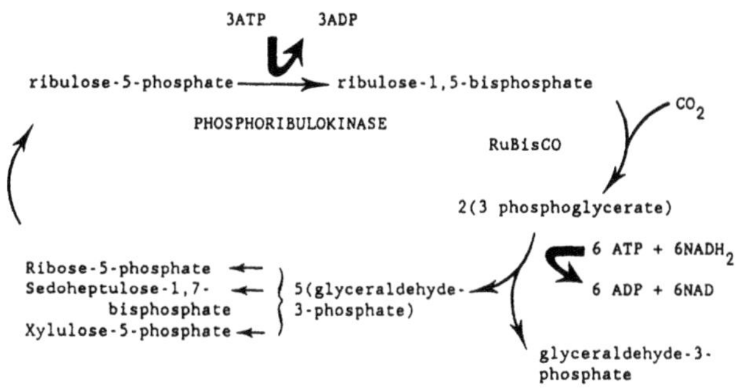

$$3CO_2 + 9ATP + 6NADH_2 \rightarrow \text{glyceraldehyde-3-phosphate} + 9ADP + 6NAD + 8Pi$$

Figure 1(a) The Calvin Cycle

RuBisCO - D-ribulose 1,5 bisphosphate carboxylase-oxygenase.

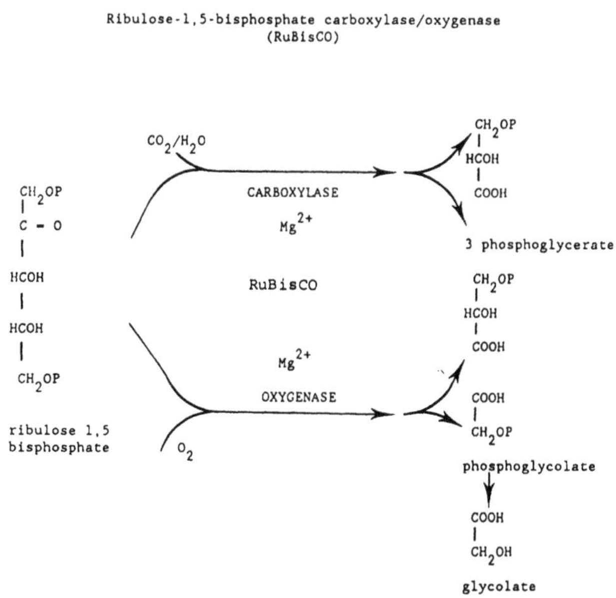

Figure 1(b)

Ribulose 1,5 bisphosphate carboxylase-oxygenase.

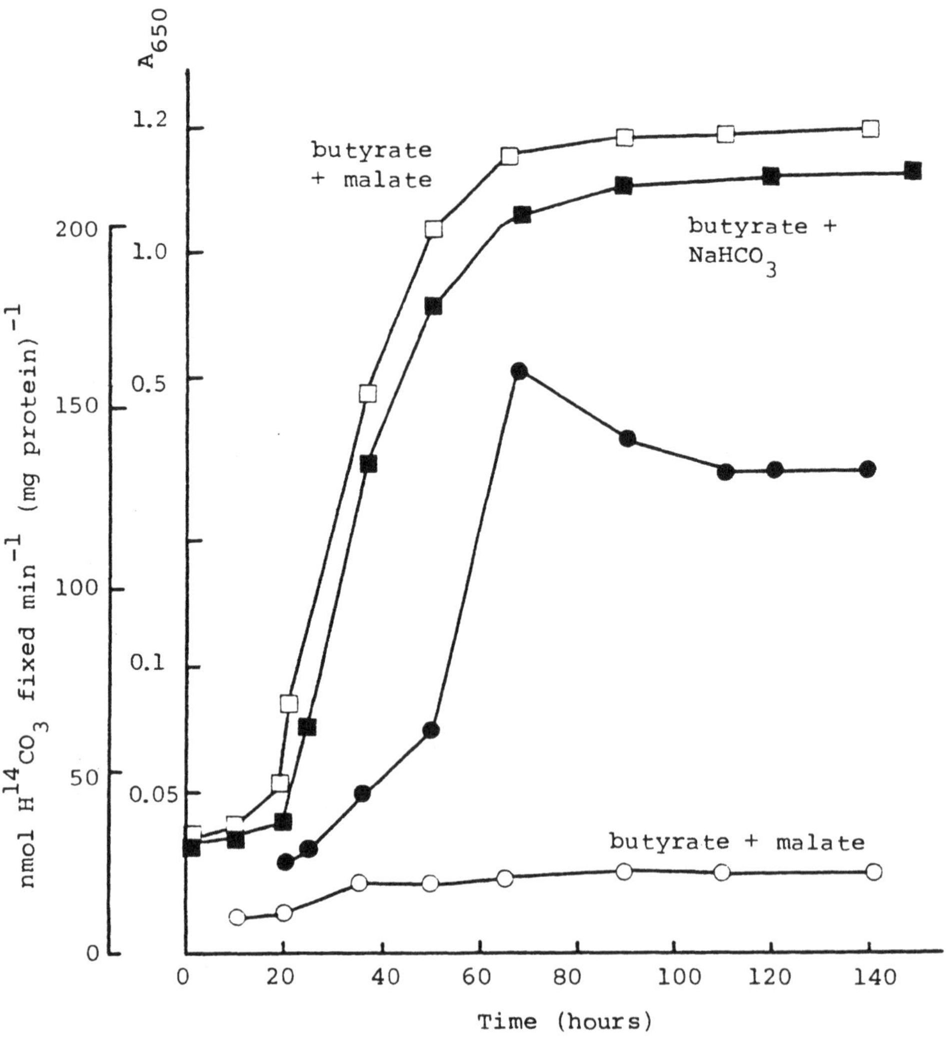

Figure 2. RubBisCo activity in soluble protein extracts prepared from *R. blastica* grown phototrophically on mixed carbon substrates.

(□ ■) A_{650}, (○ ●) RuBisCO activity.

The basal medium contained NH_4Cl 0.5, $MgSO_4.7H_2O$ 0.4, NaCl 0.4, $CaCl_2.2H_2O$ 0.05, carbon source(s) 1.0 gl^{-1}. Thiamine hydrochloride 250 µg l^{-1}, nicotinic acid 0.1 mg l^{-1}, biotin 250 mg l^{-1}. The medium was made 5 mM with respect to phosphate.

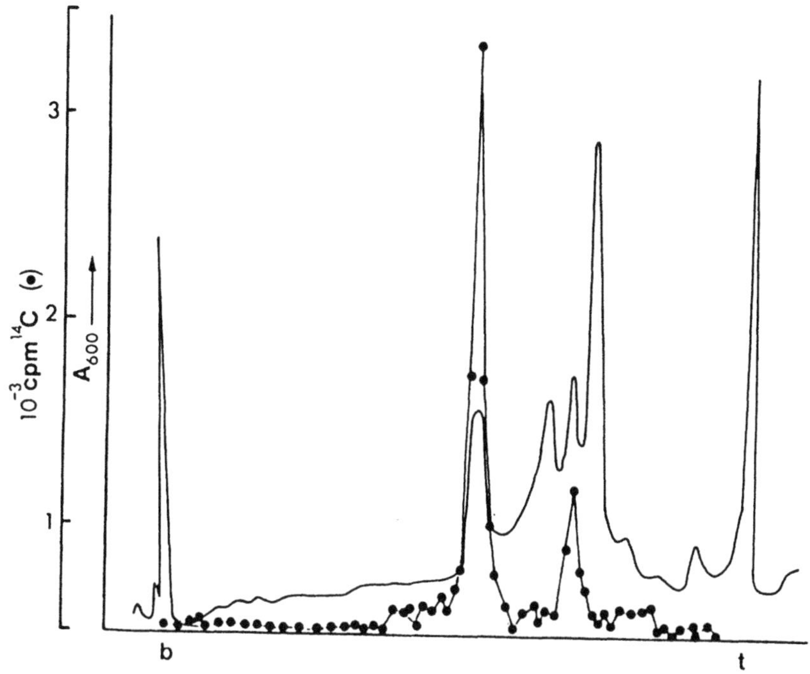

Figure 3.

The active RuBisCO fractions from a 0.2-0.8 M step sucrose gradient were pooled and applied to duplicate 5% (w/v) non-denaturing polyacrylamide gels. On completion of electrophoresis one gel was stained with Coomassie blue and scanned at 600 nm. The duplicate gel was cut into 1 mm slices and each slice assayed for CO_2 fixing activity ( ). The figure shows the presence of both form I and form II ruBisCOs from *R. blastica* in the late exponential phase of photoheterotrophic growth on pyruvate-malate medium. The form I enzyme although present in a higher concentration (as judged by the protein concentration) is less active than the form II enzyme.

t = top, b = bottom of the polyacrylamide gel.

systems for genetic exchange and few present obstacles to the application of molecular biological and genetic engineering techniques.

The quaternary structure of the majority of RuBisCOs is that of L_8S_8 with the L subunit possessing the active site for both carboxylation and oxygenation whilst the S subunit is of unknown function [2]. However, the most recent evidence suggests that although the active site(s) reside on the L subunit catalytically active enzyme is not only obtained in the presence of the S subunit. Clearly this cannot be the case with the *Rhodospirillum rubrum* (L_2) enzyme or the form II enzyme (L_6) of *R. sphaeroides*. The data available on the form I and II enzymes from *R. sphaeroides* are indicative of the L subunits being distinct in that they do not showimmunological cross-reaction [15] and proteolytic cleavage yields different peptide fragments [16, 8]. These data support the conclusion that these subunits are different gene products (B) - their physiological role(s), however, are as yet unclear.

This paper describes the form I and II enzymes from *R. blastica* and presents data on their function and expression during growth.

Consistent with previous studies on the Rhodospirillaceae, *R. blastica* can repress or derepress RuBisCO as a function of the growth conditions (figure 2). When grown phototrophically in batch culture on butyrate + malate, or butyrate + $NaCO_3$ although the growth rates and cell yield are similar there is very marked repression of RuBisCO in the former - a consequence of the high CO_2 concentration generated from the metabolism of malate. Similarly, strong repression is observed during photoheterotrophic growth on various substrates under differing conditions in continuous culture eg. under malate-limitation the RuBisCO activity was 2.2 nmoles/min/mg protein whilst CO_2 limited growth with butyrate saw an increase to 176 nmoles/min/mg protein (table 1). The primary questions addressed here relate to the expression and synthesis of the form I and II enzymes in respose to substrate availability and growth phase - are they coordinately or differentially expressed?

Purification of the form I and form II enzymes from R. blastica was achieved by centrifugation (120,000 g for 1.5 hours at 4° C) of a cell lysate, followed by ammonium sulphate precipitation of the supernatant, gel filtration (Biogel-A-5M column) and ion exchange chromatography (DEAE Trisacryl). Sucrose gradient centrifugation was the final stage of the procedure. The characteristics of both enzymes are given in table 2. As with *R. sphaeroides* the L subunits yield different peptides on proteolysis and are immunologically non cross-reactive. The oxygenase activity (µmoles O_2 consumed/min/mg protein) showed little difference - 0.13 and 0.19. However, the K_m CO_2 at pH 7.4 for the form I enzyme was 40 µM, whilst that for the form II enzyme was 102 µM ie. form I has a significantly higher affinity for CO_2. The K_m for RuBP was between 48 - 51 uM in both instances, hence the K_m CO_2 was the primary kinetic distinction between the two enzymes.

To facilitate physiological studies of the expression and synthesis of the two enzymes a rapid quantitative assay procedure was devised the principal steps of which were sucrose gradient centrifugation followed by non denaturing polyacrylamide gel electrophoresis. This permitted the qualitative and quantitative determination of both form I and II RuBisCOs (figure 3).

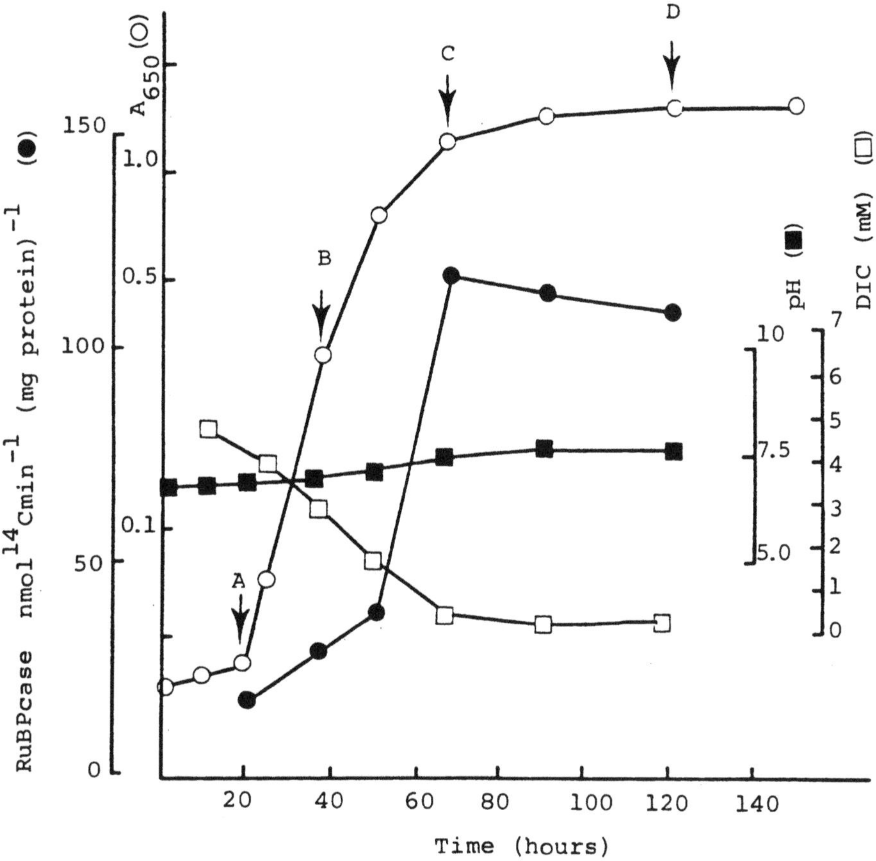

Figure 4.

Time course of synthesis of form I and form II RuBisCOs during phototrophic batch growth of *R. blastica* on 10 mM butyrate supplemented with 5 mM $NaHCO_3$. Samples were removed at the times indicated and pulse labelled (2 minutes) with ^{35}S-methionine. These were subsequently analysed for RuBisCO form I and II (Table 3) and also analysed by SDS polyacrylamide gel electrophoresis (10-30% w/v) (Figure 5).

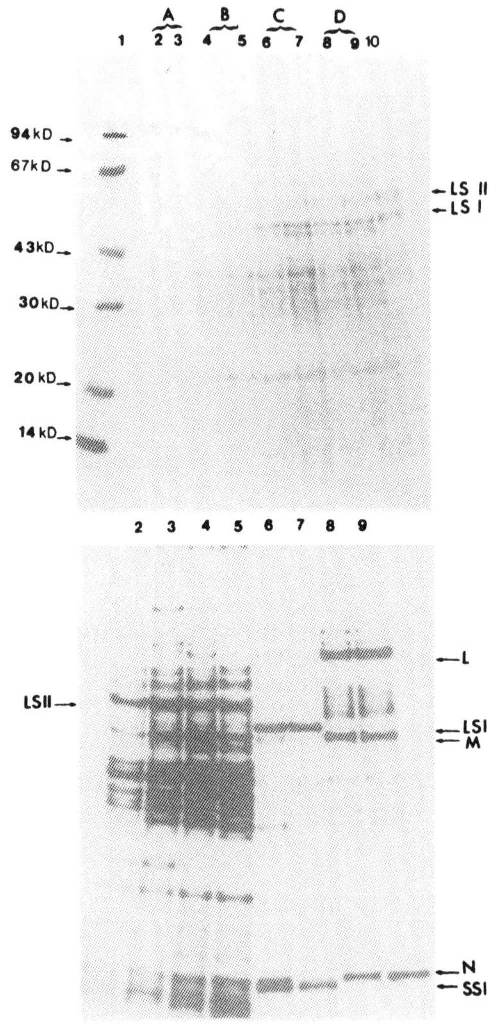

Figure 5.

(a) SDS PAGE (10-30% w/v acrylamide) of the soluble protein extracts prepared from the ^{35}S-methionine pulse labelled *R. blastica* cells from the experiment reported in Figure 4. Lane 1, molecular weight standards, 2 and 3 = A (early exponential), 4 and 5 = B (mid exponential phase), 6 and 7 = C (late exponential phase), 8 and 9 = D (stationary phase). Lane 10 is the partially purified RuBisCOs. The even numbered lanes contained equal counts of the soluble protein extract pulsed for 2 minutes, odd numbered lanes the soluble extracts after a 10 minute chase with cold methionine. Coomassie blue stained gel.

(b) Fluorograph of the above.

Analysis of the synthesis and expression of the two forms of RuBisCO during photoheterotrophic batch growth on pyruvate/malate medium showed that only the form I (L_8S_8) was present in the early exponential phase. Form II (L_6) was only detected in the mid-late growth phase and became the more abundant form in the stationary phase. This differential expression was examined further in relation to the CO_2 concentration (figure 4). A similar pattern was apparent, however, in this experiment the samples were pulse labelled with ^{35}S-methionine to ascertain the expression of both L subunits (figure 5). From this series of experiments it is apparent that the form II is synthesized when the CO_2 concentration is of the order of 1mM with the form I enzyme being derepressed when the CO_2 concentration falls below 0.05mM. Interestingly, synthesis of both enzymes is repressed in the stationary phase. Table 3 gives the levels of both enzymes relative to the total soluble protein synthesized at different growth phases in batch culture on butyrate-bicarbonate medium. There is very clear derepression of the form I enzyme at limiting CO_2 concentrations and equally apparent that form II is not under the same control ie. there would appear to be differential regulation of saynthesis of form I and II. This differential regulation was examined further by following the synthesis of the L subunits in a switch from low (0.005 mM) to high (5 mM) CO_2 concentration. The evidence is indicative that the signal for repression/derepression of form I is mediated by the CO_2 concentration. The form II L subunit was expressed only at high CO_2 concentrations. This latter series of experiments also appeared to indicate uncoupled synthesis of the form I L and S subunits.

In conclusion the available data shows that the form I and II RuBisCOs from *R. blastica* are distinctly different enzymes. The L subunits differ both immunologically and in their response to proteolytic cleavage and are most probably different gene products. From growth studies the two enzymes are differentially expressed and show variable CO_2 fixing activity dependent on the CO_2 concentration.

References

[1] Lorimer, GH. (1981) Annu. Rev. Plant Physiol. 32, 349-383
[2] Miziorko, HM and Lorimer, GH. (1983) Annu. Rev. Biochem. 52, 507-535
[3] Ellis, RJ. (1979) Trends in Biochem. Sci. 4, 241-244
[4] Ellis, RJ. and Gatenby AA. (1984) Annu. Proc. Phytochem.Soc. Eur. 23, 41-87
[5] Dijkhuizen, L and Harder, W. (1984) Ant. van Leeuwenhoek 50, 473-484
[6] Muller, ED, Chory, J. and Kaplan, S. (1985) Bacteriol. 161, 469-472
[7] Weaver KE. and Tabita, FR. (1985) J. Bacteriol. 164, 147-154
[8] Gibson, JL. and Tabita, FR. (1985) J. Bacteriol. 164, 1188-1193
[9] Jounneau, Y. and Tabita, FR. (1986) J. Bacteriol. 165, 620-624
[10] Chory, J, Muller, ED. and Kaplan, S. (1985) J. Bacteriol. 161, 307-313
[11] McFadden, BA, Torres-Ruiz, J, Daniell, H. and Sarjuni, G. (1986) Phil. Trans. R. Soc. Lond. B313, 347-358
[12] Gibson, JL. and Tabita, FR. (1977) J. Biol. Chem. 252, 943-949
[13] Gibson, JL. and Tabita, FR. (1977) J. Bacteriol. 132, 818-823
[14] Madigan, MT. and Gest, H. (1979) J. Bacteriol. 137, 524-530
[15] Gibson, JL. and Tabita, FR. (1977) J. Bacteriol. 131, 1020-1022
[16] Gibson, JL. and Tabita, FR. (1985) J. Bacteriol. 164, 1188-1193

Table 1

Growth of *Rhodopseudomonas blastica* in continuous culture under different cultural conditions

Growth conditions ($D = 0.05\ h^{-1}$)	Protein[a]	Enzyme activity[b]	$^{14}CO_2$ fixation by whole cells[c]	Dry weight ($mg.l^{-1}$)
Light anaerobic (malate limited)	5.50	9.8	43.3	408
Light aerobic (malate limited)	7.0	2.2	4.2	376
Dark aerobic (malate limited)	6.0	2.7	0.0	378
Light anaerobic (light limited)	4.7	12.7	134.0	380
Light anaerobic CO_2 limited + butyrate as substrate	5.0	176.0	546.0	409

[a] $mg.ml^{-1}$; [b] nmol $^{14}CO_2$ fixed $min^{-1}.mg\ protein^{-1}$;
[c] nmol $H^{14}CO_3$ fixed.mg dry $weight^{-1}.h^{-1}$

The basal medium was as given in Figure 2 with the stated variations.

Table 2: Characteristics of the form I and form II RuBisCos from *R. blastica*

	Form I (L_8S_8)	Form II (L_6)
Holoenzyme Mr	555,000	350,000
LS Mr	53,000	56,000
SS Mr	13,500	
Phosphogluconate (2.0 mM)	sensitive	insensitive
pH optimum	7.4	7.8 - 8.2
K_m CO_2	40 μm	102 μM

Table 3

Levels of the Form I and Form II RuBisCO relative to the total soluble protein synthesised in *R. blastica* at different growth phases in batch on butyrate-bicarbonate medium

Growth Phase	$[^{35}S]$ L-methionine incorporation into Form I and Form II RuBisCO (% of total counts in protein bands from the gel slices)	
	Form I	Form II
(A) Early log	0.01	0.06
(B) Mid-log	0.01	0.08
(C) Late log	60.8	0
(D) Stationary phase	0	0.001

AUTOTROPHIC CO_2 FIXATION IN CHEMOTROPHIC ANAEROBIC BACTERIA

Georg FUCHS, Siegfried LÄNGE, Elisabeth RUDE, Sigrid SCHÄFER, Rolf SCHAUDER, Rudolf SCHOLTZ AND Erhard STUPPERICH
Abteilung Angewandte Mikrobiologie, Universität Ulm Oberer Eselberg, D-7900 Ulm, Federal Republic of Germany

1. INTRODUCTION

An increasing number of strictly anaerobic bacteria are being found which are able to grow autotrophically with CO_2 as the only or the main source of cell carbon. They belong to different systematic groups [1], including methanogenic archaebacteria [2], gram-positive and gram-negative sulfate reducing eubacteria [3], sulfur-dependent archaebacteria [4], gram-positive and gram-negative acetogenic eubacteria [5-7], and possibly others. These anaerobes have in common that they do not use the reactions of the reductive pentose phosphate cycle (Calvin cycle) for CO_2 assimilation. Rather, two alternative pathways were found. Reviews on this subject have recently appeared [5-10].

2. AUTOTROPHIC CO_2 FIXATION VIA AN ACETYL CoA PATHWAY

Most of the autotrophic chemotrophic anaerobic bacteria were found to use an acetyl CoA pathway for CO_2 assimilation.The best studied organisms are the methanogenic *Methanobacterium thermoautotrophicum* [6, 7, 11-18] (archaebacteria), the acetogenic *Clostridium thermoaceticum* [5, 8, 19-24] (gram-positive eubacteria), and the sulfate reducing *Desulfobacterium autotrophicum* [unpublished, see also 25-27] (gram-negative eubacteria). This non-cyclic pathway appears to be used by all autotrophic species tested of the methanogenic bacteria, the acetogenic bacteria, and the sulfate reducing bacteria except for the genus *Desulfobacter*. So far, it has not been found in aerobic or facultative anaerobic microorganisms.

Acetyl CoA is formed from 2 CO_2 via two different routes leading to reduced, bound C_1-compounds at the oxidation state of carbon monoxide and methanol, respectively (Table 1). They are condensed to an activated acetyl residue which is thiolyzed with coenzyme A, rather than hydrolyzed, to give acetyl-S-Coenzyme A. Acetyl CoA synthesis from 2 CO_2, 4 H_2 and CoASH requires~1 ATP. Acetyl CoA is assimilated into cellular pyruvate synthase. Further assimilation of pyruvate requires additional carboxylation reactions. The details of the pathway vary considerably in different groups of bacteria. A unifying scheme is given in Fig. 1.

CO_2 reduction to the methanol level is brought about by several enzymes. Free formate is an intermediate in acetogenic and sulfate reducing bacteria only, but not in methanogenic bacteria. Formate is activated in an ATP-dependent reaction with different tetrahydropteridins as C_1-carriers and is then reduced to the methyl tetrahydropteridin (Table 1). In methanogenic bacteria, CO_2 reduction to methane and autotrophic CO_2 assimilation have in common the reactions and intermediates from CO_2 to

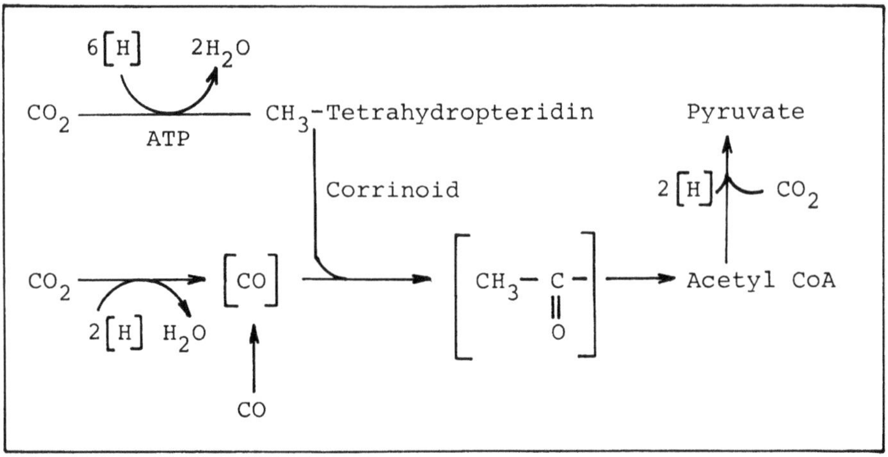

Figure 1. CO_2 fixation via the acetyl CoA pathway.

Table 1. Reactions of the acetyl CoA pathway and their thermodynamic values [7, 34]. The reactions occurring in homoacetogenic bacteria are described. The table includes some partial reactions (6a, 6b). For details see the text.

Reaction	$\Delta G_0'$ (kJ/reaction) 2[H] = H_2	$\Delta G_0'$ (kJ/reaction) 2[H] = NAD(P)H+H^+	E_0' [ox/red] (mV)
(1) $CO_2 + 2[H] \rightleftharpoons HCO_2^- + H^+$	+ 3.4	+21.5	-421
(2) $HCO_2^- + H_4folate + ATP^{4-} \rightleftharpoons N^{10}$-formyl $H_4folate + ADP^{3-} + P^{2-}$	- 8.4	- 8.4	
(3) N^{10}-formyl $H_4folate + H^+ \rightleftharpoons N^5,N^{10}$-methenyl $FH_4^+ + H_2O$	- 4.0	- 4.0	
(4) N^5,N^{10}-methenyl $H_4folate^+ + 2[H] \rightleftharpoons N^5,N^{10}$-methylene $H_4folate + H^+$	-23.0	- 4.9	-295
(5) N^5,N^{10}-methylene $H_4folate + 2[H] \rightleftharpoons N^5$-methyl $H_4folate$	-57.3	-39.2	-117
(6a) $CO_2 + 2[H] \rightleftharpoons CO + H_2O$	+20.1	+38.2	-524
(6b) N^5-methyl $H_4folate + CO + CoASH \rightleftharpoons CH_3COSCoA + H_4folate$	-21.8	-21.8	
(6) N^5-methyl $H_4folate + CO_2 + 2[H] + CoASH \rightleftharpoons CH_3COSCoA + H_4folate + H_2O$	- 1.7	+16.4	-405
(7) $2CO_2 + 8[H] + CoASH + ATP^{4-} \rightleftharpoons CH_3COSCoA + 2H_2O + ADP^{3-} + P^{2-} + H^+$	-91.0	-18.6	-296

CO_2 reduction to enzyme-bound carbon monoxide is catalyzed by the nickel enzyme CO dehydrogenase. This enzyme probably catalyzes also the C_1 addition to the acetyl intermediate and, in combination with other protein(s), its thiolysis to acetyl CoA. This enzyme complex is considered the key enzyme of this pathway. However, the reactions catalyzed by CO dehydrogenase in principle are reversible, and in some organisms CO dehydrogenase or its isoenzymes may fulfil other functions, e.g. acetyl CoA cleavage or CO oxidation. The transfer of the methyl is catalyzed by a methyltransferase and a corrinoid enzyme. Therefore, as a second characteristic feature these autotrophs exhibit high levels of different corrinoids (0.1-0.5 μmol per g dry weight), approximately 1000 times more than does *Escherichia coli.* Most of the enzymes have only been studied in acetogenic bacteria and remain to be demonstrated in the methanogenic and sulfate reducing bacteria.

3. AUTOTROPHIC CO_2 FIXATION VIA A REDUCTIVE TRICARBOXYLIC ACID PATHWAY

A few autotrophic chemotrophic anaerobic bacteria were found to use a reductive tricarboxylic acid pathway for CO_2 assimilation. This cyclic pathway has first been described in the green sulfur bacterium *Chlorobium limicola* [28-30], an anaerobic phototrophic eubacterium. It was also reported for the thermophilic Knallgasbacterium *Hydrogenobacter thermophilus* [31], an aerobic chemotrophic eubacterium. In chemotrophic anaerobes this pathway has firmly been established only in the sulfate reducing bacterium *Desulfobacter hydrogenophilus*, a gram-negative eubacterium (unpublished). Among the sulfur-dependent archaebacteria, autotrophic species occur in the aerobic, sulfur oxidizing genus *Sulfolobus*, e.g. *S. brierleyi* [32], and in the anaerobic sulfur reducing genus *Thermoproteus* [33], e.g. *T. neutrophilus.* These archaebacteria use a yet to be defined reductive carboxylic acid pathway involving reactions of the citric acid cycle. A simplified scheme of the reductive tricarboxylic acid cycle is given in Fig. 2. The reactions of the reductive tricarboxylic acid cycle are given in Table 2.

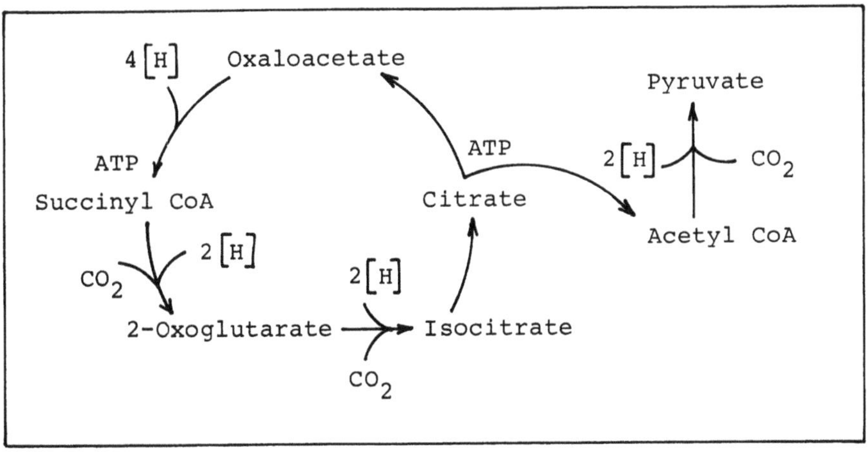

Figure 2. CO_2 fixation via the reductive tricarboxylic acid cycle.

Table 2. Reactions of the reductive tricarboxylic acid cycle and their thermodynamic values [32]. The table includes some partial reactions (8a,8b).

Reaction	$\Delta G_0'$ (kJ/reaction) $2[H] = H_2$	$\Delta G_0'$ (kJ/reaction) $2[H]$ = NAD(P)H+H$^+$	E_0' [ox/red] (mV)
(1) $Oxaloacetate^{2-} + 2[H] \rightleftharpoons Malate^{2-}$	-47.9	-29.8	-166
(2) $Malate^{2-} \rightleftharpoons Fumarate^{2-} + H_2O$	+ 3.7	+ 3.7	-
(3) $Fumarate^{2-} + 2[H] \rightleftharpoons Succinate^{2-}$	-86.0	-67.9	+ 32
(4) $Succinate^{2-} + ATP^{4-} + CoASH \rightleftharpoons Succinyl\ CoA^{-} + ADP^{3-} + P^{2-}$	+ 3.3	+ 3.3	-
(5) $Succinyl\ CoA^{-} + 2[H] + CO_2 \rightleftharpoons 2\text{-}Oxoglutarate^{2-} + CoASH + H^{+}$	+14.8	+32.9	-491
(6) $2\text{-}Oxoglutarate^{2-} + 2[H] + CO_2 \rightleftharpoons Isocitrate^{3-} + H^{+}$	- 9.6	+ 8.5	-364
(7) $Isocitrate^{3-} \rightleftharpoons cis\text{-}Aconitate^{3-} \rightleftharpoons Citrate^{3-}$	- 6.7	- 6.7	-
(8) $Citrate^{3-} + ATP^{4-} + CoASH \rightleftharpoons Acetyl\ CoA + Oxaloacetate^{2-} + ADP^{3-} + P^{2-}$	+ 5.6	+ 5.6	-
(8a) $Citrate^{3-} \rightleftharpoons Acetate^{-} + Oxaloacetate^{2-}$	+ 1.7	+ 1.7	-
(8b) $Citrate^{3-} + CoASH + H^{+} \rightleftharpoons Acetyl\ CoA + Oxaloacetate^{2-} + H_2O$	+37.4	+37.4	-
(9) $2CO_2 + 8[H] + CoASH + 2ATP^{4-} \rightleftharpoons CH_3COSCoA + H_2O + 2ADP^{3-} + 2P^{2-} + 2H^{+}$	-122.8	-50.4	-255

Citrate is formed, starting from oxaloacetate, 2 CO_2 and 8 reducing equivalents (4 H_2). A characteristic enzyme is 2-oxoglutarate synthase, catalyzing the reductive carboxylation of succinyl CoA to 2-oxoglutarate. Citrate is cleaved by ATP citrate lyase to acetyl CoA and oxaloacetate thus regenerating the starting molecule of the cycle. This reaction requires ATP cleavage to ADP + P. ATP citrate lyase may be considered a key enzyme of this pathway. Acetyl CoA synthesis from 2 CO_2, 4 H_2, and CoA requires ~2 ATP. Further assimilation of acetyl CoA requires additional carboxylation reactions and proceeds via reductive carboxylation to pyruvate.

Acknowledgements: The research of our group was supported by a grant from the Deutsche Forschungsgemeinschaft and by the Fonds der Chemischen Industrie.

REFERENCES

[1] Schleifer KH, Stackebrandt E (eds). (1985) Evolution of Prokaryotes. London. Academic Press.
[2] Balch WE, Fox GE, Magrum LJ, Woese CR, Wolfe Rs. (1979). Microbiol. Rev. 43, 260-296.
[3] Widdel F, Pfennig N. (1984) *in* Bergey's Manual of Systematic Bacteriology (Krieg NR, Holt JG, eds), Vol. 1, pp. 663-679. Williams and Wilkins, Baltimore.
[4] Fischer F, Zillig W, Stetter KO, Schreiber G. (1983) Nature 301, 511-513.
[5] Ljungdahl LG, Wood HG. (1982) *in* Vitamin B_{12} (Dolphin C, ed) pp. 165-202. Wiley, New York.
[6] Zeikus JG. (1983) Adv. Microb. Physiol. 24, 215-299.
[7] Fuchs G. (1986) FEMS Microbiol. Rev. 39, 181-213

[8] Wood HG, Ragsdale SW, Pezacka E. (1986) Trends Biochem. Sci. 11, 14-18.
[9] Fuchs G, Stupperich E. (1986) Syst. Appl. Microbiol. 7, 364-369.
[10] Diekert G, Fuchs G, Thauer RK. (1985) *in* Microbial Gas Metabolism: Mechanistic, Metabolic and Biotechnological Aspects (Poole RK, Daw CS, eds) pp. 115-13f0. Society for General Microbiology.
[11] Taylor GT, Kelly DP, Pirt SJ. (1976) *in* Proc. Symp. Microbial Production and Utilization of Gases (H_2, CH_4, CO) (Schlegel HG, Gottschalk G, Pfennig N , eds) pp. 173-180. Goltze Verlag, Göttingen.
[12] Zeikus JG, Fuchs G, Kenealy W, Thauer RK, (1977) J. Bacteriol. 132, 604-613.
[13] Fuchs G,j Stupperich E, Thauer RK. (1978) Arch. Microbiol. 117, 61-66.
[14] Daniels L, Zeikus JG. (1978) J. Bacteriol. 136, 75-84.
[15] Stupperich E, Fuchs G. (1984) Arch. Microbiol. 139, 8-13.
[16] Stupperich E, Fuchs G. (1984) Arch. Microbiol. 139, 14-20.
[17] Holder U, Schmidt DE, Stupperich E, Fuchs G. (1985) Arch. Microbiol. 141, 229-238.
[18] Länge S, Fuchs G. (1986) Eur. J. Biochem. Submitted.
[19] Diekert GB, Thauer RK, (1978) J. Bacteriol. 136, 597-606.
[20] Drake HL, Hu SI, Wood HG. (1981) J. Biol. Chem. 256, 11137-11144.
[21] Hu SI, Drake HL, Wood HG. (1982) J. Bacteriol. 149, 440-448.
[22] Ragsdale SW, Wood HG. (1985) J. Biol. Chem. 260, 3970-3977.
[23] Ragsdale SW, Clark JE, Ljungdahl L.Lundie LL, Drake HL. (1983) J. Biol. Chem. 258, 2364-2369.
[24] Pezacka E, Wood HG. (1986) J. Biol. Chem. 261, 1609-1615.
[25] Jansen K, Thauer RK, Widdel F, Fuchs G. (1984) Arch. Microbiol. 138, 257-262.
[26] Jansen K, Fuchs G, Thauer RK. (1985) FEMS Microbiol. Lett. 28, 311-315.
[27] Schauder R, Eikmanns B, Thauer RK, Widdel F, Fuchs G. (1986) Arch. Microbiol. in press.
[28] Evans MCW,Buchanan BB, Arnon DI. (1966) Proc. Natl. Acad. Sci. USA, 55, 928-934.
[29] Sirevåg R. (1975) Arch. Microbiol. 104, 105-111.
[30] Ivanovsky RN, Sintsov MV, Kondratieva EM. (1980) Arch. Microbiol. 128, 239-241.
[31] Shiba H, Kawasumi T, Igarashi Y, Kodama T, Minoda Y. (1985) Arch. Microbiol. 141, 198-203.
[32] Kandler OK, Stetter KO. (1981) Zbl. Bakt. Hyg. I Abt.Orig. C2, 111-121.
[33] Schäfer S, Barkowski C, Fuchs G. (1986) Arch. Microbiol. submitted.
[34] Thauer RK, Jungermann K, Decker K, (1977) Bacteriol. Rev. 41, 100-180.

AEROBIC AND ANAEROBIC EXTREMELY THERMOPHILIC AUTOTROPHS

R. HUBER, G. HUBER, A. SEGERER, J. SEGER, and K.O. STETTER
Lehrstuhl für Mikrobiologie, Universität Regensburg, 8400 Regensburg, Federal Republic of Germany.

1. INTRODUCTION

During the last years, various extremely thermophilic bacteria with optical growth temperatures above 80°C have been isolated from geo- and hydrothermally heated areas [14, 15, 17]. All of these organisms belong to the archaebacteria, the third kingdom of life [19]. The mode of nutrition is lithoautotrophic or organotrophic, depending on the isolates. In this paper, lithoautotrophic extreme thermophiles including some very recent isolates are described which are the primary producers of organic matter at high temperatures.

2. BIOTOPES

Up till now, all extremely thermophilic autotrophic bacteria have been isolated from terrestrial solfataric fields and submarine hydrothermal systems. Both biotopes are situated usually above magma chambers about 400 to 2000 m within the floor [10]. From there, steam consisting of water and volcanic gasses, mainly CO_2, H_2, SO_2 and H_2S, escape and are heat the soil and surface water. The examination of soil profiles within solfataric fields showed that these soils consist of two layers with different properties: On the top, there is an oxygen-containing strongly acidic layer of about 15 to 30 cm in thickness, which shows an ochre colour due to the presence of ferric iron. Below, there is an oxygen-free, slightly acidic to neutral bluish-black zone, which contains ferrous sulfide. Submarine hydrothermal systems contain seawater of neutral to slightly acidic pH which remains liquid even above 100°C due to the prevention of boiling by the hydrostatic pressure. Geothermal as well as hydrothermal areas are abundant in elemental sulfur which is formed by the oxidation of H_2S and by the reaction of H_2S with SO_2. Seawater contains about 3 mmoles of sulfate per litre.

3. ACIDOPHILIC AEROBIC AND FACULTATIVELY ANAEROBIC AUTOTROPHS

The upper oxidized layer within solfataric fields including solfataric mudholes contains coccoid lobe-shaped strictly aerobic organisms which belong to the genus *Sulfolobus* which was described by T.D. Brock [4]. The type species *Sulfolobus acidocaldarius* and most other isolates can gain their energy

Table 1: Properties and energy sources of aerobic and facultatively aerobic S-metabolizing archaebacteria

Species	DNA mol% GC	T_{max} (^0C)	Ore mixture G1	Ore mixture G1 + YE	S^0	S^0 + YE	YE	Sucrose	Anaerobic growth $H_2 + S^0$
Sulfolobus acidocaldarius DSM 639	37	85	-	-	-	+	+	+	-
Sulfolobus solfataricus DSM 1616	36	87	-	+ (no acidification !)	-	+	+	+	-
TH 2	45	80	+	n.d.	+	+	+	-	-
Kra 23	38	70	+	n.d.	+	+	-	-	-
VE 2	37	85	+	n.d.	-	-	-	-	-
VE 6	31	95	+	n.d.	-	+	+	-	-
HV 5	30	95	+	n.d.	-	-	-	n.d.	-
Acidianus brierleyii DSM 1651	31	75	-	+	-	+	+	-	+
Acidianus infernus DSM 3191	31	95	-	-	-	+	-	-	+

n.d. = not determined

chemolithotrophically by oxidation of S^o, forming H_2SO_4. Some isolates (Tab. 1) are able to grow lithoautotrophically in S^o-containing mineral medium in the presence of CO_2 and oxygen, others require the addition of traces of yeast extract (e.g. 0.01%) , the function of which is not yet clear (source of minerals or of organic carbon). Alternatively both *Sulfolobus*-species and many *Sulfolobus*-shaped isolates (Tab. 1) grow without S^o, organotrophically on yeast extract, sugars, peptone and amino acids [3, 4, 15]. Many *Sulfolobus*-isolates are able to oxidize ferrous iron [2, 5]. However, due to extensive precipitates of jarosite, even at pH 2 and in uninoculated controls, no significant growth could be observed [3, Huber and Stetter, unpublished]. *Sulfolobus brierleyi*, which has now been renamed*Acidianus brierleyi* [12] and some still unnamed *Sulfolobus*-shaped isolates [8] are moderate thermophiles, growing at temperatures up to 70°C. They are able to grow on sulfidic ores, forming sulfuric acid and solubilizing heavy metals. New ore-leaching bacteria could be isolated aerobically in mineral medium in the presence of ore mixture G1, consisting of pyrite, chalcopyrite, sphalerite and pitchblende [Tab. 1; 6]. Isolates HV5 and VE2 from samples of Hveragerthi and Hveravellir in Iceland are strictly dependent on ores and can not grow on S^o. Strains HV5 and VE6 are the most extremely ore-leaching bacteria known, growing at temperatures up to 95°C (Tab. 1). In contrast, *Sulfolobus acidocaldarius* and *Sulfolobus solfataricus* are unable to grow by the oxidation of sulfidic ores.

Recently, the new genus *Acidianus* was recognized (tentatively named *'Acidothermus"*, [14, 17]. Members have been isolated from acidic solfatara fields in Italy, Iceland, the Azores and the United States, and from a submarine hydrothermal vent in Italy [11, 12]. Although the cells of these thermoacidophiles have a coccoid, lobed shape similar to *Sulfolobus* (Fig. 1), they are different in DNA composition (G+C-content; DNA homology) and in metabolic properties.

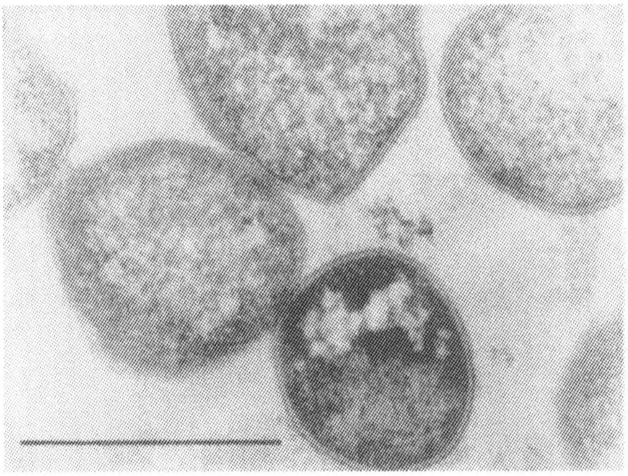

Figure 1. Thin section of *Acidianus infernus* contrasted with lead citrate and uranyl acetate.Bar 1 μm.

Table. 2: Anaerobic growth of autotrophic S-dependent archaebacteria on H_2, yeast extract and different sulfur compounds

Mode of growth	Electron donor	Electron acceptor (?)	H_2S *) (µmoles/10^8 cells)	Species
			54	*Acidianus (S.) brierleyi*
			68	*Acidianus infernus*
			122	*Thermoproteus neutrophilus*
lithoautotrophic	H_2	S^0	54	*Thermoproteus tenax*
			63	*"Pyrobaculum islandicum"*
			158	*Pyrodictium occultum*
			237	*Pyrodictium brockii*
		S^0	2	
		$S_2O_3^{2-}$	0	*"Pyrobaculum islandicum*
		SO_3^{2-}	2	
organotrophic	YE			
		S^0	2	
		$S_2O_3^{2-}$	5	*Thermoproteus tenax*
		SO_3^{2-}	2	

*) corrected by the uninoculated control

Acidianus infernus, the first species to be recognized [12], is able to grow aerobically by S^o-oxidation, forming sulfuric acid. Surprisingly, this organism is also able to grow under extremely anaerobic conditions by S^o-reduction: H_2S is formed from H_2 and S^o (Tab, 1, 2). *Sulfolobus brierleyi* also processes this previously unrecognized metabolic versatility, and was therefore considered to also belong to the genus *Acidianus* [11, 12]. A further isolate, which contains a prophage (W. Zillig, pers. comm.) and had been tentatively named "*Sulfolobus ambivalens*" [21] exhibits 60% DNA homology with the type species, *Acidianus infernus* , indicating that it is another species of the genus *Acidianus*.

4. NEUTROPHILIC AND SLIGHTLY ACIDOPHILIC ANAEROBIC AUTOTROPHS

From the lower reduced zone of solfataric fields, extremely thermophilic methanogens and S^o-dependent autotrophs could be isolated. Previously, the methanogens could be only obtained from Icelandic solfataric fields. They are rod-shaped, Gram-positive lithoautotrophs, growing by formation of methane from H_2 and CO_2 at temperatures up to 97°C. Until now, the two species *Methanothermus fervidus* and *Methanothermus sociabilis* (Fig. 2) are the only known representatives of *Methanoathermaceae* [7,8].

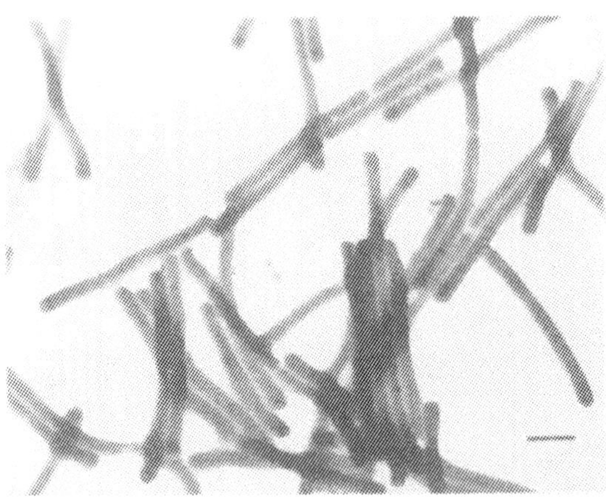

Figure 2. Electron micrograph of *Methanothermus sociabilis*, platinum-shadowed. Bar 1 µm.

The S^o-dependent autotrophs are rod-shaped organisms which divide by budding [20]. Until now, members of the two new genera *Thermoproteus* and "*Pyrobaculum*" (Fig. 3) could be isolated growing lithoautotrophically on S^o, H_2 and CO_2, gaining energy by H_2S formation (Tab. 2). "*Pyrobaculum*" has a G+C-content of only 46 mol% which is 10% lower than that of *Thermoproteus* [20]. In addition, isolates belonging to "*Pyrobaculum*" show an optimal growth temperature of 100°C which is 12°C higher than that of the *Thermoproteus*-isolates [17, 20]. Alternatively, "*Pyrobaculum islandicum*" and *Thermoproteus tenax* are able to grow without H_2 organotrophically on yeast extract in the presence of S^o, $S_2O_3^{2-}$ and SO_3^{2-} (Tab. 2).

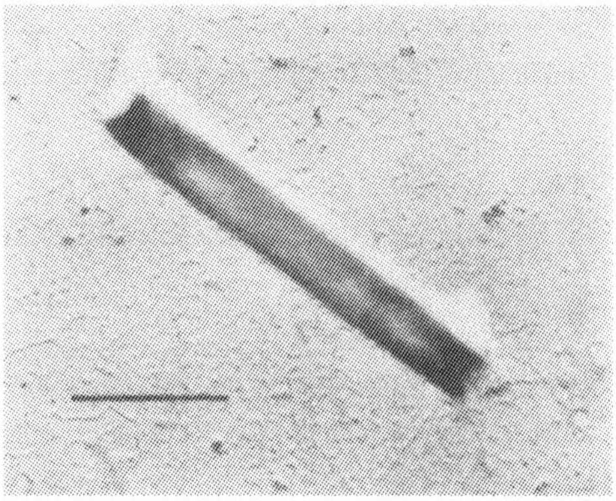

Figure 3. Electron micrograph of "*Pyrobaculum islandicum*" Geo 3, platinum-shadowed. Bar 1 µm.

During organotrophic growth, only low amounts of H_2S are formed compared with lithotrophic growth of these organisms. In the case of "*Pyrobaculum Islandicum*", organotrophic growth is also obtained when S^o is replaced by cysteine. It is most likely, these bacteria are growing organotrophically by an

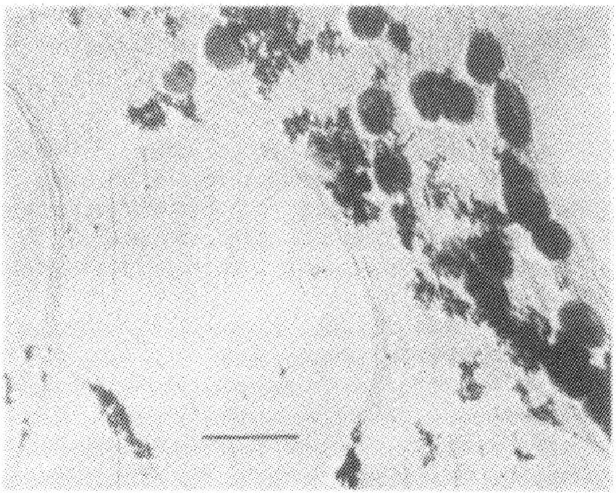

Figure 4. Electron micrograph of *Pyrodictium occultum*, platinum-shadowed. Bar 5 µm.

unknown fermentation rather than by sulfur respiration.

From the submarine hydrothermal vents of Vulcano, Italy, the most extremely thermophilic organisms known until now have been isolated. They have an optimal growth temperature of 105°C, and a maximum at around 110°C [under slight overpressure, 13]. These disc- to dish-shaped cells form unusual networks of fibers connecting the cells (Fig.4).

The new genus "*Pyrodictium*" currently contains the species *Pyrodictium occultum* and *Pyrodictium brockii*, which are strict H_2/S-autotrophs (Tab. 2). During growth, the fiber networks cover the sulfur like cobwebs [13, 16].

5. ISOLATION OF EXTREMELY THERMOPHILIC SULFATE-REDUCING ARCHAEBACTERIA

Until now, no archaebacteria thriving by dissimilatory sulfate reduction were known [9]. During our recent in situ enrichments within the hydrothermal systems close to Vulcano, Italy, coccoid to disc-shaped irregular cells could be isolated. These grew anaerobically without S^o at temperatures between 65 and 95°C with an optimum at around 83°C. They use formate, L(+) and D(-) lactate and, less efficiently, glycogen, starch, yeast extract, meat extract and bacterial homogenates as substrates. Sulfate, thiosulfate and sulfite are used as electron acceptors, while S^o is inhibitory for growth. As end products, H_2S, CO_2, and traces of methane (about 4 mmoles/l) could be detected. Growth is inhibited by 0.1 mmoles/l of Na-molybdate. Under the UV microscope, the cells show a blue-green fluorescence due to the presence of F_{420} [1]. In addition, methanopterin, but not F_{430} [A. Pfaltz, pers. comm.]or coenzyme M could be detected. The RNA polymerase of isolate VC-16 shows homology to subunits B' and B" of the methanogens, but is unique in serological cross-reaction of the heaviest subunit with components A and C of methanogens and S-dependent archaebacteria, indicating an outstanding position for the new isolates within the phylogenetic tree of the archaebacteria.

References

[1] Balch, W.E., G.E. Fox, L.J. Magrum, C.R. Woese, and R.S. Wolfe. (1979) Methanogens: Reevaluation of a unique biological group. Microbiol. Rev. 43: 260-296.

[2] Brierley, C.L., and J.A. Brierley. (1973) A chemolithoautotrophic and thermophilic microorganismisolated from an acidic hot spring. Can. J. Microbiol. 19: 183-188.

[3] Brock, T.D. (1981) Thermophilic microorganisms and life at high temperatures. Springer, Berlin, Heidelberg, New York.

[4] Brock, T.D., K.M. Brock, R.T. Belly, and R.L. Weiss. (1972) *Sulfolobus*: a new genus of sulfur-oxidizing bacteria living at low pH and high temperature. Arch. Microbiol. 84: 54-68.

[5] Brock, T. D., S. Cook, S. Peterson, and J.L. Mosser. (1976) Biochemistry and bacteriology of ferrous iron oxidation in geothermal habitats. Geochim. Cosmocim. Acta 40: 493-500.

[6] Huber, G., H. Huber, and K.O. Stetter. (1986) Isolation and characterization of new metal-mobilizing bacteria. Biotechnology and Bioengineering Symp. No. 16. J. Wiley & Sons, Inc., New York, London, Sydney, Toronto.

[7] Lauerer, G., J.K. Kristjansson, T.A. Langworthy, H. König, and K. O. Stetter. (1986) *Methanothermus sociabilis* sp. nov., a second species within the *Methanothermaceae* growing at 97°C. System. Appl. Microbiol. 8: 100-105.

[8] Marsh, R.M., P.R. Norris, and N.W. Le Roux. (1983) Growth and mineral oxidation studies with *Sulfolobus*, p. 71-81. In: G. Rossi, and A.E. Torma (eds.), Associazione Mineraria Sarda, Iglesias, Italy.

[9] Pfennig, N., and F. Widdel. (1982) The bacteria of the sulfur cycle. Phil. Trans. R. Soc. Lond. B 298: 433-441.

[10] Pichler, H. (1981) Italienische Vulkan-Gebiete III. Sammlung Geologischer Führer. Vol. 69. Gebrüder Borntraeger, Berlin, Stuttgart.

[11] Segerer, A., K.O. Stetter, and F. Klink. (1985) Two contrary modes of chemolithotrophy in the same archaebacterium. Nature 313: 787-789.

[12] Segerer, A., A. Neuner, J.K. Kristjansson, and K.O. Stetter. (1986) *Acidianus infernus* gen. nov., sp. nov., and *Acidianus brierleyi* comb. nov.: facultatively aerobic, extremely acidophilic thermophilic sulfur-metabolizing archaebacteria. Int. J. Syst. Microbiol. (in press).

[13] Stetter, K.O. (1982) Ultrathin mycelia-forming organisms from submarine volcanic areas having an optimum growth temperature of 105°C. Nature 300: 258-260.

[14] Stetter, K.O. (1986) Diversity of extremely thermophilic archaebacteria, p. 39-74. In: T.D. Brock (ed.), Thermophiles: General, Molecular and Applied Microbiology. J. Wiley & Sons, Inc., New York, London, Sydney, Toronto.

[15] Stetter, K.O., and W. Zillig. (1985) *Thermoplasma* and the thermophilic sulfur-dependent archaebacteria, p. 85-170. In: R.S. Wolfe and C.R. Woese (eds.), The Bacteria, Vol. VIII. Academic Press, New York.

[16] Stetter, K.O., H. König, and E. Stackebrandt. (1983) *Pyrodictium* gen. nov., a new genus of submarine disc-shaped sulfur-reducing archaebacteria growing optimally at 105°C. Syst. Appl. Microbiol. 4: 535-551.

[17] Stetter, K.O., A. Segerer, W. Zillig, G. Huber, G. Fiala, R. Huber, and H. König. (1986) Extremely thermophilic sulfur-metabolizing archaebacteria. System. Appl. Microbiol. 7: 393-397.

[18] Stetter, K.O., M. Thomm, J. Winter, G. Wildgruber, H. Huber, W. Zillig, D. Janecovic, H. König, P. Palm, and S. Wunderl. (1981) *Methanothermus fervidus*, sp. nov., a novel extremely thermophilic methanogen isolated from an Icelandic hot spring. Zbl. Bakt. Hyg., I. Abt. Orig. C 2: 166-178.

[19] Woese, C.R., L.J. Magrum, and G.E. Fox. (1978) Archaebacteria. J. Mol. Evol. 11: 245-252.

[20] Zillig, W., K.O. Stetter, W. Schäfer, D. Janekovic, S. Wunderl, I. Holz, and P. Palm. (1981) *Thermoproteales*: a novel type of extremely thermoacidophilic anaerobic archaebacteria isolated from Icelandic solfataras. Zbl. Bakt. Hyg., I. Abt. Orig. C 2: 205-227.

[21] Zillig, W., S. Yeats, I. Holz, A. Böck, F. Gropp, M. Rettenberger, and S. Lutz. (1985) Plasmid-related anaerobic autotrophy of the novel archaebacterium *Sulfolobus ambivalens*. Nature 313: 789-791.

PHYSIOLOGY AND BIOCHEMISTRY OF METHANOGENS AND FERMENTATIVE C_1 UTILIZERS

BIOCHEMICAL ASPECTS OF METHANE FORMATION IN *METHANOBACTERIUM THERMOAUTOTROPHICUM*

Kenneth M. NOLL, Mark I. DONNELLY*, AND Ralph S. WOLFE
Department of Microbiology, University of Illinois, Urbana, Illinois 61801, USA
*Present address: Amoco Corporation, Amoco Research Center, Naperville, IL 60566

1. INTRODUCTION

The biochemistry of methanogenesis is characterized by the involvement of unique coenzymes [1]. Consequently, investigation of the enzymology of the pathway by which carbon dioxide is reduced to methane has been severely constrained. Until recently, only a single substrate of the pathway was known, 2-(methylthio)ethanesulfonate (CH_3-S-CoM).The methylreductase system which catalyzes the reduction of CH_3-S-CoM to methane has been extensively studied, but here, too, lack of knowledge of the components involved has hampered progress. The enzyme system consists of at least four proteins and displays an absolute requirement for a heat-stable cofactor, component B [2]. We have recently established that component B is 7-mercaptoheptanoylthreonine phosphate (HS-HTP) [3]. This structure was proposed based on analysis of the authentic cofactor and was confirmed by its complete chemical synthesis.

With respect to the remainder of the pathway, the identification of methanofuran (MFR, 4), tetrahydromethanopterin (H_4MPT, 5), and C_1 derivatives of these coenzymes [6, 7] has provided the missing substrates. It is now possible to formulate a complete pathway for the reduction of carbon dioxide to methane via specific intermediates (Fig. 1). We have purified and partially characterized three enzymes involved in the interconversion of H_4MPT derivatives. We have also established another derivative, 5-formyl-H_4MPT, as an essential intermediate.

2. ENZYMOLOGY OF C_1 INTERCONVERSIONS

The model of carbon flow to methane as originally formulated was based entirely on the biological accumulation of two of the C_1 intermediates, formylmethanofuran and methenyl-H_4MPT [6, 8] and the conversion of synthesized derivatives to methane [7]. Only cursory enzymological data was available to support any of the proposed reactions. Over the past year, we have pursued the enzymology of the C_1 cycle. These investigations have led to the purification to homogeneity of three new enzymes and the characterization of the reactions they carry out [9, 10, 11]. A major consequence of this work has been the identification of another intermediate of the pathway, 5-formyl-H_4MPT, a compound not originally recognized as an intermediate of methanogenesis. The reactions catalyzed by these three enzymes effect the overall conversion of formyl-MFR to methylene-H_4MPT (Reactions 2,3, and 4 of Fig. 1).

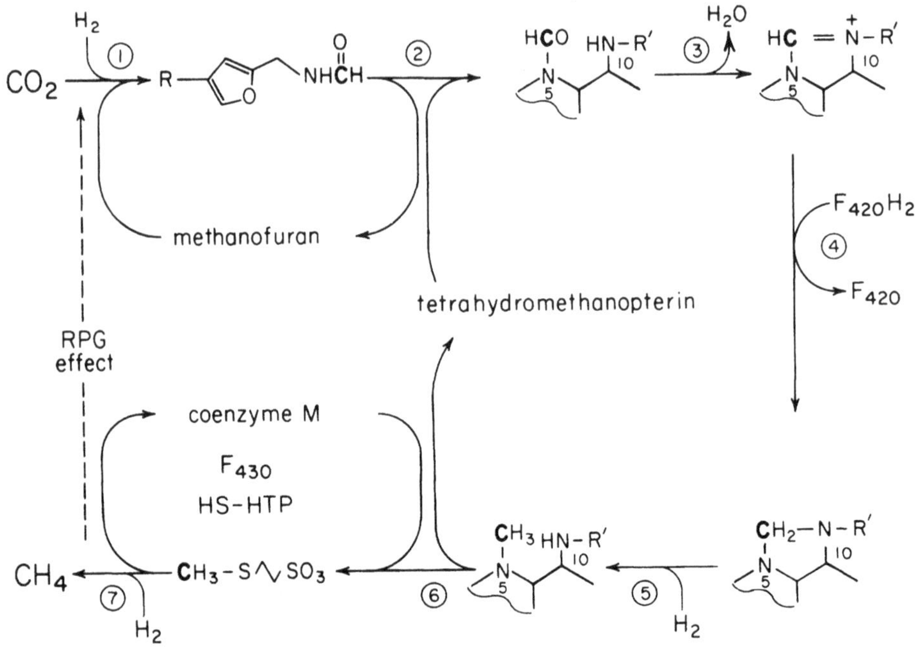

Figure 1. Proposed conversions of C_1 units during methane formation from carbon dioxide. The numbered reactions are catalyzed by 1, an uncharacterized formyl-MFR synthtase; 2, formyl-MFR:H_4MPT formyltransferase; 3, methenyl-H_4MPT cyclohydrolase; 4, methylene-H_4MPT:F_{420} oxidoreductase; 5, methylene-H_4MPT reductase; 6, methyl-H_4MPT:HS-CoM methyltransferase; and , 7, CH_3-S-CoM methylreductase.

Table 1. Reconstitution of methanogenesis from CO_2.

Fraction	Methane from CO_2 (nmol)
Cell extract	2865
Phenyl-Sepharose pass-through[a]	157
Pass-through plus purified formyltransferase[b]	1205

[a]1.0 ml cell extract was passed through a Phenyl-Sepharose column(1 x 7 cm) equilibrated with 25mM potassium phosphate buffer, pH 7.0, containing 10% ethylene glycol and 10mM 2-mercaptoethanol. Protein not retained was concentrated to a final volume of 1.0 ml.
[b]Homogeneous formyltransferase was added to the concentrated pass-through material to give a fractional concentration of protein equal to that estimated to be present in the initial cell extract (0.4%:) based on the overall purification (256-fold).

Formyl-MFR:H_4MPT formyltransferase catalyzes the transfer of a formyl group to give 5-formyl-H_4MPT [10]. The enzyme is relatively hydrophobic, and the homogeneous enzyme appears to consist of a tetramer of subunits of M_r 41,000. Due to its hydrophobicity, it was possible to bind the formyltransferase selectively to a Phenyl-Sepharose column. The pass-through material, which comprised the vast majority of the protein applied to the column, converted CO_2 to methane very poorly. Addition of the homogeneous formyltransferase to that fraction restored to a significant extent the ability to reduce CO_2 to methane (Table 1), thereby establishing the essentiality of this reaction to methanogenesis from CO_2.

The subsequent reaction of the pathway, dehydration of 5-formyl-H_4MPT to give methenyl-H_4MP, was most conveniently assayed in the reverse direction, hydrolysis of methenyl-H_4MPT [12]. The product of the formyltransferase (Fig. 1, reaction 2) was shown to be identical to that formed by the methenyl-H_4MPT cyclohydrolase (Fig. 1, reaction 3 functioning in the opposite direction from that shown) by a number of criteria [10]. The products of both reactions co-eluted using HPLC, and the spectra of the isolated derivatives were identical . Both were stable to exposure to air, unlike free H_4MPT or its 10-formyl derivative [12, 13] Further, the product of the reaction catalyzed by the homogeneous formyltransferase served as a substrate for a highly purified preparation of the cyclohydrolase [10]. The cyclohydrolase has recently been purified to homogeneity and shown to consist of a dimer of subunits of M_r 41,000. The reaction was shown to be energy-independent, in contrast to the analogous conversion of 5-formyl-H_4folate to methenyl-H_4folate [11].

The assignment of the product of these reactions as the 5-formyl as opposed to 10-formyl derivative of H_4MPT was based on analogy to the corresponding derivates of H_4folate [12]. The derivative was similar to 5-formyl-H_4folate in its UV spectrum and its lack of high sensitivity to oxygen. ^{13}C NMR data also supported the assignment. A derivative of methenyl-H_4MPT was prepared with ^{13}C in the methenyl position. The product of enzymatic hydrolysis of this compound possessed an NMR spectrum virtually identical to that reported for 5-formyl-H_4folate. Like methenyl-H_4folate, methenyl-H_4MPT is susceptible to non-enzymatic hydrolysis, but in this case the product is the 10-formyl derivative as it is in the case of the analogous folate reaction [12].

Methylene-H_4MPT dehydrogenase catalyzes the oxidation of the methylene to the methenyl derivative (Fig. 1, reaction 4 functioning in the reverse direction). In crude extracts the reaction proceeded without the addition of an electron donor. By purification of the enzyme to homogeneity, it was possible to establish that coenzyme F_{420} was the physiological electron acceptor [9]. In the absence of added acceptor, no activity was detected. Alternative electron acceptors gave less than two percent of the rate obtained with coenzyme F_{420} (Table 2).

We further demonstrated that the reaction stoichiometry was 1 : 1, as expected since coenzyme F_{420} functions as a two electron donor [14]. Also, we showed that the reaction was reversible; reduced coenzyme F_{420} ($F_{420}H_2$) served as electron donor for the reduction of methenyl-H_4MPT. At physiological pH, the equilibrium favors methenyl-H_4MPT [9]. Due to its abundance and low potential,

Table 2. Activity of electron acceptors of the methylene-H_4MPT dehydrogenase.

Electron acceptor	Relative activity (%)
Coenzyme F_{420}	100.0
NAD	1.0
NADP	0.9
FAD	1.7
FMN	1.3

coenzyme F_{420} has long been attributed a role as an electron donor in methanogenesis, but to date the above data constitute the only rigorous demonstration that it can serve as an electron donor in the reduction of a C_1 derivative of the methanogenic pathway. Reactions 5 and 6 (Fig. 1) are speculative.

3. STRUCTURE AND SYNTHESIS OF HS-HTP

The reduction of methylcoenzyme M to methane is carried out by a complex enzymic system composed of at least four enzymes. Two of these, components C and A2, have been purified to homogeneity [15, 16]. Component C has been proposed to be the site of the methyl reduction to methane [17]. The role of A2 has not been established. Coenzyme F_{420} reductase activity is present in the A1 fraction [2]. The final fraction A3, is required for activity, but no role has been assigned to it yet.

In addition to these enzyme fractions, a number of cofactors are required for this seemingly simple reaction. Although many cofactors have been identified, with the exception of CH_3-S-CoM, their roles in the methyıreductase reaction are unclear. ATP and Mg(II) are essential, but in catalytic amounts. These are apparently required for activation [18]. CH_3-S-CoM acts as the methyl donor [19]. FAD was found to be required in the more resol ved enzymic system [2]. Coenzyme F_{420} can stimulate the rate of methane formation, but reduced F_{420} was not found to act as an electron source [2]. Cyanocobalamin also stimulates the rate of methane formation (20). Finally, a unique nickel tetrapyrrole (F_{430}) was found to be an integral part of component C [21].

When the methylreductase system was first resolved into three components, a new cofactor requirement was revealed. Component B was found to be a heat-stable, low molecular weight cofactor required when cell extracts were fractionated by ion-exchange chromatography [22]. Cell extracts stripped of soluble cofacters by gel filtration were also dependent upon this cofactor for methanogenesis from CH_3-S-CoM. Subsequent investigation has revealed that component B is 7-mercaptoheptanoylthreonine phosphate (HS-HTP). Details of this work have been presented elsewhere [3, 23] .

Component B was purified by a series of anion-exchange and reversed-phase HPLC chromatography

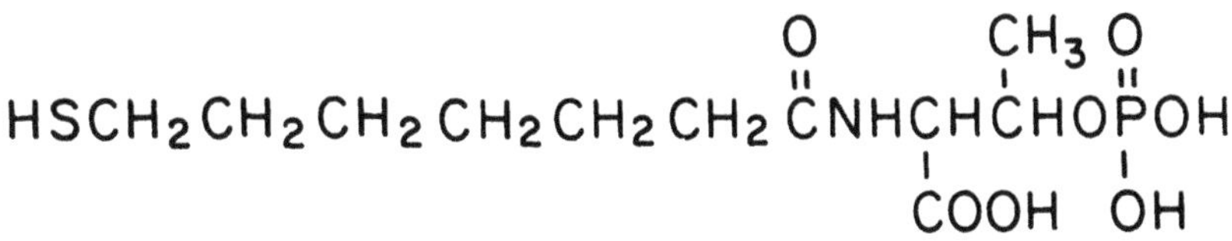

7-mercaptoheptanoylthreonine phosphate
(HS-HTP)

steps (3). This procedure gave a compound that was judged pure since it gave a single peak upon elution from two types of HPLC columns each using a different means of detection. This purification yielded 9 mg of pure material from 1 kg wet weight of cells.

Chemical analysis of component B suggested one phosphate per molecule if its molecular weight were 575 (triethlammonium salt). Fast-atom bombardment mass spectrometry (FABMS) revealed the compound had a molecular weight of 343 as the free acid. It also indicated the presence of sulfur in the molecule. There appeared to be losses of both dithiotreitol (from the matrix) and 2-mercaptoethanol (HSEtOH), providing further evidence for the presence of sulfur and suggesting it was present as a mercaptan. The elemental formula, as determined by high resolution FABMS was $C_{11}H_{22}NO_7PS$.

The ^{1}H NMR spectrum proved to be unexpectedly simple (Fig 2). The two triplets at 2.89 and 3.87 ppm were consistent with the presence of HSEtOH. The couplings of the signals at 1.37, 1.44, 1.64, 1.70, 2.35 and 2.78 ppm indicated a chain of six methylenes. The shift of the 2.78 ppm triplet and the shift and splitting of the 2.35 ppm multiplet were consistent with methylenes bound to a disulfide and a carbonyl group, respectively. The methyl at 1.31 ppm and the methines at 4.2 and 4.7 ppm were found to be related through coupling and were identical to those observed for N-acetylphosphothreonine (24). This is presumably bound to the mercaptopheptanoyl moiety by an amide linkage.This was supported by a 1640 cm^{-1} signal observed in the IR spectrum of component B.

Based on these data, we proposed that component B as isolated was the mixed disulfide of HSEtOH and 7-mercaptoheptanoylthreonine phosphate. This structure was verified by chemical synthesis. 7-mercaptoheptanoate was prepared from diethylpimelate [23] using a method employed for the synthesis of lipoic acid [25]. The diethylpimelate was converted to the mono ester, mono acid chloride, which was reduced to ethyl 7-hydroxheptanoate. 7-Mercaptoheptanoate was synthesized by refluxing the hydroxy acid with thiourea in hydrobromic acid. The resulting mercaptan was oxidized with iodine in the presence of HSEtOH to give the mixed disulfide of HSEtOH and 7-mercaptoheptanoate. This mixed disulfide was condensed with phosphothreonine using dicyclohexylcarbodiimide and N-hydroxy-succinimide to give the mixed disulfide of HSEtOH and HS-HTP. This was found to be active in the *in vitro* methanogenic assay and it gave the same ^{1}H NMR spectrum as authentic component B. It was further demonstrated that the HSEtOH moiety could be removed from the compound with full retention of the activity. Thus, the HSEtOH moiety was probaly an artifact of purification and the authentic structure is HS-HTP.

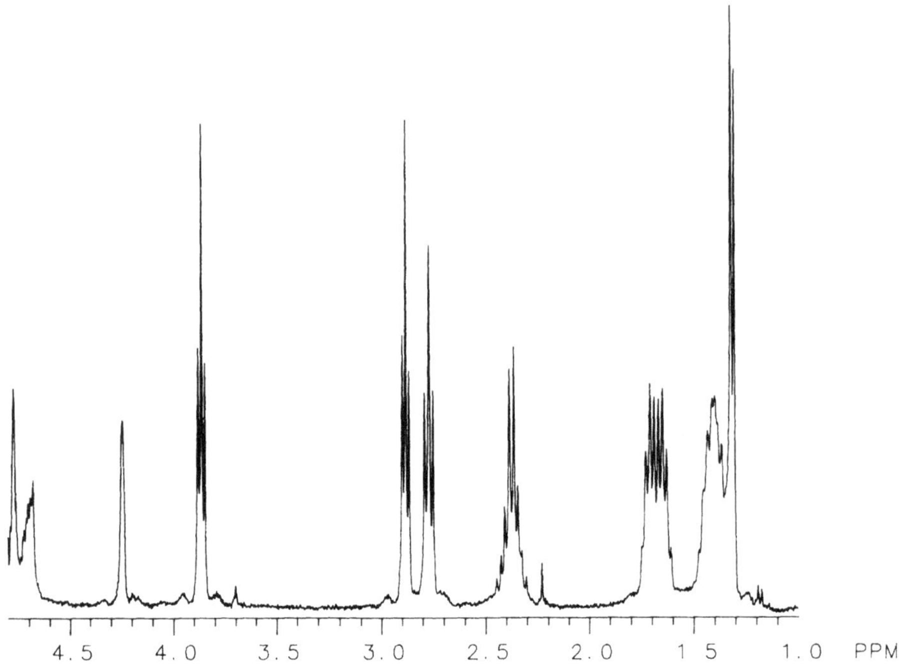

Figure 2. 360-MHz ^{1}H NMR spectrum of authentic HS-HTP. The sample was dissolved in D_2O. TSP was the chemical shift standard [0.0 ppm]

4. PROSPECTS

The purification to homogeneity of the transformylase, cyclohydrolase, and dehydrogenase establishes the enzymology of a major portion of the cycle (Fig. 1). Preliminary investigation of the remaining reactions, with the exception of the terminal methylreductase, has yielded little definitive data. The enzymatic reduction of methylene-H_4MPT has not been demonstrated *in vivo* or *in vitro,* and the methyltransferase (reaction 6 of Fig. 1) appears to be complex. Finally, characterization of the initial reduction of CO_2 to give formyl-MFR may be quite difficult. All evidence indicates that this reaction is obligately coupled to the final step of the pathway, the reduction of CH_3-S-CoM to methane. This phenomenon, the "RPG effect", was originally observed in crude extracts [25] and has been substantiated with extracts depleted of cofactors by G-25 chromatography. Attempts to uncouple these systems by inhibition of the methylreductase or by omission of any one of the components essential for the methylreductase reaction have been unsuccessful.

Several possible roles for HS-HTP seem apparent. First, the mercaptan may serve as a methyl carrier, perhaps accepting the methyl group from CH_3-S-CoM or donating it to HS-CoM. It may function as both a mercaptan and a disulfide (with another HS-HTP or HS-CoM) much the way lipoic acid does in fatty acid synthetase. Finally, it may be active in an adenylated form, thus having a structure analogous to coenzyme A. In this form it may be an effector of the methylreductase and its actual role may be to coordinate methanogenesis and acetate biosynthesis. Acetate is the primary product of autotrophic carbon assimilation and its biosynthesis shares at least one intermediate with the methanogenic pathway [26]. With the structure of HS-HTP now in hand, we are now in a position to begin meaningful studies of its role in methane formation by the methylreductase system.

The work reported here was supported by grants from the National Science Foundation (DMB 81 18178) to R.S.W. K.M.N. was the recipient of a U.S. Public Health Service Cell Biology Traineeship (GM 07283-11)

REFERENCES

[1] Wolfe, R.S. (1985) Trends Biochem. Sci. 10, 396-399.
[2] Nagle, D.P., Jr. and Wolfe, R. S. (1983) Proc. Natl. Acad. Sci. USA 80, 2151-2155.
[3] Noll, K,M., Rinehart, K.L., Jr., Tanner, R.S., and Wolfe, R.S. (1986) Proc. Natl. Acad. Sci. USA 83, 4238-4242.
[4] Leigh, J.A., Rinehart, K.L., Jr., and Wolfe, R. S. (1984) J. Amer. Chem. Soc. 106, 3636-3640.
[5] Van Beelen, P., Stassen, A.P.M., Bosch, J.W.G., Vogels, G.D., Guijt, W., and Haasnoot, C.A.G. (1984) Eur.J.Biochem. 138, 563-571.
[6] Leigh, J.A., Rinehart, K.L., Jr., and Wolfe, R.S. (1985) Biochemistry 24, 995-999.
[7] Escalante-Semerena, J.C., Rinehart, K.L., Jr., and Wolfe, R.S. (1984) J. Biol. Chem. 259, 9447-9455.
[8] Van Beelen, P., Van Neck, J.W., de Cock, R.M., Vogels, G.D., Guijt, W., and Haasnoot, C.A.G. (1984) Biochemistry 23, 4448-4454.
[9] Hartzell, P.L., Escalante-Semerena, J.C., Zvilius, G., and Donnelly, M.I. (1985) Biochem. Biophys. Res.Comm. 133, 884-890.
[10] Donnelly, M.I. and wolfe, R.S. (1986) Abstr. Ann. Meet. Amer. Soc. Microbiol., I-56, 174.
[11] DiMarco, A.A., Donnelly, M.I., and Wolfe, R.S. (1986) Submitted to J. Bacteriol.
[12] Donnelly, M.I., Escalante-Semerena, J.C., Rinehart, K.L. Jr., and Wolfe, R.S. (1985) Arch. Biochem. Biophys. 242, 430-439.
[13] Escalante-Semerena, J.C., Leigh, J.A., Rinehart, K.L., Jr., and Wolfe, R.S. (1984) Proc. Natl. Acad. Sci. USA 81, 1976-1980.
[14] Jacobson, F. and Walsh, C. (1984) Biochemistry 23, 979-988.
[15] Ellefson, W. L. and Wolfe, R.S. (1981) J. Biol. Chem. 256, 4259-4262.
[16] Rouviere, P. E., Escalante-Semerena, J.C., and Wolfe, R.S. J.Bacteriol. 162, 61-66.
[17] Ellefson, W. L. and Wolfe, R.S. (1980) J. Biol. Chem. 255, 8388-8389.
[18] Roberton, A.M. and Wolfe, R.S. (1969) Biochim. Biophys. Acta. 192, 420-429.
[19] Taylor, C.D. and Wolfe, R.S. (1974) J. Biol. Chem. 249, 4879-4885.
[20] Whitman,W.B. and Wolfe, R.S. (1982) Fed.Proc.Fed.Amer.Soc.Exp.Biol. 41, 5157 (abstr.).
[21] Ellefson, W.L., Whitman, W.B., and Wolfe, R.S. (1982) Proc. Natl. Acad. Sci. USA 79, 3707-3710.
[22] Gunsalus, R.P. and Wolfe, R.S. (1980) J. Biol. Chem. 255, 1891-1895.
[23] Noll, K.M., Donnelly, M.I., and Wolfe, R.S. (1986) Abstr. Ann. Meet. Amer. Soc. Biol. Chem., Fed. Proc. 45, 1543.
[24] Pogliani, L., Ziessow, D., and Kruger, C. (1979) Tetrahedron 35 2867-2873.
[25] Walton, E., Wagner, A.F., Bachelor, F.W., Peterson, L.H., Holly, F.W., and Folkers, K. (1955) J. Amer.Chem.Soc. 77, 5144-5149
[26] Lange, S. and Fuchs G. (1985) FEBS Lett. 181, 303-307..

INTERMEDIARY STEPS IN METHANOGENESIS

C. van der DRIFT, J.T. KELTJENS and G.D. VOGELS

Department of Microbiology, University of Nijmegen, Nijmegen, The Netherlands

1. INTRODUCTION

The eight electron reduction of CO_2 to methane proceeds through the oxidation levels formate, formaldehyde and methanol. In this process the C_1-unit remains bound to three unique C_1-carriers, methanofuran, 5,6,7,8-tetrahydromethanopterin (H_4MPT) and coenzyme M (HS-CoM, 2-mercaptoethane-sulfonic acid) [1,2,3] (Fig. 1). The pathway shown in Fig. 1 was established for *Methanobacterium thermoautotrophicum*, but the presence of the typical C_1-carriers in all species tested suggests that this CO_2 reduction route is common to all hydrogenotrophic methanogenic bacteria [4].

On the basis of data available from the literature concerning reactions involved in methanogenesis and analogous reactions occurring in non-methanogenic organisms we have calculated [3] the free energy changes of the partial reactions and the standard midpoint potentials of the varuous reduction steps; these values are included in Fig. 1. The first step is a thermodynamically unfavorable reaction. In accordance herewith, the reverse reaction, the oxidation of formyl-methanofuran proceeds readily, even under H_2 atmosphere [5]. On the other hand the reduction of CH_3S-CoM is exergonic to such an extent as to allow ATP synthesis. The exergonic character was confirmed by the demonstration that ATP is synthesized during the reduction of methanol by cell suspensions of *Methanosarcina barkeri* [6] and by the isolation of *Methanosphaera stadtmaniae* which only uses methanol and H_2 as energy sources [7]. Hence, a coupling of the primary and the terminal reduction steps may be expected. This is elegantly demonstrated in the RPG effect found in many but not all methanogens [8, 9]: the CO_2 reduction under

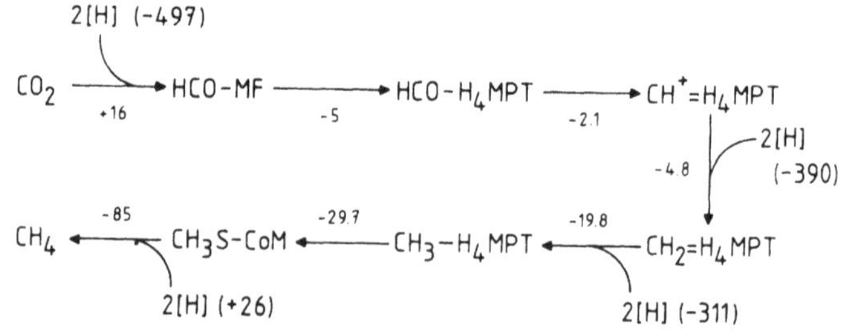

Figure 1. Pathway of CO_2 reduction to methane. The standard free-energy changes (kJ/mol)J and the standard reduction potentials (mV, values between brackets) are given for the partial reactions. MF, methanofuran.

H_2 is drastically stimulated by the addition of CH_3S-CoM (2-methylthioethanesulfonic acid) to the reaction mixture. The stmulatory effect is only transient and can be repeated by subsequent addititons of CH_3S-CoM. The nature of the coupling is unclear. From an energetic point of view one could envisage several possibilities : (i) ATP produced in the last step is (partly) used in the first step, (ii) ATP is used for reversed electron transport to provide the first step with low-redox potential electrons, (iii) energy or an energy-rich intermediate formed in the terminal step is transferred directly to the first step by an intimate coupling of both enzyme systems [9], (iv) the coupling is less specific: an active CH_3S-CoM reductase reaction pulls the first step by actively removing the products of the primary and intermediate steps. In addition, the role of CH_3S-CoM in the stimulation needs a further explanation. It was suggested [9] that the compound either acts as a positive effector of (one of the) methanogenic enzymes, or is involved in the synthesis of an energy-rich intermediate required for the coupling of the first and terminal steps. We have tested the different possibilities and the results of the investigations will be discussed below.

2. MATERIAL AND METHODS

M. thermoautotrophicum strain ΔH (DSM 1053) was mass-cultured in 12-l or 300-l fermenters and cell-free extracts of the organism were prepared as described previously [10]. Coenzyme-free cell extracts (G-25 treated extract) and crude mixture of heat-stable cofactors (boiled cell-fee extract, BCFE) were

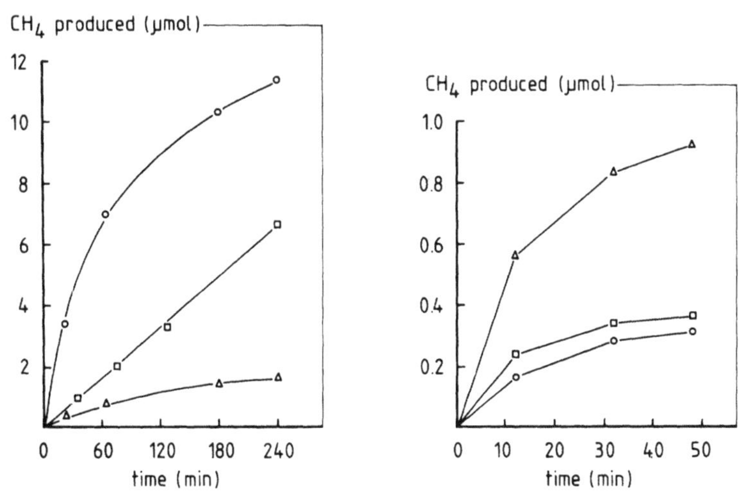

Figure 2. A) (left) Stimulation of the CO_2 reduction under H_2 by formaldehyde. Symbols:□, 80% H_2/20% CO_2; 0, 80% H_2/20% CO_2 plus formaldehyde (1500 nmol) ; Δ,100% H_2 plus formaldehyde (1500 nmol). B) (right) Stimulation of the formaldehyde reduction under H_2 by CH_3S-CoM. Symbols: o, formaldehyde (360 nmol);□ CH_3S-CoM (500 nmol), Δ, formaldehyde (360 nmol) plus CH_3S-CoM (500 nmol).

obtained according to Leigh and Wolfe [11]. Routine assays were performed in 10-ml crimped-sealed serum vials. Reaction mixtures contained 60 to 100 μl cell extract (1.5 to 2.5 mg of protein), 22 mM $MgCl_2$, ATP (2.5mM) if desired and other reactants as described in the text, supplemented with K^+-TES buffer (100 mM, pH 7) to a final volume of 200μl. After pregassing with the desired gasphase, the reactions were started by placing the vials at 60°C; CH_4 was measured as described before [12].

3. RESULTS AND DISCUSSION

3.1 Stimulation of CO_2 reduction

CO_2 reduction is not only stimulated by CH_3S-CoM, but a stimulation is also brought about by other intermediates of methanogenesis: methenyl-H_4MPT, methylene-H_4MPT and methyl-H_4MPT [13]. Moreover formaldehyde (Fig. 2A) and serine are easily converted into methylene-H_4MPT and act as stimulants [9]. The above-mentioned compounds all are converted to CH_3S-CoM (Fig. 1). In addition, a series of cell-carbon intermediates enhance CO_2 reduction to methane, viz. pyruvate, 2-oxoglutarate, isocitrate, malate, oxaloacetate [9, 10] and carbon monoxide (Fig. 3) These compounds have in common that they are oxidized in reactions that show low standard redox potentials; they might stimulate by providing reducing equivalents on oxidation.

Addition of CH_3S-CoM not only stimulated the CO_2 reduction, but also the conversion of formyl-methanofuran [5] and formaldehyde (Fig. 2B) to methane. This indicates that the RPG effect does not imply a specific coupling between the first and the last step of methanogensis.

3.2 Stimulation of CO_2 reduction by carbon monxide

CO_2 reduction under H_2 atmosphere was stimulated by CO (Fig. 3). The stimulation was inhibited by cyanide, which did not affect the rate of methanogenesis from CO_2 (Fig. 3). Cyanide is known to be an inhibitor of carbon monxide dehydrogenase which catalyzes the following reaction:

$$CO + H_2O \rightarrow CO_2 + 2H^+ + 2e \quad (E'_0 = -518\ mV)$$

Thus it appears that CO may stimulate CO_2 reduction by provididng the system with reducing equivalents.

When the CH_3S-CoM reduction was inhibited by 0.25 mM 2-bromoethane-sulfonic acid (BES) and HS-CoM was added to trap the C_1-units, CO_2 became reduced to CH_3S-CoM. Addition of CH_3S-CoM did not stimulate the reaction, eliminating the possibility that the compound acts as a positive effector on the enzymes of the primary steps of reduction. CO_2 conversion to CH_3S-CoM was stimulated by CO and CN^- inhibited the stimulation. Moreover, this experiment proves that the first step may proceed independently from the terminal step, which pleads against a direct coupling.

3.3. Role of ATP in the process of methanogenesis from H_2 and CO_2

The CH_3S-CoM -stimulated CO_2 reduction was dependent on the presence of ATP [8, 9] (Fig. 4.). In the absence of ATP the reaction showed a sigmoid flexure suggesting that during the reaction a net amount of ATP is produced which enables an acceleration of the velocity. This was confirmed by the inhibition of methanogenesis on adddition of the uncoupler tetrachlorosalicylanilide (TCS) (10 μM) to

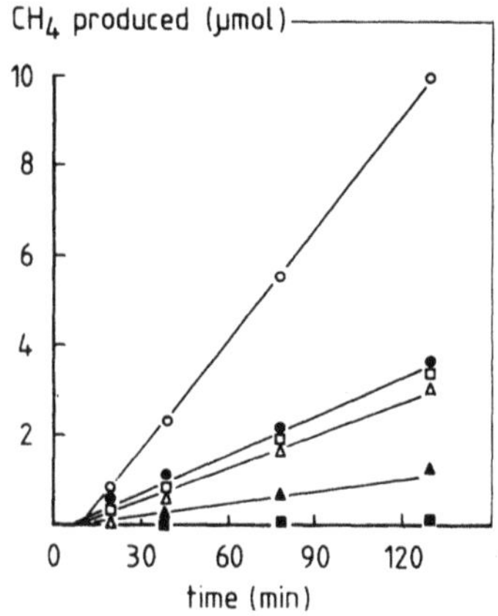

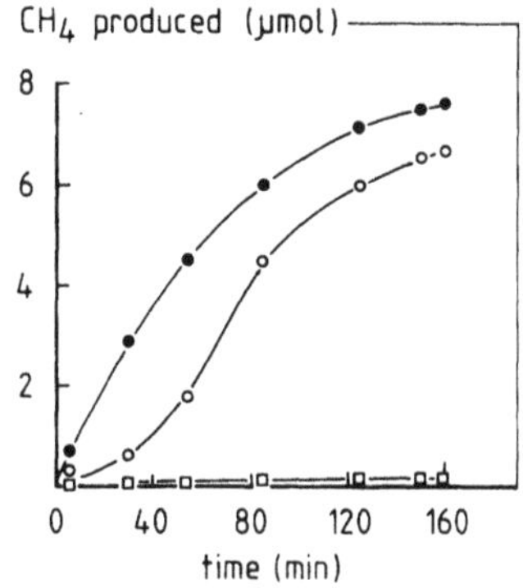

Figure 3. (left) Stimulation of the CO_2 reduction under H_2 by carbon monoxide. Symbols: O, ▲, 40% H_2/10% CO_2/50% CO; ●, Δ, 40% H_2/10% CO_2/50% N_2; □, ■, 50% H_2/50% CO. ▲, Δ, ■ were measured in the presence of KCN (1 mM).

Fligure 4. (right) Effect of ATP on the CH_3S-CoM - stimulated CO_2 reduction. The time course of methane formation under 80% H_2/20% CO_2 plus CH_3S-CoM (1000 nmol) was measured in the presence (2.5 mM) (●) or absence (O, □) of ATP; □ contained 10 μM Tetrachlorosalicylanilide.

Table 1. ATP use in the process of methanogenesis[a,b]

Substrates[c]		ATP added[d] (nmol)	Total CH_4 formed (nmol)	Ratio CH_4/ATP[e]
CH_3S-CoM	+ H_2	30-480 (30)	180-1520 (180)	3.4-6.0 (6.0)
CH_3S-CoM	+ H_2/CO_2	30-480 (60)	800-15000 (8000)	26.7-133 (133)
formaldehyde	+ H_2	30-480 (30)	1300-1900 (1300)	3.9-43.3 (43.3)
formaldehyde	+ H_2/CO_2	30-480 (30)	8260-11250 (8260)	23.5-275 (275)

[a]Reactions were measured in the presence of 10μM tetrachlorosalicylanilide. [b]Values between brackets indicate the amount of ATP added and the total CH_4 formed resulting from it at which an optimal ratio CH_4/ATP was obtained. [c]CH_3S-CoM and formaldehyde were 10 μM.
[d]For each series 5 reaction mixtures were prepared containing ATP varying between the values indicated. [e]The ratio CH_4/ATP was calculated from each individual reaction.

the reaction mixture lacking ATP (Fig. 4). When ATP was added to reaction mixtures containing TCS, methanogenesis was the same as in controls that did not contain TCS. Thus, TCS does not inhibit the methanogenic enzymes. The efffect of TCS opposes the possibility that ATP is used to drive a reversed electron flow ; such a reversed electron transport should be uncoupler-sensitve even in the presence of ATP. By adding varying amounts of ATP to TCS-containing reaction mixtures the stoichiometry between ATP consumption and methane formation can be determined (Table 1). It follows that ATP is consumed only in catalyctic amounts. Remarkably, more ATP is used per mol methane formed in the reduction of CH_3S-CoM- or formaldehyde under H_2 than in the CH_3S-CoM or formaldehyde-stimulated CO_2 reduction. The data in Table 1 permit the conclusion that the primary steps of CO_2 reduction do not require an appreciable amount of ATP and that ATP is consumed in the terminal and penultimate steps.

3.4 Consumption of ATP in the conversion of formaldehyde and HS-CoM to CH_3S-CoM

When we studied the conversion of formaldehyde and HS-CoM under H_2 to CH_3S-CoM in cell-free extracts of *M. thermoautotrophicum* in which the terminal reaction was inhibited by BES, it was found that this reaction required catalytic amounts of ATP: the presence of 400 nmol ATP allowed the conversion of 900 nmol formaldehyde to CH_3S-CoM. The ratio will be an overestimate, since the particular cell extract possessed a rather high ATPase activity. Since it was found that the reduction of methylene-H_4MPT to methyl-H_4MPT readily proceeds in the absence of ATP [14], ATP must be consumed in the methyltransferase reaction from methyl-H_4MPT to HS-CoM. The activity of the formaldehyde conversion varied considerably among the various batches of cell extracts but it appeared that the activity was stimulated by preincubation of cell extract under H_2 atmosphere. After preincubation the activities of the various extracts were about equal. This treatment did not affect the methylene-H_4MPT reduction to methyl-H_4MPT. This ATP-dependent reductive activation is reminiscent of the activation of the methanol: HS-CoM methyltransferase system of *M. barkeri* [15-20], which is composed of two enzymes, *viz.* methanol: 5-hydroxybenzimidazolylcobamide (B_{12}-HBI) methyltransferase (MT_1) and Co-methyl-5-hydroxybenzimidazolylcobamide (CH_3-B_{12}-HBI): HS-CoM methyltransferase (MT_2) [17-20]. In *M. barkeri* ATP and H_2 are required to activate MT_1 [19,20].

Crude cell extracts of *M. thermoautotrophicum* contained various B_{12}-HBI derivatives as determined by HPLC, but CH_3-B_{12}-HBI was absent [C. Poirot, unpublished results]. However, after incubation of extract with formaldehyde and ATP under H_2 CH_3-B_{12}-HBI was the only B_{12}-HBI derivative extractable from the mixture. In the presence of excess HS-CoM only demethylated B_{12}-HBI derivatives were found. These results suggest the presence of a methyl-H_4MPT: B_{12}-HBI methyltransferase. Such an enzyme would catalyze the methyl group transfer from methyl-H_4MPT to HS-CoM in collaboration with Co-methyl-B_{12}: HS-CoM methyltransferase analogously to the methanol: HS-CoM methyltransferase system in *M. barkeri*. The presence of Co-methyl-B_{12}: HS-CoM methyltransferase was already demonstrated in *Methanobacterium* species [21] but no specific function could yet be attributed to this enzyme.

3.5 Consumption of ATP in the conversion of CH_3S-CoM to methane

ATP is required for the reductive demethylation of CH_3S-CoM both in crude extracts [22] (Table 1) and in the resolved CH_3S-CoM reductase system [23]. It was suggested that ATP activates the system by modifying an enzyme or enzyme-bound cofactor [24]. A simplified reductase system was recently described by Ankel-Fuchs and Thauer [25]. Their system is composed of CH_3S-CoM reductase, component B, a heat-stable cofactor of the reaction [23], and corrinoid and dithiothreitol which serve as the electron donors; ATP is not needed. In agreement herewith, we found that ATP was not per se required for the reduction of CH_3S-CoM in a system containing a crude coenzyme-free enzyme mixture (Sephadex G25-treated extract) supplemented with varying amounts of boiled cell-free extract (BCFE) (Fig.5). Treatment of BCFE with hexokinase-glucose and addition of TCS (10 μM) to the reaction mixture did not affect methanogenesis. However, in the absence of BCFE ATP stimulated the conversion of CH_3S-CoMunder H_2 atmosphere (Fig.6). The results suggest that BCFE contains an activating compound which is derived from ATP. Interestingly, ATP (2.5 mM) could be replaced by low concentrations (1 μM of S(±)-adenosyl-methylcoenzyme M (SAMCOM), which contains an activated methyl group. Stimulation was maximal when both ATP and SAMCOM were present in the reaction mixture (Fig. 6). SAMCOM could be derived from ATP and CH_3S-CoM in a way analogous to the formation of S(+)-adenosyl-methionine, the activator of B_{12} containing 5-methyltetrahydrofolate: homocysteine methyltransferase (methionine synthetase), from ATP and methionine. The presence of SAMCOM in *M. thermoautotrophicum* remains to be established.

3.6 Conclusions

In *M. thermoautotrophicum* and some other methanogenic bacteria [8, 9] the primary and terminal steps of methanogenesis are coupled as expressed in the RPG effect. The effect implies a drastic stimulation of CO_2 reduction by the addition of ATP and CH_3S-CoM or a metabolic precursor thereof to the reaction mixtures. In addition, a number of compounds like pyruvate and carbon monoxide stimulate the CO_2 reduction. From the results described above it can be concluded that no appreciable amounts of ATP are used in the primary steps of the reaction sequence. ATP is neither used in driving reversed electron transport to provide the first step of CO_2 reduction with low-redox potential electrons, since TCS does not inhibit CO_2 reduction when ATP is present. CO_2 reduction proceeds also when the CH_3S-CoM reduction is inhibited by BES. Moreover, CH_3S-CoM not only stimulates the reduction of CO_2, but also of reduced C_1 intermediates, *viz.* formyl-methanofuran and formaldehyde.Furthermore, a direct coupling between the first and the last step was excluded: ATP appears to be used for the activation of the methylreductase and methyltransferase reactions. Apart from ATP CH_3S-CoM stimulates methanogenesis from CO_2. The presence of CH_3S-CoM does not stimulate CO_2 conversion upto the level of this methylated compound opposing the possibility that CH_3S-CoM acts as a positive effector on the steps involved in its production. Since a direct coupling between the terminal and first step was excluded CH_3S-CoM cannot be involved in the formation of an energy-rich intermediate required for such a coupling. Either CH_3S-CoM acts as a positive effector of the CH_3S-CoM reductase system, or the compound is functioning in the activation of this system. The latter possibility is corroborated by the finding that SAMCOM stimulates the CH_3S-CoM reduction. An explanation of the RPG effect thus could be that methylreductase, activated by ATP and CH_3S-CoM, performs a rapid removal of the products of the primary and intermediate steps.

Many details of the activation mechanism of methylreductase remain to be established. With respect to this activation some remarkable parallels exist with the methyl-H_4MPT: HS-CoM and the methanol: HS-CoM methyl-transferase systems of *M. thermoautotrophicum* and *M. barkeri*, respectively. All three

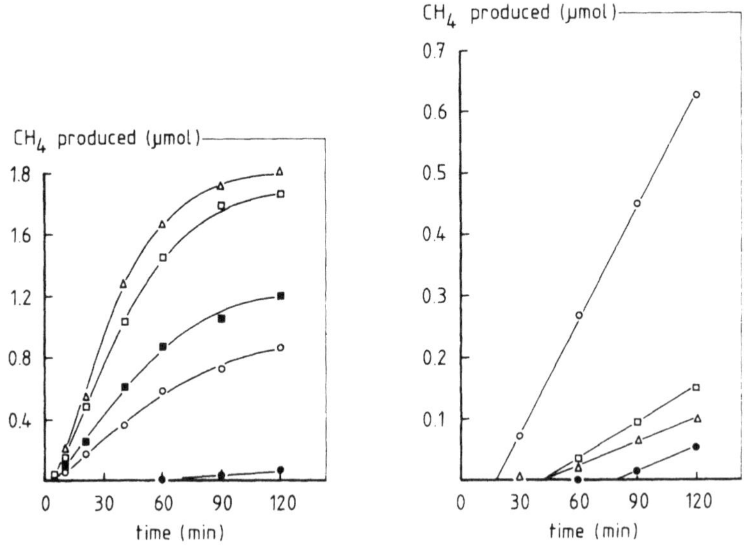

Figure 5. (left) Stimulation of CH_3S-CoM reduction under H_2 by boiled cell-free extract. The reaction mixture contained 80 μl Sephadex G25-treated extract (1.6 mg protein). $MgCl_2$ (22 mM), CH_3S-CoM (2 μmol) and 0 (●), 10 (O), 20 (■), 40 (□) or 80 (Δ) μl boiled cell-free extract.
Figure 6. (right) Stimulation of CH_3S-CoM reduction under H_2 by ATP and S(±)-adenosyl-methylcoenzyme M (SAMCOM). The reaction mixtures were the same as in Fig. 5; no BCFE was added. Symbols: ●, no further additions; Δ, plus ATP (2.5 mM); □, plus SAMCOM (1 μM); O, plus ATP (2.5 mM) and SAMCOM (1 μM).

reactions require catalytic amounts of ATP and reductive preactivation with H_2 increases the activity (data not shown for methylreductase). The key enzymes of the three systems contain tetrapyrroles as their prosthetic groups: B_{12}-HBI and the nickel (II) tetrapyrrole factor F_{430} are present in the methyltransferases [18, 20, this paper] and in methylreductase [26], respectively. During the activation of MT_1 the central Co atom is converted to its most reduced state (Co^I) [19] and active CH_3S-CoM reductase was recently proposed to contain Ni^I [27]. Electrons for the reduction of Co^{III} to Co^I can be provided by pyruvate [16] or carbon monoxide [20]. Remarkably, both compounds stimulate CO_2 reduction in *M. thermoautotrophicum.* Hence, pyruvate, carbon monoxide, 2-oxoglutarate and others could act as stimulants by generating reducing equivalents for the reductive activation of the methylreductase and methyltransferase systems.

References

[1] Escalante-Semerena JC, Leigh JA, Wofe RS. (1984) In Microbial Growth on C_1 compounds (Crawford RL, Hanson RS, eds), 191-198. Am. Soc. Microbiol., Washington, D.C.
[2] Wolfe RS. (1985) Trends Biochem. Sci. 10, 396-399.
[3] Keltjens JT, Van der Drift C. (1986) FEMS Microbiol. Rev., in press.
[4] Jones WJ, Donelly MI, Wolfe RS. (1985) J. Bacteriol. 163, 126-131.
[5] Leigh JA, Rinehart KL Jr., Wolfe RS. (1985) Biochemistry 24, 995-999.
[6] Blaut M, Gottschalk G. (1985) Trends Biochem. Sci. 10, 486-489.
[7] Miller TL, Wolin MJ. (1985) Arch. Microbiol. 141, 116-122.
[8] Gunsalus RP, Wolfe RS. (1977) Biochem. Biophys. Res. Commun. 76, 790-795.
[9] Romesser JA, Wolfe RS. (1982) J. Bacteriol. 152, 840-847.
[10] Van Beelen P, Thiemessen HL, De Cock RM, Vogels GD. (1983) FEMS Microbiol. Lett. 18, 135-138.
[11] Leigh JA, Wolfe RS. (1983) J. Biol. Chem. 258, 7536-7540.
[12] Hutten TJ, De Jong MH, Peeters BPH, Van der Drift C, Vogels GD. (1981) J. Bacteriol. 145, 27-34.
[13] Escalante-Semerena JC, Rinehart KL Jr., Wolfe RS. (1984) J. Biol. Chem. 259, 9447-9455.
[14] Keltjens JT, Caerteling GC, Van der Drift C, Vogels GD. (1986) Syst. Appl. Microbiol. 7, 370-375.
[15] Shapiro S. (1982) Can. J. Microbiol. 28, 1409-1411.
[16] Van der Meijden P, Heythuysen HJ, Sliepenbeek HT, Houwen FP, Van der Drift C, Vogels GD.(1983) J. Bacteriol. 153, 6-11.
[17] Van der Meijden P, Heythuysen HJ, Pouwels A, Houwen F, Van der Drift C, Vogels GD. (1983) Arch. Microbiol. 134, 238-242.
[18] Van der Meijden P, Jansen LPJM, Van der Drift C, Vogels GD. (1983) FEMS Microbiol. Lett. 19, 247-251.
[19] Van der Meijden P, Van der Lest C, Van der Drift C, Vogels GD. (1984) Biochem. Biophys. Res. Commun. 118, 760-766.
[20] Van der Meijden P, te Brömmelstroet BW, Poirot CM, Van der Drift C, Vogels GD. (1984) J. Bacteriol. 160, 629-639.
[21] Taylor CD, Wolfe RS. (1974) J. Biol. Chem. 249, 4886-4890.
[22] Gunsalus RP, Wolfe RS. (1978) J. Bacteriol. 135, 851-857.
[23] Gunsalus RP, Wolfe RS. (1980) J. Biol. Chem. 255, 1891-1895.
[24] Whitman WB, Wolfe RS. (1983) J. Bacteriol. 154, 640-649.
[25] Ankel-Fuchs D, Thauer RK. (1986) Eur. J. Biochem. 156, 171-177.
[26] Thauer RK, Ankel-Fuchs D, Diekert G, Gilles H-H, Graf E-G, Jaenchen R, Moll J, Schönheit P. (1984) In Microbial Growth on C_1 compounds (Crawford RL, Hanson RS, eds), 188-190. Am. Soc. Microbiol., Washington, D.C.
[27] Albracht SPJ, Ankel-Fuchs D, Van der Zwaan JW, Fontijn RD, Thauer RK. (1986) Biochim. Biophys. Acta 870, 50-57.

RECENT DEVELOPMENTS ON THE BIOCHEMISTRY OF METHANOGENESIS FROM ACETATE

Michael J. K. NELSON, Katherine C. TERLESKY, and James G. FERRY
Department of anaerobic Microbiology, Virginia Polytechnic Institute and State University, Blacksburg, Virginia, 24061, USA

1. INTRODUCTION

Athough acetate is the major precursor for methanogenesis in nature, biochemical studies on this pathway have lagged behind studies on the pathway of carbon dioxide reduction to methane. In 1955, it was demonstrated that the methyl group of acetate is metabolized intact to methane by mixed cultures [17]. Research was stimulated 23 years later when Mah et al. [12] showed that *Methanosarcina barkeri* utilizes acetate as a sole growth substrate and produces methane at rates comparable to that observed in nature. Methods for large-scale culture of these slow growing organisms have been improved by use of the pH auxostat which delivers the large amounts of acetic acid required for growth while maintaining a constant pH [21]. Methanogenesis from acetate in cell exracts [1, 9] is an important development. Coenzyme M was shown to be a methyl carrier in the pathway by demonstrating deuterated methyl coenzyme M in *Methanosarcina thermophila* cells utilizing deuterated methyl acetate for methanogenesis [11]. Recently, the involvement of coenzyme M was also demonstrated in *Methanosarcina barkeri* using ^{14}C-methyl labelled acetate [6]. The terminal step in methanogenesis from acetate is the reductive demethylation of methyl coenzyme M [6, 14]; thus, a common mechanism exists for the terminal step in methanogenesis from all known substrates. However, the mechanism of acetate cleavage and the steps leading to methyl coenzyme M in acetotrophic methanogens are unknown.

2. ACTIVATION OF ACETATE

Eikmanns and Thauer [2] have pointed out, from thermodynamic considerations, that cleavage of acetate may require an activation step. The high levels of acetate thiokinase activity in acetate-grown *Methanothrix soehngenii* [5] and *M. thermophila* [unpublished results] indicate acetyl-CoA as a possible activated form in these organisms. However, the ability of acetyl-phosphate to support methanogenesis from acetate in cell extracts of *M. barkeri* [6] indicates this compound as a potential activated form of acetate in this organism. A third possibility, component B, must also be considered since this cofactor may be structurally analogous to coenzyme A [15] and therefore may be acetylated.

3. THE INVOLVEMENT OF CARBON MONOXIDE DEHYDROGENASE

A hypothesis is proposed in which the carbon-carbon bond of activated acetate is cleaved by an enzyme with CO dehydrogenase activity followed by transfer of the carbonyl carbon to a bound reduced one-carbon intermediate, possibly CO. This hypothesis is based on a reversal of the pathway for acetate synthesis in the acetogenic clostridia. A purified CO dehydrogenase from *Clostridium thermoaceticum* catalyzes the synthesis of acetyl-CoA from CoA-SH, methyl corrinoid, and an enzyme-bound one-carbon intermediate [18]; the enzyme forms a nickel (III)-iron-carbon species with a unique electron paramagnetic resonance signal when reduced with CO [19].

Several lines of evidence support an involvement of CO dehydrogenase in the pathway of methanogenesis from acetate. The CO dehydrogenase activities of *Methanosarcina* spp. are greater in acetate-grown cells compared to cells grown on alternate substrates [7, 14]. Acetate-grown cells of *M. thermophila* also contain CO-dependent methyl coenzyme M methylreductase activity [14]. Furthermore, a CO dehydrogenase is implicated in the synthesis of acetyl-CoA for cell carbon of the nonacetotrophic carbon dioxide-reducing methanogen *Methanobacterium thermoautotrophicum* [23]. Antibodies to the nickel-containing CO dehydrogenase isolated from *M. barkeri* inhibit methanogenesis from acetate in cell extracts [6]. More recent evidence strongly implicates a CO dehydrogenase isolated from *M. thermophila* [24]: (i) synthesis of the enzyme is induced in acetate-grown cells; (ii) the enzyme comprises greater than 10% of the cellular protein suggesting involvement in a major catabolic pathway; (iii) the enzyme contains a nickel center with an electron paramagnetic resonance spectrum nearly identical to the published spectra of the clostridial CO dehydrogenase [19] catalyzing the synthesis of acetyl-CoA; and (iv) the pattern of cyanide inactivation of CO dehydrogenase activity suggests that cyanide and CO bind to a common site, possibly nickel. Although the indirect evidence is strong for the involvement of an enzyme with CO dehydrogenase activity in the cleavage of acetate, this function remains to be directly shown.

Attempts to demonstrate free CO as the one-carbon intermediate derived from the carbonyl carbon of acetate have been unsuccessful. Eikmanns and Thauer [2] have suggested that tightly bound CO may be an intermediate in *M. barkeri*. Experiments were performed with resuspended cells of *M. thermophila* to determine whether free CO was an intermediate in this organism. Unlike *M. barkeri* [2] where methanogenesis from aetate is completely inhibited by low concentrations of CO, the rate of methanogenesis from acetate by cell suspensions of *M. thermophila* was inhibited apprroximately 65% in an atmosphere of 100% CO [data not shown). Cell suspensions were incubated with 1-^{14}C-acetate in an atmosphere of N_2 or CO until ca. 10 µmol of CH_4 was produced. The head space was then assayed for ^{14}CO and $^{14}CO_2$ (Table 1). Little CO was produced form the C-1 of acetate when incubated in a N_2 atmosphere. A significant amount of CO was produced from the C-1 of acetate when the experiments were performed in a CO atmosphere; however, the majority of the C-1 of acetate was converted to CO_2. Control experiments demonstrated CO may also be produced from CO_2. The results of the experiment conclusively demonstrate that CO in an unbound form is not a direct intermediate in the conversion of the C-1 of acetate to CO_2 during acetate-dependent methanogenesis. If this were the case, dilution with unlabelled CO should have resulted in most of the label trapped in the CO atmosphere with very little converted completely to CO_2. The results of the experiments, however, do not rule out the possiblility of CO in a bound form which is able to exchange with free CO at a rate which is slower than the rate of conversion of the C-1 acetate to CO_2. Whether such an exchange is responsible for production of some of the CO from the C-1 of acetate or whether all of this CO is produced indirectly by conversion of CO_2 to CO cannot be established from the present data. Formate has been ruled out as a potential intermediate [2, 14].

Table 1. Conversion of 1-^{14}C-acetate to ^{14}CO and $^{14}CO_2$ by cell suspensions of *Methanosarcina thermophila*.
Methanosarcina thermophila cells (0.2 g wet weight) were resuspended in 5.0 ml sealed Wheaton vials containing 0.5 ml 0.1 M potassium phosphate buffer (pH 7.0), 10 mM $MgCl_2$ and either 25 μmol acetate or 10μmol Na_2CO_3 and incubated at 50°C. After termination of the reaction, N_2 was passed through the reaction vials and through three 20 ml vials connected in series each containing 8 ml of an ethanolamine:methanol mixture (60:40 v/v), and finally into a sealed evacuated 500 ml filtering flask containing 100 ml 6 mM $PdCl_2$. After 4 hours at 25°C the 500 ml filtering flask was sparged with N_2 through three ethanolamine:methanol traps. The first set of traps quantitatively absorbed all CO_2 as indicated by controls; aliquots (0.8 ml) of these traps were counted to determine the label present in CO_2. CO that passed through the first traps was converted to CO_2 by the $PdCl_2$ in the filtering flask. The resulting CO_2 was then absorbved in the second set of traps which were counted to determine label present as CO. Label that was not converted to CO or CO_2 (mostly present as unconsumed acetate) remained in the reaction vial, an aliquot of which was also counted for determining the nonvolatile radioactivity.

	Counts Added[(a)]		Counts Recovered[(a)]			
Gas Phase (100%)	Compound	Total	Non-volatile	CO_2	CO	Total
N_2	1-^{14}C-acetate	1,337	729	670	20	1,420
CO	1-^{14}C-acetate	1,650	913	504	103	1,520
CO	1-^{14}C-acetate	1,650	1,041	420	76	1,538
CO	$Na_2{}^{14}CO_3$	158	23	102	26	151

[(a)]Counts per minute x 10^{-3}

4. INVOLVEMENT OF CORRINOIDS

The involvement of corrinoids in the pathway of acetate conversion to methane was first proposed by Stadtman [22] basd on the known function of this methyl carrier in the pathway of acetogenesis in the clostridia. Recent evidence implicates the involvement of a corrinoid protein during methanogenesis from acetate in whole cells of *M. barkeri* [3]. The results, based on propyl iodide inhibition, suggest that the corrinoid functions in methyl transfer after carbon-carbon bond cleavage and before reductive demethylation of coenzyme M. However, no direct evidence is available showing a methyl transfer function. The CO dehydrogenase isolated from *M. thermophila* is a complex enzyme comprised of five subunits and also contains a cobamide with spectral properties similar to 5-hydroxybenzimidazolyl-cobamide common in methanogens [24]. This cobamide is methylated with methyl iodide consistent with its function as a methyl carrier [unpublished results]. It is tempting to speculate that the CO dehydrogenase complex cleaves acetate transferring the methyl group to a subunit containing the cobamide. Corrinoids have also been implicated in the synthesis of acetyl-CoA by *M. thermoautotrophicum* [4]. Curiously, a corrinoid-containing membrane protein complex has recently been identified in *M. thermoautotrophicum* [20].

5. METHYL COENZYME M METHYLREDUCTASE

The terminal step in methanogenesis from acetate is the reductive demethylation of methyl coenzyme M with electrons derived from the proposed one-carbon intermediate. Demonstratioon of a CO-dependent methylreductase in acetate-grown *M. thermophila* is consistent with CO dehydrogenase as the electron donor [14]. However, the involvement of H_2 and hydrogenase cannot be excluded since H_2 also supports methylreductase activity in acetate-grown *M. thermophila* although this organism is unable to utilize H_2 and carbon dioxide as substrates for methanogenesis when grown on acetate [14]. Furthermore, rapid H_2 production from CO occurs in extracts of *M. thermophila* (Figure 1). The addition of methyl coenzyme M to extracts decreased the rate of H_2 production possibly by utilization of H_2 for reductive demethylation to methane. Thus, the participation of H_2 as an intermediate in electron transfer to methylreductase must be considered. Both the production and consumption of H_2 in growing cultures of *Methanosarcina* spp. [10] is consistent with this idea, although other functions for H_2 metabolism can also be envisioned.

6. ATP SYNTHESIS

Several divers hypotheses are proposed for the mechanism of ATP synthesis in acetotrophic methanogens. Wolfe [25] has proposed a mechanism in which an electron pair from the cleavage of acetate traverses the membrane where protons are consumed and the $2e^- + 2H^+$ are transported back across the membrane to reduce the methyl group (of methyl coenzyme M) to methane thereby establishing a proton gradient. Zehnder and Brock [26] have proposed two mechanisms that do not involve electron transport but instead results in a pH gradient driven by the energy released from decarboxylation. Krzycki and Zeikus [9] have suggested a mechanism similar to that of Wolfe but that incorporates the involvement of CO dehydrogenase as the electron donor to the membrane-bound electron transport chain ultimately reducing methyl coenzyme M. Lovley and Ferry [10] have proposed a

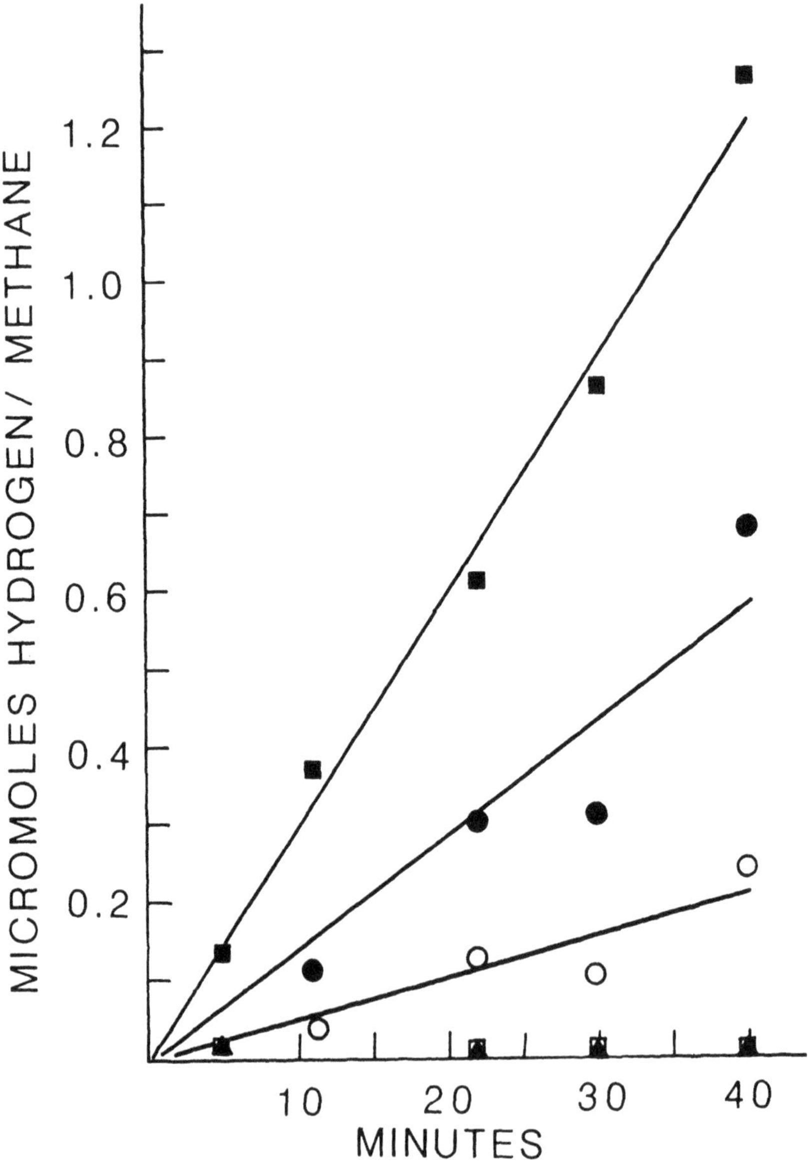

Figure 1. Hydrogen production from carbon monoxide, and methane production from methyl coenzyme M in cell extracts of *Methanosarcina thermophila*. Experimental conditions were as described in [14] for the standard methyl coenzyme M methylreductase assays. Symbols: (■) H_2 produced, 1:9 CO:N_2 atmosphere; (●) H_2 produced, 1:9 CO: N_2 atmosphere + methyl coenzyme M; (▲) H_2 produced, 100% N_2 atmosphere + methyl coenzyme M; (□) CH_4 produced, 1:9 CO:N_2 atmosphere; (o) CH_4 produced, 1:9 CO: N_2 atmosphere + methyl coenzyme M.

hypothetical mechanism in which H_2 is an intermediate between oxidation of the proposed reduced one-carbon compound (possibly CO) and reduction of methyl coenzyme M; this hypothesis is analogous to the H_2 cycling mechanism for ATP synthesis proposed for the sulfate-reducing bacteria [16]. The production of H_2 from bound CO is an interesting hypothesis since it provides a potential reaction for energy conservation (CO[free] + H_2O = CO_2 + H_2, $\Delta G^{o\prime}$ = 20 kJ/mol) in addition to the reductive demethylation of methyl coenzyme M.

7. SUMMARY

Based on the present understanding, a new model of methanogenesis form acetate can be proposed (Figure 2) in which an activated form of acetate is cleaved by an enzyme with CO dehydrogenase activity with transfer of the carbonyl carbon as CO bound to a nickel site on the enzyme, and donation of the methyl group to a corrinoid protein. In the reactions that follow, the model proposes that the methyl group is transferred to coenzyme M and the bound CO is oxidized to carbon dioxide with electron flow to methyl coenzyme M through an electron transport chain coupled to ATP synthesis. Proof of the proposed reactions await reconstitution of an enzyme system converting the methyl group of acetate to coenzyme M. Studies on the electron transport chain that couples the oxidation of CO and reduction of methyl coenzyme M will be important to understanding the mechanism of energy conservation in the acetotrophic methanogens.

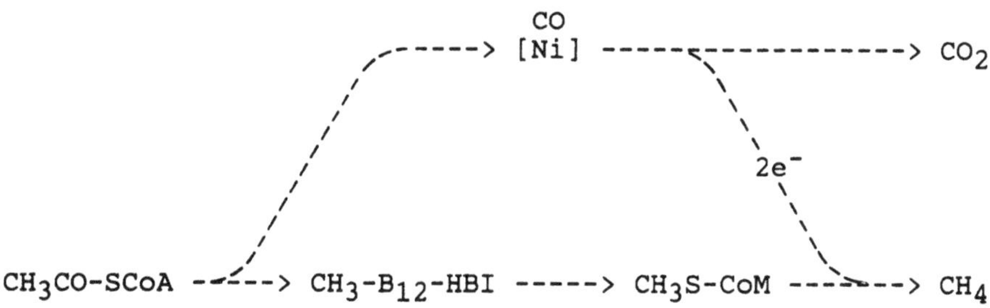

Figure 2. A new model for methanogenesis from acetate. Methyl coenzyme M (CH_3-S-CoM) is a known intermediate [11]; only indirect evidence supports the other proposed intermediates. CH_3-B_{12}-HBI is CH_3-B_{12}-5-hydroxybenzimidazolylcobamide.

References

[1] Baresi, L. (1984) J. Bacteriol. 160, 365-370.
[2] Eikmanns, B, and RK Thauer. (1984) Arch. Microbiol. 138, 365-370.
[3] Eikmanns, B, and R.K Thauer. (1985) Arch. Microbiol. 142,175-179.
[4] Holder, U, DE Schmidt, E Stupperich, G Fuchs. (1985) Arch. Microbiol. 141, 229-238.
[5] Kohler, HPE, AJB Zehnder. (1984) FEMS Microbiol. Lett. 21, 287-292.
[6] Krzycki, JA, LJ Lehman, and JG Zeikus (1985) J. Bacteriol. 163, 1000-1006.
[7] Krzycki, JA, RH Wolkin, and JG Zeikus. (1982) J. Bacteriol. 149, 247-254.
[8] Krzycki, JA, JG Zeikus. (1984) J. Bacteriol. 158,231-237.

[9] Krzycki, JA, JG Zeikus. (1984) FEMS Microbiol. Lett. 25, 27-32.
[10] Lovley, DR, and JG Ferry. (1985) Appl. Environ. Microbiol. 49, 247-249.
[11] Lovley, DR, RH White, JG Ferry. (1984) J. Bacteriol. 160,521-525.
[12] Mah, RH, MR Smith, L Baresi, (1978) Appl. Environ. Microbiol. 35 1174-1184.
[13] Nagle, Jr., DP, and RS Wolfe. (1983) Proc. Natl. Acad. Sci. USA. 80,2151-2155.
[14] Nelson, MJK, JG Ferry. (1984) J. Bacteriol. 160, 526-532.
[15] Noll, KM, KL Rinehart, Jr., RS Tanner, RS, Wolfe. (1986) Proc. Natl. Acad. Sci., USA. 83, 4238-4242.
[16] Odom, JM, H D Peck. (1981) FEMS Microbiol. Lett. 12, 47-50.
[17] Pine, MJ and HA Barker. (1955) J. Bacteriol. 71, 644-648.
[18] Ragsdale, SW and HG Wood. (1985) J. Biol. Chem. 260, 3970-3977.
[19] Ragsdale, SW, HG Wood, and WE Antholine. (1985) Proc. Natl. Acad. Sci. USA. 82, 6811-6814.
[20] Schultz, H,G Fuchs. (1986) FEBS Lett. 198, 279-282.
[21] Sowers, KR, MJ Nelson, JG Ferrry. (1984) Curr. Microbiol. 11, 227-230.
[22] Stadtman, TC. (1967) Ann. Rev. Microbiol. 21, 121-142.
[23] Stupperich, EG.Fuchs. (1984) Arch. Microbiol. 139, 14-20.
[24] Terlesky, KC, MJK Nelson,JG Ferry. (1986) J. Bacteriol. (In Press.)
[25] Wolfe, RS. IJ Higgins. (1979) in: Microbial Biochemistry (Quale, JR. ed.) 21, 267-299.
[26] Zehnder, JB, T.D.Brock. (1979) FEBS Lett. 107, 1-3.

C_1-UTILIZING ACETOGENS

G. DIEKERT
Fachbereich Biologie, Philipps-Universität Marburg, Federal Republic of Germany

1. INTRODUCTION

Acetogenic bacteria are strictly anaerobic eubacteria which catalyze the formation of acetate from CO_2 in their energy metabolism. The CO_2 reduction to acetate can serve as an electron sink for reducing equivalents generated during oxidative degradation of carbobydrates and various other substrates. However, acetate synthesis from CO_2 is also the major catabolic pathway for growth of the acetogens on C_1-compounds as energy substrates such as H_2 plus CO_2 and carbon monoxide. From the finding that most acetogens can grow on H_2 plus CO_2 and form acetate as the major end product of metabolism it can be deduced that CO_2 reduction to acetate must be coupled to the generation of ATP - probably via electron transport phosphorylation. This communication will mainly deal with the pathway of CO_2 reduction to acetate and with the key enzyme involved, namely the carbon monoxide dehydrogenase. The physiological role of this enzyme is discussed as well as some energetical aspects of the carbon monoxide dehydrogenase reaction. For recent reviews on the subject see [1-4].

2. CO_2 REDUCTION TO ACETATE

The synthesis of acetate from 2 CO_2 according to

$$\text{(a)} \quad 2\,CO_2 + 8\,[H] \rightarrow CH_3COO^- + H^+ + 2\,H_2O$$

requires (i) the reduction of 1 CO_2 with 6 electrons to the methyl group and (ii) the reduction of 1 CO_2 with 2 electrons to the carboxyl group. The pathway of the methyl group formation has been thoroughly investigated by L. Ljungdahl, H. G. Wood, and coworkers [1, 5]. CO_2 is first reduced to formate by CO_2 reductase. The formate is then bound to tetrahydrofolate at the expense of ATP. The pteridine-bound formate is subsequently dehydrated to methenyl tetrahydrofolate, which is then reduced via methylene terahydrofolate to methyl tetrahydrofolate as the precursor of the methyl group. Early investigations have indicated that the methyl is transferred to a corrinoid [5]. Although the synthesis of the methyl group of acetate has been elucidated since many years, almost nothing was known about the formation of the carboxyl group until recently. Early evidence was presented by Wood and coworkers, that the carboxyl group of pyruvate rather than free CO_2 is the precursor of the carboxyl group of acetate [6]. A reductive transcarboxylation from pyruvate would make sense for acetogens which degrade carbohydrates via pyruvate as intermediate; it is, however, unlikely for acetogens growing on C_1-compounds (see also [7]). Thus the synthesis of the carboxyl group of acetate remained unclear until the discovery of carbon monoxide dehydrogenase in the acetogenic bacteria [8]. This enzyme, which contains nickel as a cofactor [9-11], catalyzes in vitro the oxidation of CO to CO_2 with artificial electron acceptors such as methyl viologen (MV):

(b) $CO + 2\,MV^{2+} + H_2O \rightarrow CO_2 + 2\,MV^{+} + 2\,H^{+}$

Evidence was presented that CO dehydrogenase is somehow involved in CO_2 reduction to acetate - probably in one of the final steps of acetate synthesis [8]. This was later confirmed and specified by investigations of Wood and coworkers. The authors demonstrated with an enzyme system purified from *Clostridium thermoaceticum* an isotopic exchange of CO and the carboxyl group of acetyl CoA, which is the primary product of the carboxyl-methyl "addition" [12].

Later it was demonstrated with cultures of *Acetobacterium woodii* growing on H_2 and CO_2 that CO also in vivo can be the precursor of the carboxyl of acetate [13] and that cell suspensions of this organism and of *Clostridium thermoaceticum* formed traces of free CO from H_2 and CO_2 [14, 15]. Inhibition studies indicated that CO dehydrogenase catalyzes the reduction of CO_2 to CO [14]. Wood and coworkers then reported that purified CO dehydrogenase alone mediated the isotopic exchange of ^{14}CO and the carboxyl group of acetyl CoA [16]. The authors concluded that CO dehydrogenase in vivo mediates the formation of acetyl CoA from CO, the methyl group, and CoA. However, the activity of the isotopic exchange was very low - even with the purified enzyme. Moreover, it was demonstrated that cyanide, an inhibitor of CO oxidation [8] as well as of CO formation, did not effect CO incorporation into the carboxyl group of acetate [14]. Thus the role of CO dehydrogenase in acetyl CoA synthesis from CO and the methyl group is still unclear and requires further investigations. Probably *C. thermoaceticum* is not an appropriate organism to study the carbonylation reaction in detail, and it might be favourable to

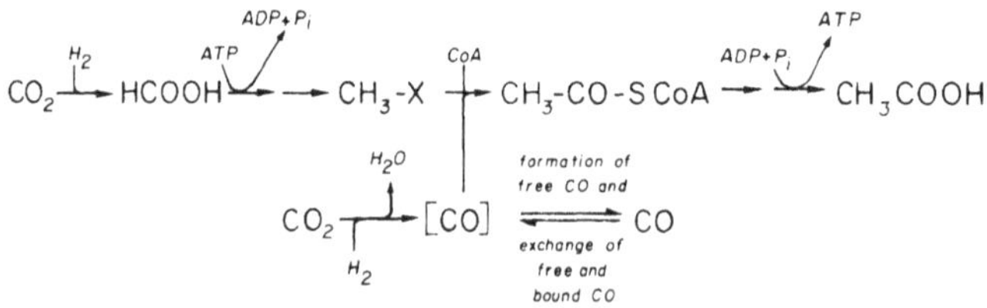

Figure 1. Tentative scheme for the acetate synthesis from 2 CO_2 in acetogenic bacteria. [CO] = bound carbon monoxide.

use an acetogen, which catalyzes a carbonylation at catabolic activities with free carbon monoxide as a substrate. A tentative scheme for the CO_2 reduction to acetate is given in Fig. 1.

3. ENERGETICS OF CO FORMATION AND CO OXIDATION

As mentioned above, one of the physiological functions of CO dehydrogenase is the reduction of CO_2 to CO. Since the standard redox potential of the CO_2/CO couple (-550 mV) is lower than that of the H^+/H_2 couple (-420 mV), CO formation from H_2 and CO_2 is an endergonic reation:

$$\text{(c)} \quad CO_2 + H_2 \rightarrow CO + H_2O \qquad (\Delta G^{0\prime} = +20 \text{ kJ/mol})$$

The equilibrium of reaction (c) is far on the side of CO_2 and H_2. Therefore it was tested with *A. woodii*, whether CO formation from CO_2 and H_2 requires the input of metabolic energy [15]. The suspensions were incubated under defined conditions with H_2/CO_2 (80%/20%) in the headspace. CO formation was determined until an apparent equilibrium was reached. The equilibrium concentration of CO in the gas phase exceeded by far the expected value calculated for the conditions used. This finding clearly showed that CO formation from CO_2 and H_2 is energy dependent. Investigations on the influence of ionophores or the ATP synthase inhibitor dicyclohexylcarbodiimide (DCCD) indicated, that this reaction could be driven by the electrochemical proton potential; unambiguous evidence, however, is still lacking. Surprisingly the ionophore nigericin, which mediates a non-electrogenic antiport of H^+ and K^+, strongly inhibited CO formation without effecting the membrane potential [15]. The possibility, that somehow the driving force for CO_2 reduction to CO could be the ΔpH-component of the electrochemical proton potential, has to be seen.

Cell suspensions of *A. woodii* also mediated the reverse reaction, i.e. the conversionof CO to CO_2 and H_2 according to

$$\text{(d)} \quad CO + H_2O \rightarrow CO_2 + H_2 \qquad (\Delta G^{0\prime} = -20 \text{ kJ/mol})$$

If CO formation from CO_2 and H_2 requires metabolic energy, it could be that the reverse reaction is coupled to the generation of ATP. This was confirmed by experiments with cell suspensions of *A. woodii* [15]. It was shown that CO oxidation could drive the energy dependent uptake of histidine into the cells. The influence of different ionophores and of DCCD on CO oxidation and on the ATP levels in cells incubated in the presence of carbon monoxide indicated that ATP formation from CO oxidation could be by way of electron transport phosphorylation.

Various acetogenic bacteria are able to grow on carbon monoxide as the sole energy source (for the most recent report see [17]). During growth on CO acetate and CO_2 are formed as the major end products:

(e) $4\,CO + 2\,H_2O \rightarrow 1\,CH_3COO^- + 1\,H^+ + 2\,CO_2$ ($\Delta G^{o\prime}$ = -165 kJ/mol acetate)

This pathway includes the oxidation of CO to CO_2. Since the latter reaction can be coupled to the generation of ATP, acetate formation from CO can be energetically more favourable than from H_2 plus CO_2.

Recently we isolated a strain of *Peptostreptococcus productus* (see also [17]), which grows very well on carbon monoxide as sole energy source. It seemed tempting to find out with this organism whether free carbon monoxide was directly incorporated into the carboxyl of acetate or whether CO is all oxidized to CO_2 prior to CO_2 reduction to acetate according to Fig. 1. Preliminary experiments with cell suspensions of this organism incubated in the presence of labelled CO and unlabelled CO_2 or vice versa indicated that *P. productus* mediates the incorporation of carbon monoxide in free form into acetate (data not shown). Therefore it is advantageous to carry on experiments with this organism to study the carbonylation reaction in detail.

References

[1] Wood, H.G., Drake, H.L., Hu, S.-I. (1982). In: Amino acids, fermentations, and nucleic acids: A symposium (E:E: Snell, Ed.), pp. 29-56. Ann. Rev. Inc., Palo Alto, California.
[2] Fuchs, G. (1986). FEMS Microbiol. Rev., in press.
[3] Diekert, G. 1986. In: Bioinorganic chemistry of nickel. VCH Publishers Inc., Deerfield Beach, in press.
[4] Diekert, G., Fuchs, G., Thauer, R.K. (1985) In: Microbial gas metabolism: Mechanistic, metabolic, and biotechnological aspects (R.K. Poole, C.S. Dow, Eds.), pp. 115-130. Academic Press Inc., New York.]
[5] Ljungdahl, L.G., Wood, H.G. (1982). In: B_{12}, Vol. 2 (D.Dolphin, Ed.), pp. 165-202. Wiley, New York.
[6] Schulman, M., Ghambeer, R.K., Ljungdahl, L.G., Wood, H.G. (1973). J. Biol. Chem. 248, 6255-6261.
[7] Eden. G., Fuchs, G. (1982). Arch. Microbiol. 133, 66-74.
[8] Diekert, G.B., Thauer, R.K. (1978). J. Bacteriol. 136, 597-606.
[9] Diekert , G., Graf, E.G., Thauer, R.K. (1979). Arch. Microbiol. 122, 117-120.
[10] Ragsdale, S.W., Clark, J.E., Ljungdahl, L. G. , Lundie, L.L., Drake, H.L. (1983). J. Biol. Chem. 258, 2364-2369.
[11] Diekert, G., Ritter, M. (1983). FEBS Lett. 151, 41-44.
[12] Hu, S.-I., Drake, H.L., Wood, H.G. (1982). J. Bacteriol. 149, 440-448.
[13] Diekert, G., Ritter, M. (1983). FEMS Microbiol. Lett. 17, 299-302.
[14] Diekert, G., Hansch, M., Conrad, R. (1984). Arch. Microbiol. 138, 224-228.
[15] Diekert, G., Schrader, E., Harder, W. (1986). Arch. Microbiol. 144, 386-392.
[16] Wood, G.H., Ragsdale, S.W., Pezacka, E. (1986). TIBS 11, 14-18.
[17] Lorowitz, W.H., Bryant, M.P. (1984). Appl. Environ. Microbiol. 47, 961-964.

ECOLOGY OF C_1-METABOLIZING ANAEROBES

B. SCHINK

Fachbereich Biologie, Mikrobiologie, Philipps-Universität, S-3550 Marburg/Lahn. West Germany

1. INTRODUCTION

C_1-Compounds occur in nature at all redox states of carbon with oxidation numbers from - 4 to + 4. Aerobic and anaerobic microorganisms are involved in the transformation of these compounds in natural and man-made environments. Whereas methane and carbon dioxide are comparably inert metabolites which accumulate especially in anoxic environments to high amounts, the other C_1-compounds are rather reactive and are transformed quickly. Usually, they are hardly detectable in anoxic sediments or sludges. In the boundary zones between oxic and anoxic layers in water bodies and sediments or in heterogeneous environments such as soils or biofilms, also a high exchange of such metabolites between oxic and anoxic compartments has to be expected. It has to be emphasized that also the digestive tract of many animals and man represents an anoxic environment, and that a considerable part of the human diet is digested without the participation of molecular oxygen.

Among the anaerobic bacteria metabolizing C_1-compounds, homoacetogenic bacteria are perhaps the most versatile ones with respect to the range of substrates utilized, followed by methanogenic and sulfate-reducing bacteria. Competition for C_1-compounds between these groups of anaerobes, therefore, often yields results very different from those obtained in similar experiments with hydrogen or acetate as substrates.

The present contribution gives a survey on recent issues in the metabolism of C_1-compounds by anaerobic bacteria. Emphasis is placed on the mutual relationships of the various bacteria and on their interaction with the surrounding abiotic environment. A more extensive review on anaerobic C_1-metabolism was published recently [1].

2. TRANSFORMATION OF C_1-COMPOUNDS

2.1 Anaerobic methane oxidation

Methane is the terminal reduced product formed from organic matter in the apparent absence of other inorganic electron acceptors than CO_2. Its reactivation to serve as a source of carbon or electrons usually required the action of a monooxygenase enzyme in the presence of molecular oxygen. The question has never been answered satisfactorily whether methane could also been activated in the absence of oxygen. Indication of a sulfate-dependent methane oxidation came mainly from distribution curves of methane, sulfate, and sulfide in anoxic marine sediments and water bodies [2] as well as from radiotracer studies with $^{14}CH_4$ [3]. Experiments with pure cultures demonstrated that methanogenic bacteria catalyze a reoxidation of CH_4 to CO_2. This oxidation only occurred with simultaneous methane formation and the rates of methane formation were always 3-5 orders of magnitude higher than methane oxidation [4]. Labelling studies provided preliminary evidence that methane oxidation was not simply a back reaction of methane production, and it was assumed that a separate mechanism of methane oxidation exists which bypasses the methyl CoM-reductase reaction [4]. Thus, there is a mechanism of anaerobic methane activation, but the necessity of simultaneous methane formation in the methanogenic bacteria studied rendered this mechanism insufficient to explain net methane oxidation with concomitant sulfate reduction. Mixed culture experiments with sulfate-reducing and methanogenic bacteria were carried out in our laboratory together with Dr. Fritz Widdel to check whether acetate or hydrogen consumption by sulfate reducers could allow net methane oxidation by methanogenic bacteria. The results were never convincing, and a significant difference between sulfide formation in the presence or in the absence of methane could not be demonstrated reproducibly.

The recent observation by G. Fuchs and his coworkers [5] that several acetate-oxidizing sulfate reducers exhibit considerable methane formation (mini-methane formation) may provide a new key to this long-discussed problem. If the mini-methane formation reaction is reversible, methane can be oxidized by these bacteria, and the electrons derived would flow to sulfate reduction. Although the overall reaction is slightly exergonic (ΔG_0 = -13 kJ mol^{-1}) it would represent only a cometabolic activity for these bacteria, probably without further impact on their energy budget. This could explain why nobody so far succeeded in the enrichment or isolation of methane-oxidizing sulfate reducers.

2.2. Utilization of methyl compounds

Free methanol is formed by enzymatic cleavage of pectin methyl esters by aerobic and anaerobic pectinolytic bactera. Apparently none of these bacteria utilizes the methanol itself. [6]. A further source of methanol is methane oxidation by aerobic methylotrophic bacteria which always form traces of methanol as by product [7]. Some of this methanol may also enter anoxic environments.

The methyl groups of methoxylated aromatic compounds, e.g. lignin monomers, are aerobically released as formaldehyde by monooxygenase action [8]. Methanol formation is only possible if the ether linkage has been transformed to an ester linkage by action of a dioxygenase [9]. Anaerobic demethylation of methoxylated aromatic compounds by homoacetogenic [10-13] and facultatively anaerobic bacteria [14] does not release methanol at measurable concentrations. Studies with *Acetobacterium woodii* revealed that growth rates and yields were considerably higher with methoxylated aromatics than with free methanol indicating that methanol is not an intermediate [15]. Probably the methyl group is directly transferred to a (B_{12}?) carrier, but biochemical studies on this anaerobic ether cleavage reaction are still lacking.

Methanol is utilized by methanogenic and homoacetogenic bacteria. Some of the latter ones which can be considered as the C_1 metabolizers *par excellence* form also butyrate besides acetate at high substrate concentrations. Recently homoacetogenic bacteria were isolated which grow also with methylamines [16], a group of substrates which was so far reserved for methanogenic bacteria [17].

Utilization of methanol by sulfate reducers has only recently been proven unequivocally. One *Desulfovibrio* strain oxidizes methanol only cometabolically without concomitant growth [18] wheras another *Desulfovibrio* strain grows at the expense of methanol oxidation [19]. *Desulfotomaculum orientis* can oxidize methanol with concomitant sulfate reduction. In the absence of sulfate, it can grow with methanol or formate and carry out a homoacetate fermentation [20]. It is amazing which metabolic versatility *Desulfotomaculum* now exhibits after it had for long a time been considered a minor and metabolically inferior cousin of *Desulfovibrio*.

Homoacetogenic, methanogenic, and sulfate-reducing bacteria produce hydrogen at a low partial pressure during degradation of methyl compounds and other organic substrates. This phenomenon led to the establishment of the"hydrogen-cycling" hypothesis [21-22], but can also be explained simply by an exchange of the redox state of the cell-internal electron carriers with a free hydrogen pool via hydrogenase enzymes. Today, there is no reliable evidence any more in support of the "hydrogen-cycling" hypothesis. Nonetheless, the pool of free hydrogen produced can be used up by hydrogen-oxidizing anaerobes which thus exert a severe influence in the fermentation balance of the primary organism [23-25]. This interspecies hydrogen transfer between methylotrophic and non-methylotrophic hydrogen-oxidizing anaerobes could help to understand how e.g. methanol oxidation could stimulate sulfate reduction observed in intertidal sediments [26]. The recent isolation of methanol-oxidizing sulfate reducers provides another explanation for this phenomenon. However, the cooperation of the three major groups of C_1-metabolizing anaerobes appears to be more complex: In salt marshes and freshwater sediments, methanol and methylamines were nearly exclusively transformed to methane while acetate and hydrogen metabbolism led primarily to sulfide production [27-28]. These contradictory results demonstrate that more environmental factors than simply substrate affinities and utilization spectra govern the mutual relationships between bacteria cooperating in the anaerobic food chain.

2.3. Formaldehyde metabolism

Formaldehyde is extremely reactive and very toxic to living cells. Nonetheless, it is formed in nature by oxygenase-dependent demethylation of lignin constituents [8], and traces may also enter anoxic environments. In our laboratory, we used hexamethylene tetramine as a formaldehyde substitute which is slowly hydrolyzed in aqueous solution to formaldehyde and ammonia.

From all enrichment cultures with hexamethylene tetramine, we isolated homoacetogenic bacteria resembling *Acetobacterium* sp., and the known species of this genus also grew well with this substrate. We also isolated some new strains of spore-forming anaerobes which are still being characterized. It remains to be established whether formaldehyde or its dismutation products methanol and formate are the actual growth substrates for these bacteria.

2.4 Decarboxylations and carboxylations

Dicarboxylic acids with 2 to 4 C-atoms are degraded by decarboxylation to the respective monocarboxylic acids. The free energy change of these reactions (15-25 kJ mol^{-1}) is sufficient to cover the energy needs for growth of these bacteria [29,30, Schink, unpublished results].

However, also carboxylation reactions can play important roles in anaerobic degradation. Methanogenic degradation of propionate proceeds via carboxylation to methylmalonyl CoA as was demonstrated by tracer experiments with enrichment cultures and sediment samples [31,32]. Also C_2-compounds can be converted via two carboxylations to propionate, and its reoxidation again includes carboxylation to a C_4 dicarboxylic acid [33,34]. The formation of propionate represents the third major anaerobic process by which carbon dioxide is employed as anaerobic electron acceptor. It also provides so far the only pathway for producing odd-numbered fatty acids as products of the fermentation of C_1 and C_2 compounds [34,35].

Carboxylations appear also to be involved in the anaerobic degradation of isovalerate [36], branched-chain hydrocarbons [37], and aliphatic ketones [Schink and Platen, unpublished results]. Although some of these reaction sequences are known from aerobic biodegradation systems, the energetical aspects of these processes are highly interesting because the bacteria often have to carry out the carboxylation steps at the expense of less than one ATP equivalent.

It remains to be examined whether also interspecies transfers of intermediates other than hydrogen (e.g. acetate) help to balance the energy budgets of the bacteria involved.

2.5 Influence of acidity on C_1 metabolizing anaerobes

The CO_2/bicarbonate system is the natural buffer in neutral to slightly alkaline sediments and lakes. Comparison of a mildly acidic lake (Knaacke Lake; pH 6.1) with a neutral lake (Lake Mendota; pH 7.1) revealed that at lower pH homoacetogenic bacteria outcompete methanogenic bacteria for hydrogen and hydrogen equivalents [38]. As a consequence, the organic matter is channelled to methane and CO_2 nearly exclusively through acetate (95%), contrary to the situation in neutral environments in which acetate cleavage accounts only for 60-70% of total methane formation. Obviously the homoacetogens are more tolerant to acidic conditions than the methanogens. It is, however, also not clear yet whether the pH itself or the concomitant decrease in the CO_2 pool size (about 90%) is primarily responsible for this effect. It is interesting to note that in the slightly acidic Knaack Lake sediment also the "homopropionate" fermenting bacteria were by far more important than in a neutral sediment [34]. Both processes " harbour" bound CO_2 in the form of carboxylic groups, and it is tempting to speculate whether this effect is related to the low CO_2 pool size in this environment.

3. CONCLUSIONS

Isolation and thorough characterization of new anaerobes has broadened our understanding of anaerobic transformation of C_1-compounds considerably in the recent past. It is amazing to note that more than ten new types of homoacetogenic bacteria have recently been isolated, wheras Wieringa's first isolation of *Clostridium aceticum* could not be reproduced for nearly 40 years. Perhaps this success is due to the use of bicarbonate-rich media which mimic the situation in a natural anoxic environment by far better than did the phosphate-buffered media used before. The search for more environmentally representative isolates in hand with studies on substrate transformations in complex anoxic environments will further help to elucidate the mutual relationship among C_1-metabolizing anaerobes to study their intermediary metabolism, and to employ their biochemical capacities for treatment of waste materials and for the production of suited secondary products.

References

[1] Zeikus JG. (1983) Adv. Microbial Physiol. 24, 215-299.
[2] Reeburgh WS, Heggie DT. (1977) Limnol. Oceanogr. 22, 1-9.
[3] Panganiban AT, Patt TE, Hart W, Hanson RS. (1979) Appl. Environ. Microbiol. 37, 303-309.
[4] Zehnder AJB, Brock TD. (1979) J. Bacteriol. 137, 420-432.
[5] Schauder R, Eikmanns B, Thauer RK, Widdel F, Fuchs G. (1986) Arch. Microbiol. 145, 162-172.
[6] Schink B, Zeikus JG. (1980) Curr. Microbiol. 4, 387-390.
[7] Wolfe RS, Higgins IJ. (1979) Internat. Rev. Biochem. Microbial Biochemistry, Vol. 21, University Park Press, Baltimore, pp. 267-352.
[8] Bernhardt FH, Staudinger H, Ullrich V. (1970) Hoppe-Seyler's Z. Physiol. Chem. 351, 467-478.
[9] Donnelly MJ, Dagley S. (1980) J. Bacteriol. 142, 916-924.
[10] Bache R. Pfennig N. (1981) Arch. Microbiol. 130, 255-261.
[11] Krumholtz LR, Bryant MP. (1985) Int. J. Syst. Bacteriol. 35, 454-456.
[12] Frazer AC, Young LY. (1985) Appl. Environ. Microbiol. 49, 1345-1347.
[13] Mountfort DO, Asher RA. (1986) Arch. Microbiol. 144, 55-61.
[14] Grbic-Galic D. (1985) Appl. Environ. Microbiol. 50, 1052-1057.
[15] Tschech A, Pfennig N. (1984) Arch. Microbiol. 137, 163-167.
[16] Eichler B, Schink B. (1984) Arch. Microbiol. 140, 147-152.
[17] Hippe H, Caspari D, Fiebig K, Gottschalk G. (1979) Proc. Natl. Acad. Sci. USA 76, 494-498.
[18] Braun M, Stolp H. (1985) Arch. Microbiol. 142, 77-80.
[19] Nanninga HJ, Gottschal JG. (1986) FEMS Microbiol. Ecol. 38, 125-130.
[20] Klemps R, Cypionka H, Widdel F, Pfennig N. (1985) Arch. Microbiol. 143, 203-208.
[21] Odom JM, Peck HD. (1981) FEMS Microbiology Lett. 12, 47-50.
[22] Odom JM, Peck HD. (1984) Ann. Rev. Microbiol. 34, 551-592.
[23] Phelps TJ, Conrad R, Zeikus JG. (1985) Appl. Environ. Microbiol 50, 589-594.
[24] Cord-Ruwisch R, Ollivier B. (1986) Arch. Microbiol. 144, 163-165.
[25] Heijthuijsen JHFG, Hansen TA. (1986) FEMS Microbiol. Ecol. 38, 57-64.
[26] King GM, Klug MJ, Lovley, DR. (1983) Appl. Environm. Microbiol. 45, 1848-1853.
[27] Oremland RS, Marsh LM, Polcin S. (1982) Nature 296, 143-145.
[28] Lovley DR, Klug MJ. (1983) Appl. Environ. Microbiol. 45, 1310-1315.
[29] Allison MJ, Dawson KA, Mayberg WR, Foss JG. (1985) Arch. Microbiol. 141, 1-7.
[30] Schink B, Pfennig N. (1982) Arch. Microbiol. 133, 209-216.
[31] Koch M, Dolfing J, Wuhrmann K, Zehnder AJB. (1983) Appl. Environ. Microbiol. 45, 1411-1414.
[32] Schink B. (1985) J. Gen. Microbiol. 131, 643-650.
[33] Schink B. (1984) Arch. Microbiol. 137, 33-41.
[34] Schink B, Phelps TJ, Eichler B, Zeikus JG. (1985) J. Gen. Microbiol. 131, 651-660.
[35] Goldberg I, Cooney CL. (1981) Appl. Environ. Microbiol. 41, 148-154.
[36] Stieb M, Schink B. (1986) Arch. Microbiol. 144, 291-295.
[37] Schink B. (1985) FEMS Microbiol. Ecol. 31, 69-77.
[38] Phelps TJ, Zeikus JG. (1984) Appl. Environ. Microbiol. 48, 1088-1095.

PHYSIOLOGY AND BIOCHEMISTRY OF METHYLOTROPHIC BACTERIA

PHYSIOLOGY AND BIOCHEMISTRY OF METHYLOTROPHIC BACTERIA

H. DALTON AND I.J. HIGGINS
Department of Biological Sciences, University of Warwick, Coventry, U.K. and Biotechnology Centre, Cranfield Institute of Technology, U.K. and the Leicester Biocentre, University of Leicester, Leicester, U.K.

1. INTRODUCTION

The importance of a thorough understanding of the physiology and biochemistry of a microorganism before proceeding to genetic studies or industrial exploitation cannot be over-emphasised. In the case of the methylotrophic bacteria, we have a considerable body of basic biochemical information for a range of species including representatives of facultative and obligate methylotrophs and both Type I and Type II methanotrophs. However, much remains to be discovered concerning the important areas of growth physiology and metabolic regulation. The main theme of this contribution, which introduces the section on methylotrophic physiology and biochemistry, is regulation and control, the area in which in the view of the authors, the greatest progress has been made since the last symposium in this series. Table I summarises the important topics discussed herein and in the ensuing contributions.

Table 1. Recent advances in the physiology and biochemistry of methylotrophic bacteria

Growth physiology and control	- Type I and Type II obligate methanotrophs
Metabolic regulation	- Facultative methylotrophs
	- Obligate methanotrophs
Role of PQQ enzymes in methylotrophic metabolism	
Role and range of C-1 carriers in metabolism	
New metabolic findings arising from exploitation of novel techniques, especially *in vivo* NMR.	

Even at the levels of basic enzymology and metabolic mechanisms, important new findings continue to be made as exemplified in the contributions to this volume on the biochemistry of C-1 carriers [1] and the role of PQQ enzymes [2]. Our knowledge of the distribution and importance of the quinoproteins has expanded considerably in the last few years. Of particular note is progress in our understanding of their mechanism, mode of coupling to the respiratory chain and most recently in the case of methanol dehydrogenase, modulation of substrate specificity by a modifier protein [3].

There has in the past been a tendency to develop a somewhat simplistic view of methylotrophic metabolism. This applies particularly to the primary oxidative pathways which it is now clear involve multiple enzymes and complex regulatory systems. Sophisticated mechanisms are required to control the flow of C-1 metabolites between biosynthesis and energy transduction. More particularly however, there is the special problem of methylotrophic oxidative metabolism to be considered namely the involvement of a highly toxic compound, formaldehyde as a mainline intermediate. Formaldehyde occupies a unique position in methylotrophic metabolism, not only because of its toxicity, but because it is also the key intermediate at the branchpoint between anabolism and catabolism. It is only in the last few years that we have begun to understand the role of formaldehyde as a regulatory intermediate and the mechanisms involved in controlling its concentration *in vivo.* The information has arisen from a range of approaches and laboratories : A detailed metabolic study of the mechnisms involved in facultative methylotrophs, Arthrobacter P1 in particular, is described in this volume [4]. In the case of the methanotrophs, results have been obtained from studies as far removed as basic enzymology and protein chemistry [5] and whole organism NMR [6]. It has been known for many years that formaldehyde can indeed accumulate to growth-inhibitory concentrations [7]. Using carbon-13 NMR to investigate metabolite concentrations in methanotrophs, it has recently been shown that formaldehyde levels are not significant when a soluble methane monooxygenase s(MMO) system (see below) is present. In contrast, even in the same species, the presence of particulate MMO (pMMO) is associated with formaldehyde accumulation.

Over the years, there have been many problems of inconsistencies and irreproducibilities in studies with methylotrophic and especially methanotrophic bacteria. It is now becoming possible to rationalise these problems in the light of new information, particularly about multiple enzymes, modulation of enzyme activities by protein modifiers, conventional metabolic regulation studies, controlled growth techniques and in *in vivo* metabolic studies. We exemplify -ecent progress below by reference to two important areas where rapid progress has taken place, namely formaldehyde regulation and regulation of methane oxidation.

2. FORMALDEHYDE REGULATION

There are two intimately linked and poorly understood aspects of this important problem, namely regulation of formaldehyde levels and regulation of metabolism by formaldehyde. The current status of our understanding of these processes in facultative species is discussed in this volume [4]. In Arthrobacter P1, formaldehyde generated from the metabolism of methylamine or added directly, dramatically induces the enzymes specifically involved in the RuMP pathway, hexulosephosphate synthase and hexulosephosphate isomerase. It appears that formaldehyde also regulates its own rate of synthesis from methylamine by inhibiting the activities of both the methylamine transport system and the amine oxidase.

capable of such efficient regulation of formaldehyde levels. The control mechanism may well depend primarily on the ratio of oxidised to reduced coenzyme. High concentrations of reduced coenzyme, which might be expected to be associated with high formaldehyde levels would encourage utilisation of reduced coenzyme by the MMO in preference to alternative electron donors. More work is required, especially with the particulate system to test this hypothesis.

3. REGULATION OF METHANE OXIDATION

Substantial advances in our understanding of two aspects of the control of MMO activities and synthesis have occurred in the three years since the last C-1 symposium. The first is the definitive role played by copper ions in the expression of either sMMO or pMMO; the second is the role of the enigmatic protein B, a necessary component of the MMO system. It has been reported that the *M. capsulatus* (Bath) enzyme exists in two forms [10]: Firstly, the soluble enzyme comprised of three proteins (A, B and C) which uses NADH as electron donor via protein C passing reducing equivalents on to protein A; secondly, the particulate enzyme about which little is known, but which can also use NADH as electron donor and which is only expressed when the copper:biomass ratio is high. The situation appears to be closely similar in the Type II organism, *Methylosinus trichosporium* [11].

Recently, the accumulation of more information concerning the physicochemical nature of the protein components has facilitated a clearer understanding of how copper regulates MMO's. NADH interacts with protein C, which is an iron-sulphur flavoprotein of molecular weight 44,000, by first reducing the FAD centre of the protein (a two-electron reduction), which in turn passes electrons on to the iron-sulphur centre (a one-electron reduction). Reduced protein C, acting as a transformase, is then able to pass single electrons on to protein A[12]. Protein A, a non-haem iron protein of molecular weight 220,000, is the site of methane oxidation [5] and oxygen reduction [9] and accepts electrons from protein C. Unlike the iron centre of protein C which is associated with acid labile sulphur groups, the iron centre of protein A appears to be in the form of a μ-oxobridge structure in which the pair of iron atoms are antiferromagnetically-coupled through an oxygen atom [13].

In the presence of high levels of copper ions, there is a rapid inhibition of sMMO both *in vitro* and *in vivo* which is caused primairly by a rapid loss of the iron sulphur centre, thereby preventing electron transfer from protein C to protein A. Under these circumstances protein C still remains its NADH acceptor: reductase activity since it is still capable of passing electrons on to artificial acceptors via the flavin centre [14]. A secondary effect is also observed at higher copper concentrations in which the FAD centre is lost and electrons can no longer be accepted from NADH. Interestingly, at these concentrations of copper, the complex intracytoplasmic membranes sometimes observed in methanotrophs appear *in vivo* and there is a dramatic increase in the levels of pMMO. Furthermore, there appears to be a requirement for copper ions for full expression of pMMO activity in *M. capsulatus* (Bath) [15] and *Mts. trichosporium* [11]. The new proteins that appear on exposure to high levels of copper are distinct from their soluble counterparts in several respects. Their molecular weights on SDS

It has long been known that the isolated quinoprotein, methanol dehydrogenase from methanol-utilising methylotrophs rapidly oxidises formaldehyde in addition to methanol and some other alcohols. Two species, *Methylophilus methylotrophus* and *Pseudomonas* AM1 have been shown to elaborate a modifier protein [3]. This large protein (dimer of molecular weight, 140,000) decreases the affinity of the enzyme for formaldehyde by over 97%, preventing significant oxidation of the aldehyde. It also changes the substrate specificity towards higher alcohols somewhat. This protein clearly plays an important regulatory role as oxidation of formaldehyde to formate by the periplasmic methanol dehydrogenase would represent loss of carbon substrate as well as being energetically less favourable. However, it is possible that this process might be desirable under some conditons in order to protect the organism from a tendency to accumulate formaldehyde. It was shown [3] that the ratio of M protein to the dehydrogenase was much lower (1:15 compared to 1:2) in organisms grown under oxygen-limitation as opposed to carbon-limitation. This would be consistent with a release of control on the methanol dehydrogenase, allowing it to oxidise formaldehyde under conditons when this toxic intermediate might be expected to accumulate.

A very special situation pertains in the methanotrophs in that the first step in the oxidative sequence, the oxidation of methane by a monooxygenase utilises reducing equivalents, most of which are derived from the further oxidation of formaldehyde. Also a s MMO, being NADH-linked, generates oxidised coenzyme acceptor for formaldehyde dehydrogenase; this may also be true in part where a p MMO is present which can also use NADH. Whatever the details of the particular system involved, there is an intimate connection between the initial step in the metabolic process and the further metabolism of the key branch-point intermediate, formaldehyde. Regulation of the two types of MMO and control of the switch from one form to the other is discussed in the next section. Whilst we know less about the mechanisms of the control links between the MMO's and formaldehyde metabolism, there are a number of facts available which allow the construction of plausible theories:

1. It is clear that formaldehyde rarely, if ever accumulates to significant levels in methanotrophs containing s MMO [6].
2. Formaldehyde accumulation has been observed in organisms with p MMO [6].
3. Under some conditions methanotrophs are difficult to grow in pure culture or to adapt to growth on methanol, probably due to formaldehyde accumulation [8].
4. It has recently been shown that the B protein of the s MMO modulates the oxidase and oxygenase activities of this enzyme and that formaldehyde via its effect on the B protein causes stimulation of the NADH oxidase activity of the MMO [9].
5. There is evidence that formaldehyde represses synthesis of s MMO in *Methylococcus capsulatus* [Bath] [10].

It seems likely therefore that in the case of organisms containing the s MMO, formaldehyde regulates the supply of oxidised coenzyme acceptor required for its further oxidation by stimulating its generation via the oxidase function of the MMO. In this way high levels of formaldehyde are avoided. In contrast, organisms containing the energetically more efficient particulate enzyme (see below) appear not to be polyacrylamide gel systems are approximately 46,000, 35,000 and 25,000 respectively, the pMMO is readily inhibited by agents that do not affect sMMO [16,17] and there is a substantial increase in the carbon conversion efficiency of methane into biomass of 38% with the transition from sMMO to pMMO [18].

It has been argued that since the improvement in growth yield in organisms possessing particulate MMO is associated with the ability of ethanol to act as an electron donor to the particulate enzyme *in vivo,* then there may be a direct coupling of methanol dehydrogenase to MMO. Such an argument, of course, has been advanced earlier [20,21] although the yield data was not available at the time. Since one can now readily dictate the form of MMO in the cell by altering the level of copper in the medium, one can predict that demonstration of reproducible *in vitro* coupling of MDH to MMO might be forthcoming in the next few years. It has recently been shown *in vitro* that succinate is a highly effective source of reducing equivalents for the pMMO in membrane preparations of *Mts. trichosporium* via a mechanism not involving pyridine nucleotides [22]. Certainly the discovery of the switch between the two forms of the enzyme involving copper ions has helped to clarify the somewhat confused situation between a number of groups working on the different MMO's and there is now a measure of agreement on this [11].

As indicated earlier, our simplistic view of methane oxidation by MMO had belied the complex regulatory patterns that control both synthesis and activity of the enzyme. Crucial to our understanding of how the system functions is the purification of the components involved in catalysis. For many years it had been known that 3 proteins were necessary for activity, but only recently has the role of protein B been understood in the overall scheme of methane oxidation. The purified protein has a molecular weight of about 16,000 and contains no prosthetic groups. In the absence of protein B, proteins C and A catalyse the NADH-driven four electron reduction of oxygen to water only. Even though methane may be present it is not transformed. If protein B is now added to the system (in the absence of methane) electron flow to protein A is dramatically reduced to a value of zero at equivalent concentrations of protein B and protein A. If methane (or any other oxidisable hydrocarbon substrate) is now added back to the system then rapid oxygenation of substrate occurs with half the oxygen being reduced to water and the other half ending up in the organic product (methanol). Under the circumstances it is clear that protein B can act as a regulator of enzyme activity converting the enzyme from an oxidase (in the absence of B) to an oxygenase (in the presence of B). Under normal circumstances, however, one would expect protein B to be present when proteins A and C were also present, so what is its role *in vivo*? One possibility might be that protein B acts by binding to protein A ($K_{diss} = 1.4 \times 10^{-7}$M.J. Green, personal communication) thereby permitting methane to bind and be oxidised by protein A. In the absence of methane the A/B complex now prevents the wasteful oxidation of NADH which can be very important in organisms which are essentially limited for their NADH supply. Another role for protein B may be to prevent excessive levels of the toxic metabolite formaldehyde from accumulating within the organisms. It has been shown in *Mts. trichosporium,* for example, that during methanol metabolism formaldehyde can accumulate within cells but only when pMMO rather than sMMO is present [6]. One explanation for this could be that pMMO is relatively ineffective at using NADH *in vivo* as an electron donor so that NADH tends to accumulate (NADH oxidase levels are low in these organisms [23].) Consequently formaldehyde dehydrogenase, which requires NAD^+, would be operating at low activities due to the low level of NAD^+ within the cell. This would not be the case when sMMO was present because sMMO is an effective sink for NADH and readily generates NAD^+ in the presence of formaldehyde, which acts as an uncoupler of MMO (by dissociating protein B from the complex), thereby allowing NADH to be oxidised to NAD at the expense of methane oxidation.

It is quite clear form these two examples of regulation in methanotrophs that we have only scratched the surface of a fascinating control system. The reports, at this meeting, on the cloning of protein A of the sMMO from *M. capsulatus* (Bath) (Mullins, Murnell, Salmond & Dalton) and the solubilisation and purification of the pMMO from *M. capsulatus* (strain M) (Gvozdev, Akentieva, Tzuprun, Tagunava, Orlova & Kiselev) could well herald the way forward to a better understanding of the two systems. Certainly the lack of a facile genetic system in the methanotrophs has significantly hampered our advances in this area despite many attempts by various groups to develop them. We still await breakthroughs in this area.

References

[1] Ljungdahl LG. (1986) This volume
[2] Duine JA, Frank J, Dijkstra M. (1986) This volume`
[3] Page MD, Anthony C. (1986) J. Gen. Microbiol. 132, 1553-1563.
[4] Dijkhuizen L, Levering PR. (1986) This volume
[5] Green J, Dalton H. (1986) Biochem. J. 236, 155-162.
[6] Cornish A, Nicholls KM, Scott D, Hunter BK, Aston WJ, Higgins IJ, Sanders M. (1984) J. Gen. Microbiol. 130, 2565-2575.
[7] Ribbons DW, Harrison JW, Wadzinski AM. (1970) Ann. Rev. Microbiol. 24, 135-158.
[8] Linton JD, Vokes J. (1978) FEMS Microbiol. Letts. 4, 125-128.
[9] Green J, Dalton H. (1985) J. Biol. Chem. 260, 15795-15801.
[10] Dalton H, Prior SD, Leak DJ, Stanley SH. (1984) Microbial Growth on C-1 Compounds (RL Crawford, RS Hanson eds.) American Society for Microbiology pp 75-83.
[11] Burrows JK, Cornish A, Scott D, Higgins IJ. (1984) J. Gen. Microbiol. 130, 3327-3333.
[12] Lund J, Woodland MP, Dalton H. (1985) Eur J. Biochem. 147, 297-305.
[13] Woodland MP, Cammack R. (1985) Microbial Gas Metabolism (RK Poole, CS Dow eds.) Academic Press pp 209-213.
[14] Green J, Prior SD, DAlton H. (1985) Eur. J. Biochem. 153, 137-144.
[15] Prior SD, Dalton H. (1985) J. Gen. Microbiol. 131, 155-163.
[16] Scott D, Brannan J, Higgins IJ. (1981) J. Gen. Microbiol. 125, 63-72.
[17] Stanley SH, Prior SD, Leak DJ, Dalton H. (1983) Biotechnol. Lett. 5, 487-492.
[18] Leak DJ, Dalton H. (1986) Appl. Microbiol. Biotechnol. 23, 470-476.
[19] Leak DJ, Dalton H. (1986) Appl. Microbiol. Biotechnol. 23, 477-481.
[20] Tonge GM, Harrison DEF, Higgins IJ. (1977) Biochem. J. 161, 333-344.
[21] Duine JA, Frank J, Jongejan JA, Dijkstra M. (1984) Microbial Growth on C-1 Compounds (RL Crawford, RS Hanson eds.) American Society for Microbiology pp 91-97.
[22] Cornish A, MacDonald J, Burrows KJ, King TS, Scott D, Higgins IJ. (1985) Biotechnol. Lett. 7, 319-324.
[23] Anthony C. (1982) Biochemistry of the Methylotrophs. Academic Press.

METABOLIC REGULATION IN FACULTATIVE METHYLOTROPHS

L. DIJKHUIZEN and P. R. LEVERING
Department of Microbiology, University of Groningen and Microbiological R&D Laboratories, Organon International B.V., The Netherlands

1. INTRODUCTION

In aerobic methylotrophs carbon assimilation proceeds initially by conversion of three C_1 units into a C3 compound via a cyclic pathway.In bacteria three of these pathways are now firmly established, namely the ribulose bisphosphate (RuBP) or Calvin cycle (e.g. in *Paracoccus denitrificans, Xanthobacter autotrophicus*), the ribulose monophosphate (RuMP) cycle (e.g. in *Methylophilus methylotrophus*) and the serine pathway (e.g. in *Methylobacterium* AM1, *Hyphomicrobium* X). A fourth pathway, the xylulose monophosphate (XuMP) cycle, uniquely operates in methylotrophic yeasts.

The three C_1 assimilation pathways involving sugar phosphate molecules as intermediates are not only functionally but also structurally very similar. They all employ condensation of a C_1 unit, CO_2 or formaldehyde, with a C_5 sugar phosphate as the key mechanism of C_1 carbon assimilation (fixation phase). The main differences are found in the identity of the enzymes specifically involved in these reactions. Further steps in these cycles are catalyzed by enzymes from the glycolytic or Entner-Doudoroff pathways (cleavage phase) and the oxidative pentose phosphate pathway (rearrangement phase) [1].

Over the years it has become evident that the synthesis of enzymes specifically involved in the utilization of C_1 compounds is under metabolic control. This raises the question which regulatory mechanisms control the levels of these enzymes. A clear understanding of these processes is not only important when considering the role of these organisms in nature, but also with regard to the relation between energy generation and carbon assimilation, and from an applied point of view [2]. Metabolic regulation manifests itself most clearly in facultative methylotrophs. These organisms are able to synthesize these C_1 enzymes at high rates, or to completely switch-off their synthesis, depending on the availability of C_1 compounds in the environment. Facultative methylotrophs are also suitable hosts for genetic studies since they allow the isolation and characterization of mutants blocked in C_1 specific functions.

In this paper recent studies on the regulation of C_1 metabolism in methylotrophs employing the RuMP cycle of formaldehyde fixation are described. Regulation of the Calvin cycle [3,4], XuMP cycle [5] and the serine pathway [6,7,8] is discussed in the literature indicated, and elsewhere in this volume.

2. REGULATION OF THE RuMP CYCLE

Most of the basic research on the physiology and biochemistry of the RuMP cycle has been performed with obligate methylotrophs. The available information on the regulation of the synthesis of enzymes specifically involved in C-1 metabolism in these organisms therefore remained relatively scarce. Since formaldehyde is a highly reactive and potentially toxic compound, it has to be expected that RuMP cycle methylotrophs have not only developed an efficient enzymatic machinery to deal with formaldehyde, but also accurately control the flux of this compound [9].

Table 1. Enzyme profiles of *Arthrobacter* P1 during growth on various substrates.

Enzyme	Growth substrate: MA	Glu	Chol	Ace	EA
^{14}C-Methylamine uptake	470	0	-	0	32
Amine oxidase	740	0	0	0	780
Catalase	139	32	139	9	200
Hexulose phosphate synthase	742	0	197	0	47
Hexulose phosphate isomerase	90	0	31	0	-
6-Phosphofructokinase	58	64	-	-	16
Fructose-1, 6-bisphosphate aldolase	410	402	250	207	145
Transketolase	836	225	167	169	126
Transaldolase	1185	437	238	282	-
Phosphoriboisomerase	1105	218	-	197	-
Ribulose-5-phosphate 3-epimerase	11300	604	-	522	-
Glucose-6-phosphate dehydrogenase (NADP)	264	198	97	94	72
6-Phosphogluconate dehydrogenase (NADP)	238	291	141	160	166

Activities are expressed as nmoles.min^{-1}.mg^{-1} of protein, except for catalase: ΔE240.min^{-1}.mg^{-1} of protein
-: not determined
Abbreviations: MA, methylamine; Glu, glucose; Chol, choline; Ace, acetate; EA, ethylamine.

2.1. Methylamine metabolism in Arthrobacter P1

For unknown reasons the isolation of methanol-utilizing facultative RuMP cycle methylotrophs has turned out to be extremely difficult. We have therefore carried out a study of the regulation of C_1 metabolism in a versatile methylotrophic bacterium, *Arthrobacter* P1, which is able to grow rapidly on methylated amines and on many "heterotrophic" substrates, but not on methanol [10]. The metabolism of methylamine in *Arthrobacter* P1 (Fig. 1; Table 1) shows a number of interesting features which make

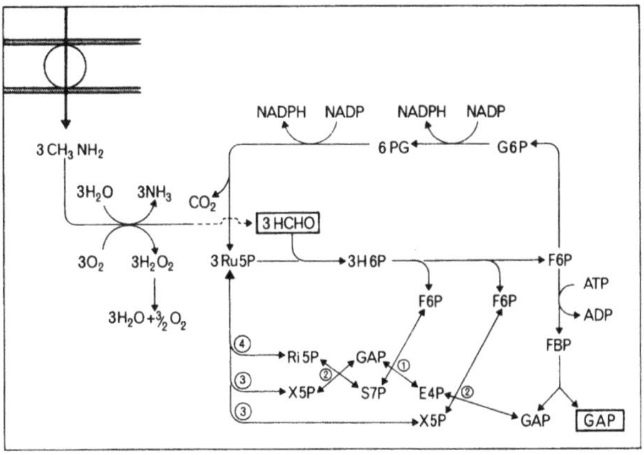

Figure 1. Metabolism of methylamine in *Arthrobacter* P1. Abbreviations: GAP, glyceraldehyde-3-phosphate; E4P, erythrose-4-phosphate; S7P, sedoheptulose-7-phosphate; X5pP, xylulose-5-phosphate; Ri5P, ribose-5-phosphate; Ru5P, ribulose-5-phosphate; H6P, hexulose-6-phosphate; F6P, fructose-6-phosphate; FBP, fructose-1, 6-bisphosphate; G6P, glucose-6-phosphate; 6PG, 6-phosphogluconate.

this organism ideal as a model system for studies on the regulation of C_1 metabolism [11]. The first step in the metabolism of methylamine is its uptake into the cell via an inducible, energy-dependent transport system [12]. Comparable mechanisms for methylamine uptake have been observed in the serine pathway methylotrophs *Methylobacterium* MA [13] and *Hyphomicrobium* X [14]. A study of the properties of this system in cytoplasmic membrane vesicles of *Arthrobacter* P1 [12] showed that it has a high affinity for methylamine ($K_{app} =$ 20 - 25 μM) and accumulates this compound in unmodified form in a process driven by the membrane potential. Interestingly, methylamine uptake by these vesicles is inhibited by ethylamine (competitively; K_i = 75 μM) and other lower primary amines, but notby ammonia. This indicates that an amine specific transport system is involved and not the ammonia carrier. Strong (non-competitive) inhibition was also observed with formaldehyde (K_i = 1.6 μM) and acetate (k_i = 0.8 μM).

Inside the cell, methylamine is oxidized to formaldehyde, hydrogen peroxide, and ammonia by a primary amine oxidase [11,15]. Cytochemical staining experiments revealed that amine oxidase activity is mainly located on invaginations of the cytoplasmic membrane [11]. Fixation of the formaldehyde involves the specific RuMP cycle enzymes hexulose phosphate synthase (HPS) and hexulose phosphate isomerase (HPI). In subsequent studies it became clear that in *Arthrobacter* P1 the FBP aldolase cleavage and TK/TA-rearrangement variant of the RuMP cycle is present [16]. Since enzymes for the linear oxidation of formaldehyde (via formate to CO_2) were undetectable in methylamine-grown cells of *Arthrobacter* P1, energy generation during growth of the organism on C_1 compounds apparently proceeds via the so-called dissimilatory RuMP cycle.

2.1.1. Regulation of the synthesis of enzymes of methylamine metabolism.

The presence of regulatory mechanisms governing the synthesis of the enzymes of methylamine metabolism became evident from an analysis of enzyme profiles of *Arthrobacter* P1 during growth on various substrates. As shown in Table 1, *Arthrobacter* P1 does not synthesize the amine transport system, amine oxidase, HPS and HPI during growth on glucose or acetate. High levels of these enzymes are present in methylamine-grown cells on the other hand. Synthesis of the amine transport system and amine oxidase, however, was also observed with ethylamine-grown cells [17], whereas HPS and HPI were present in choline-grown cells [18]. In *Arthrobacter* P1 the metabolism of choline was found to proceed via glycine, resulting in the intermediate production of three formaldehyde molecules. These results suggested that the synthesis of enzymes involved in methylamine oxidation and those of formaldehyde fixation in *Arthrobacter* P1 is regulated non-coordinately. Methylamine (and ethylamine) apparently induces the synthesis of the amine transport system and amine oxidase, whereas formaldehyde induces the synthesis of HPS and HPI, the specific enzymes of the RuMP cycle.

When *Arthrobacter* P1 was grown on an equimolar (25 mM) mixture of acetate and methylamine in batch culture [19], acetate turned out to be the preferred growth substrate (Fig. 2). Methylamine utilization only started after the acetate concentration in the culture had dropped below 5 mM. Acetate did not fully repress the synthesis of amine oxidase and HPS since activities of these enzymes were detected from t = 2 h onwards. Whereas amine oxidase activity steadily increased, HPS activity remained low until acetate had disappeared from the medium. The observed sequential utilization of these two substrates appeared to be due to an almost complete inhibition of the amine transp[ort system [12] and amine oxidase by acetate [19]. This will prevent intracellular production of formaldehyde from methylamine and therefore may explain the failure of the cells to synthesize HPS under these conditions. It remained possible, however, that acetate or its metabolism additionally controlled synthesis of the RuMP cycle enzymes via a mechanism of repression. Such a repression by "heterotrophic" substrates is generally observed in facultatively autotrophic bacteria that fix CO-2 via the Calvin cycle [3,4]. Because of the close analogy between the RuBP- and RuMP cycles, the involvement of such a control mechanism in *Arthrobacter* P1 was also possible and could not be excluded on the basis of the available evidence.

2.1.2. Growth of Arthrobacter P1 with formaldehyde

In view of the above it was decided to study growth of *Arthrobacter* P1 on formaldehyde in more detail. Such an approach first of all would enable us to obtain more direct evidence for the suggested sequential induction of the synthesis of the C_1 specific enzymes. In addition, it would open up the possibility to directly study effects of "heterotrophic" substrtates on formaldehyde metabolism without

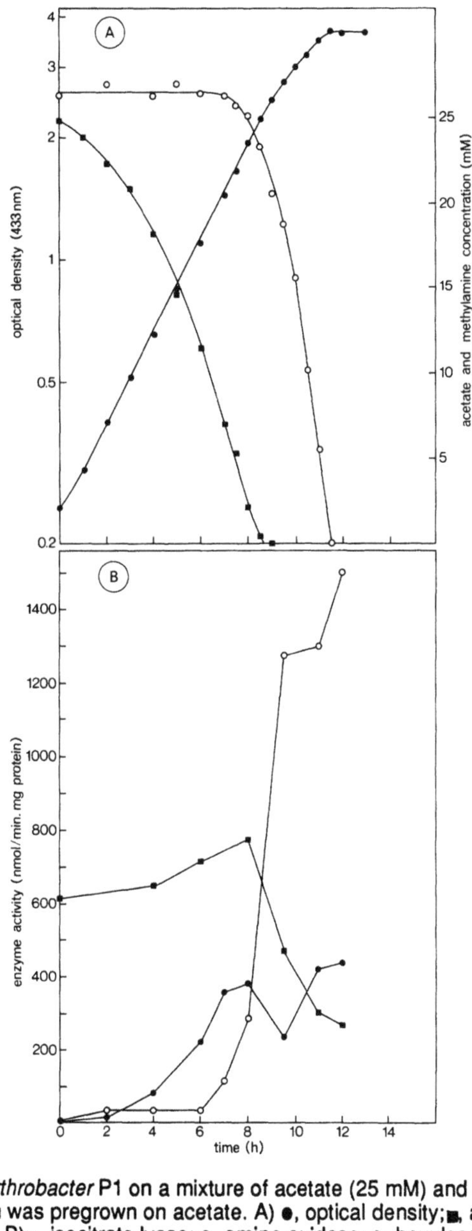

Figure 2. A,B) Growth of *Arthrobacter* P1 on a mixture of acetate (25 mM) and methylamine (25 mM) in batch culture. The inoculum was pregrown on acetate. A) ●, optical density; ■, acetate concentration; o, methylamine concentration. B) ■, isocitrate lyase; ●, amine oxidase; o, hexulose phosphate synthase.

interference of negative effects of such substrates on the enzyme systems involved in the conversion of methylamine into formaldehyde. In view of the extremely toxic nature of formaldehyde, making it virtually impossible to grow microorganisms on this compound alone in batch culture, *Arthrobacter* P1 was first of all grown in formaldehyde-limited continuous cultures [20]. These cultures were established in smooth substrate transitions, form choline-limited conditions, via choline plus formaldehyde-limitation to growth on formaldehyde as the sole carbon- and energy source. This procedure did not result in (transient) accumulation of formaldehyde and the cultures thus obtained were stable up to D = 0.20 h^{-1}, which is in the same order of magnitude as the μ_{max} on methylamine (0.28 h^{-1}; 11). Enzyme analysis subsequently showed that the synthesis of the enzymes involved in the conversion of methylamine into formaldehyde and in formaldehyde fixation is indeed induced sequentially in this organism: the amine transport system and amine oxidase are only synthesized in the presence of methylamine, whereas the RuMP cycle enzymes are apparently induced by formaldehyde, present in the medium reservoir or following its intracellular production from methylamine (Table 2). Compared to growth with methylamine the molar growth yield on formaldehyde was approximately 30% higher. This difference appears to be mainly due to the expenditure of energy for the uptake of methylamine into the cells since in both cultures formaldehyde was metabolized in an identical way and no evidence for a formaldehyde transport system was found [20].

Table 2. Enzyme activities in cell-free extracts and molar growth yields (Y_{mol}) of *Arthrobacter* P1 grown in methylamine- or formaldehyde-limited continuous cultures at a dilution rate of 0.10 h^{-1}.
Activities are expressed as $nmol.min^{-1}.mg^{-1}$ protein, except for catalase; $\Delta E240.min^{-1}.mg^{-1}$ protein. Y_{mol} is expressed as g dry weight of cells produced/mol substrate consumed.

Enzyme	Growth-limiting substrate (S_R = 25 mM) Methylamine	Formaldehyde
14C-Methylamine uptake	290	0
Amine oxidase	1365	0
Catalase	150	20
Hexulose phosphate synthase	1170	1830
Hexulose phosphate isomerase	-	+
Transketolase	800	1075
Y_{mol}	9.0-9.2	11.5-12.0

0, not detectable
-, not determined
+, present but fluctuating activities (see text)

Unequivocal evidence that formaldehyde is the inducing signal for HPS synthesis was subsequently obtained in experiments involving addition of formaldehyde to resting cell suspensions of *Arthrobacter*

P1 (Fig. 3; 21). The question to what extent this system is additionally controlled by repression exerted by "heterotrophic" substrates was subsequently studied in experiments in which formaldehyde (and for a comparison also methylamine) was added continuously at a low rate to cultures of *Arthrobacter* P1

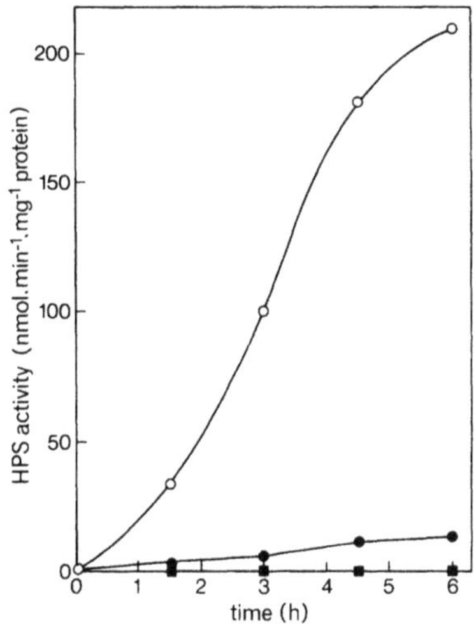

Figure 3. Induction of the synthesis of hexulose phosphate synthase in resting cell suspensions prepared from glucose-grown cells of *Arthrobacter* P1. From zero time onwards formaldehyde (final condentration 0.25 mM) was added every 30 min. o, •,▪, hexulose phosphate synthase activity in the absence and in the presence of chloramphenicol, 25 and 100 µg/ml, respectively.

growing exponentially on "heterotrophic" substrates [21]. If synthesis of the RuMP cycle enzymes were strictly subject to catabolite repression under these conditions, this addition of formaldehyde would result in accumulation of this compound in the culture and would soon reach lethal concentrations. If on the other hand solely an induction mechanism were involved then it should be possible to find a certain low rate of formaldehyde addition that would allow induction of the synthesis of HPS and the subsequent fixation of formaldehyde without causing accumulation of this compound in the culture. Since it has become clear over the years that glucose may be a strong source of catabolite repression in some organisms, but that organic acids may act as such in other organisms [22], the experiments were performed with both acetate and glucose. As expected, addition of methylamine at a rate of 1 mmol. l^{-1} h^{-1} to cells of the organism growing in batch culture on acetate resulted in accumulation of most of the methylamine. Since under these conditions hardly any methylamine will be oxidized to formaldehyde (inhibitory effects of acetate on methylamine uptake and amine oxidase) this, or the possible involvement of catabolite repression exerted by acetate or its metabolism, apparently resulted in failure of the organism to synthesize HPS (Fig. 4). Addition of formaldehyde on the other hand (Fig. 5) resulted in the immediate and very rapid synthesis of HPS and not in accumulation of significant levels of formaldehyde. Comparable results were obtained in experiments involving addition of formaldehyde to cells growing exponentially on glucose [21]. These results, obtained under carbon excess conditions, clearly show that induction by formaldehyde is the overriding control mechanism in the regulation of the synthesis of HPS in *Arthrobacter* P1.

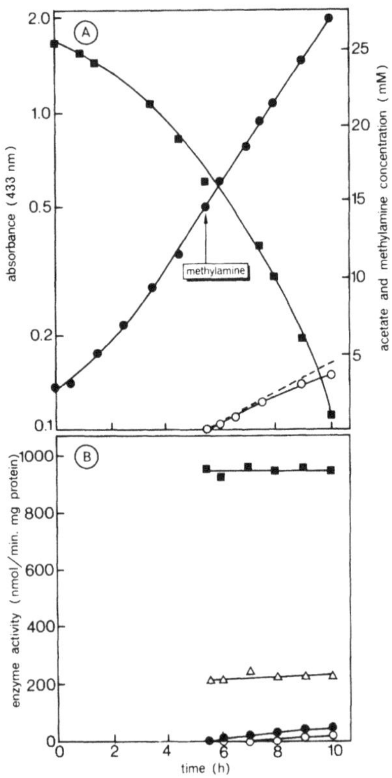

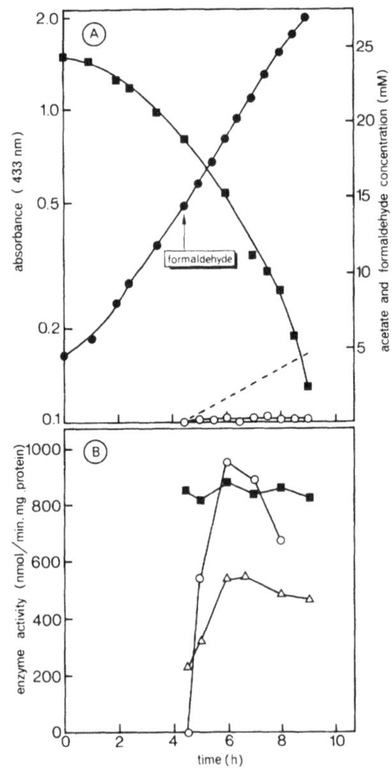

Figure 4.(left) Growth of *Arthrobacter* P1 on acetate (25 mM) and effect of a continuous addition of methylamine (1 mmol. 1^{-1}. h^{-1}). A) ●, absorbance; ■, acetate concentration; o, methylamine concentration; - - - , theoretical curve for methylamine accumulation. B) ■, isocitrate lyase; Δ, transketolase; ●, amine oxidase; o, hexulose phosphate synthase.

Figure 5.(right) Growth of Arthrobacter P1 on acetate (25 mM) and effect of a continuous addition of formaldehyde (1 mmol. 1^{-1}.h^{-1}). A) ●, absorbance; ■, acetate concentration; o, formaldehyde concentration; - - - , theoretical curve for formaldehyde accumulation. B) ■, isocitrate lyase; Δ, transketolase; o, hexulose phosphate synthase.

2.1.3. Presence of transaldolase isoenzymes in Arthrobacter P1

In *Arthrobacter* P1, and in other methylotrophs, the enzyme transaldolase may play an important role in the regeneration of the C_1 acceptor molecule ribulose monophosphate during growth on C_1 compounds [1,16]. This enzyme also functions in the metabolism of those substrates which are converted via the pentose phosphate pathway to fructose-6-phosphate and glyceraldehyde-3-phosphate, i.e. gluconate and xylose in *Arthrobacter* P1. Among gluconate-negative mutants of *Arthrobacter* P1 strains occurred which also were unable to grown on xylose because they had lost the ability to synthesize transaldolase [23]. Contrary to expectation, these mutants still grew on methylamine and were even able to use gluconate as a carbon source at a normal rate provided methylamine was supplied as a nitrogen source. Under these conditions high levels of transaldolase were observed. Partial purification of the transaldolases synthesized by gluconate-grown cells of *Arthrobacter* P1 and methylamine-grown cells of one of these mutants, strain Art 98, subsequently revealed the presence of two transaldolase isoenzymes. These enzymes displayed similar kinetics and also functionally they appear to be very similar. This followed most clearly form the observation that, in the presence of methylamine as the sole nitrogen source, Art 98 was still able to utilize gluconate, reacing the same doubling time as normally observed with wild-type *Arthrobacter* P1 (see above). The major differences are the heat sensitivities of the enzymes and the regulation of their synthesis. In substrate transition experiments it became clear that the transaldolase enzyme purified from gluconate-grown cells is synthesized constitutively whilst the other enzyme, purified from methylamine-grown cells of mutant Art 98, is specifically induced by formaldehyde. It was therefore concluded that strain Art 98 had lost the ability to synthesize the transaldolase which normally becomes expressed constitutively. The additional recruitment via induction of a "methylotrophic" transaldolase during growth on C_1 compounds may simply serve to sufficiently enhance the flow of carbon via the rearrangement reactions in order to ensure continuous availability of the C_1 acceptor molecule ribulose-5-phosphate under these conditions. The question whether the presence of only the constitutive transaldolase enzyme during growth on methylamine would cause a decreased maximal growth rate cannot be answered at the moment and has to await the isolation of mutants deficient in the "methylotrophic" transaldolase. In view of the large differences in the activities of the enzymes of the non-oxidative pentose phosphate pathway/RuMP cycle (Table 1) in methylamine- and glucose-grown cells, it might be speculated that *Arthrobacter* P1 is able to additionally induce a complete set of these enzymes during growth on C_1 compounds.

3. CONCLUSIONS

Methylamine oxidation in bacteria may occur via an amine dehydrogenase (e.g. in *Methylobacterium* AM1), an amine oxidase (e.g. in *Arthrobacter* P1) or methylated amino acids (e.g. in *Methylobacterium* MA, *Hyphomicrobium* X) [24]. Whereas amine dehydrogenase is a periplasmic enzyme, the latter two enzyme systems are located in the cytoplasm of the cells and organisms employing these mechanims for methylamine oxidation apparently additionally require the presence of an amine specific transport system [12,13,14]. In *Arthrobacter* P1 methylamine transport and amine oxidase are strongly inhibited by formaldehyde (but not by ammonia), a mechanism of feedback inhibition which may serve to prevent further accumulation of toxic formaldehyde. In this organism, however, the presence of formaldehyde, even during growth with "heterotrophic" substrates, also induces a vey rapid synthesis up to high levels of the formaldehyde-fixing enzyme HPS. It is likely that other RuMP cycle enzymes (HPI and one of the transaldolase isoenzymes) are also induced by formaldehyde. The regulation of the synthesis of RuMP cycle enzymes in *Arthrobacter* P1 (induction by formaldehyde) is therefore completely different from that generally observed for the enzymes of the Calvin cycle in facultatively autotrophic bacteria (repression/derepression mechanism; 3,4). The rationale behind this probably is that in autotrophic bacteria energy generation can proceed completely independently of CO_2 fixation, whereas in *Arthrobacter* P1 formaldehyde has to be fixed via HPS and HPI in the RuMP cycle both for energy generation and carbon assimilation. In this respect a comparative study of the regulation of RuMP cycle enzymes in *Nocardia* sp. 239, a bacterium that does use a linear sequence for the oxidation of formaldehyde, may yield interesting further information on the relation between the particular enzymology and regulation of metabolic pathways. The first results obtained in the study of metabolic regulation in this organism in fact do indicate that synthesis of RuMP cycle enzymes is not only induced by methanol (or formaldehyde) but also sensitive to repression exerted by the metabolism of "heterotrophic" substrates [25,26].

REFERENCES

[1] Anthony C. (1982) The biochemistry of methylotrophs. London, Academic Press.
[2] Dijkhuizen L, Hansen TA, Harder W. (1985) Trends Biotechnol. 3: 262-267.
[3] Dijkhuizen L, Harder W. (1984) Antonie van Leeuwenhoek 50: 473-487.
[4] Dijkhuizen L, Harder W. (1985) In: Comprehensive biotechnology, vol. 1, pp 409-423. Oxford, Pergamon Press.
[5] Veenhuis M, van Dijken JP, Harder W. (1983) Adv. Microb. Physiol. 24: 1-78.
[6] O'Connor ML. (1981) In: Proc. 3rd Int. Symp. Microb. Growth C-1 Compounds, pp. 294-300. London, Heyde;n.
[7] Brooke AG, Attwood MM. (1983) J. Gen. Microbiol. 129: 2399-2404.
[8] Brooke AG, Attwood MM. (1985) FEMS Microbiol. Lett. 29: 251-256.
[9] Attwood MM, Quayle JRI. (1984) In: Proc. 4th Int. Symp. Microb. Growth C-1 Compounds, pp. 315-323. Washington DC, ASM.
[10] Levering PR. (1985) Metabolic regulation in the facultative methylotroph *Arthrobacter* P1. PhD Thesis, University of GRoningen, The Netherlands.
[11] Levering PR, van Dijken JP, Veenhuis M, Harder W. (1981) Arch. Microbiol. 129: 72-80.
[12] Dijkhuizen L, de Boer L, Boers RH, Harder W. Konngs WN. (1982) Arch. Microbiol. 133: 261-266.
[13] Bellion E, Khan MYA, Romano MJ. (1980) J. Bacteriol. 142: 786-790.
[14] Brooke AG, Attwood MM. (1984) J. Gen. Microbiol. 130: 459-463.
[15] Van Vliet-Smits M, Harder W, van Dijken JP. (1981) FEMS Microbiol. Lett. 11: 31-35.
[16] Levering PR, Dijkkhuizen L, Harder W. (1982) FEMS Microbiol. Lett. 14: 257-261.
[17] Levering PR, Dijkhuizen L. Harder W. (1984) Arch. Microbiol. 139: 188-195.
[18] Levering PR, Binnema DJ, van Dijken JP, Harder W. (1981) FEMS Microbiol. Lett. 12: 19-25.
[19] Levering PR, Dijkhuizen L. (1985) Arch. Microbiol. 142: 113-120.
[20] Levering PR, Croes LM, Tiesma L. Dijkhuizen L. (1986) Arch. Microbiol. 144: 272-278.
[21] Levering PR, Croes LM, Dijkhuizen L. (1986) Arch. Microbiol. 144: 279-285.
[22] Harder W, Dijkhuizen L. (1982) Phil. Trans. R. Soc. London B. 297: 459-480.
[23] Levering PR, Dijkhuizen L. (1986) Arch. Microbiol. 144: 116-123.
[24] Large PJ. (1981) In: Proc. 3rd Int. Symp. Microb. Growth C-1 Compounds, pp. 55-69. London, Heyden.
[25] Hazeu W, de Bruyn JC, van Dijken JP. (1983) Arch. Microbiol. 135: 205-210.
[26] Dijkhuizen L. Clement W, Witteveen CFB, van der Woude M, de Boer L. (1986) In: Proc. 6th Int. Symp. Biol. Actinomycetes, in press.

QUINOPROTEINS IN THE DISSIMILATION OF C_1 COMPOUNDS

J.A. DUINE, J. FRANK, Jzn and M. DIJKSTRA
Laboratory of Microbiology & Enzymology, Delft University of Technology, Julianalaan 67, 2628 BC Delft, The Netherlands.

1. INTRODUCTION

The recent interest in microbial enzymology of specialized groups of bacteria has thrown up several novel cofactors which are involved in the special redox reactions catalyzed by these organisms. In particular the study of methanogens and methylotrophs has led to the discovery of unknown compounds participating in the pathways of methane formation and degradation. Some of these compounds consist of familiar cofactors fused together to one compound, others have a simpler but unfamiliar structure. PQQ (pyrrolo quinoline quinone) belongs to the latter group and is exceptional since its distribution is widespread. Although it was originally detected in methanol dehydrogenase, its functioning has now also been established in other enzymes, some of these occurring not only in micro-organisms but also in man. Information about PQQ and PQQ-containing enzymes (quinoproteins) can be found elsewhere [1,2]. The present review gives the latest results of those quinoproteins involved, or suspected to be involved, in steps of C_1 dissimilation routes.

2. METHANOL DEHYDROGENASES

2.1. Introduction

All Gram-negative methylotrophic bacteria investigated so far are able to oxidize methanol via the quinoprotein, methanol dehydrogenase (EC 1.1.99.8) (MDH). The Gram-positive, *Nocardia* sp. 239, uses a quite different enzyme for that purpose, namely NAD-linked quinoprotein methanol dehydrogenase which has also been found in the methanotroph *Methylococcus capsulatus* strain Bath [3]. Progress on MDH obtained after the last C_1 Symposium [4] will be discussed below.

2.2. Synthesis

As could be expected for a periplasmic enzyme, the protein chains of MDH are synthesized with a leading sequence [5]. Assemblage of the dimeric enzyme from protein chains and cofactor probably occurs at the outside of the cytoplasmic membrane, in line with the substantial amounts of PQQ found in the culture fluid of methylotrophic bacteria and the instantaneous reconstitution of the apoenzyme with PQQ observed in PQQ$^-$-mutants [6]. Curiously, the monomeric as well as the dimeric enzyme form can be isolated from *Pseudomonas* BB1. Both forms seem to be functional since growth conditions substantially change the ratio in which they are isolated but not the methanol-oxidizing capacity of the cells. In addition, both forms become inactivated during the conversion of the suicide substrate cyclopropanol by whole cells. Monomer and dimer have identical kinetic parameters, indicating that the subunits in the dimer do not influence each others catalytic activity [7].

Conversion of ethanol during growth of *Hyphomicrobium* X on that substrate most probably occurs via MDH since cyclopropanol treatment of these cells completely inhibits methanol as well as ethanol oxidation and it is known that MDH becomes irreversibly blocked by this compound and NAD-dependent alcohol dehydrogenases become not [8]. On the other hand, the facultative methylotroph *Acetobacter methanolicus* MB 70 synthesizes MDH as well as a dye-linked alcohol dehydrogenase [9]. This is rather surprising because ethanol is a perfect substrate for MDH in vitro and the dye-linked enzyme might also be a quinoprotein alcohol dehydrogenase since several enzymes of that type have been found in non-methylotrophs [10,11].

A quite different situation exists in *Pseudomonas* AM1. When this organism is grown on propanediol, MDH is used for oxidation, although this compound is a very poor substrate for MDH in vitro. In this case, a so-called modifier protein is synthesized, presumably binding to MDH and improving the kinetic parameters for propanediol oxidation substantially [12].

2.3. Molecular properties

One of the intriguing properties of MDH is the large difference between the spectra of its redox forms, those of other quinoproteins and those of PQQ [1,2]. An obvious explanation is that PQQ in MDH interacts with another redox group or an amino acid residue. A search for metals in the enzyme gave negative results [13] and SH/SS-group determinations of the different redox forms gave a constant ratio. It seems unlikely, therefore, that a second redox group is responsiblc for the deviating spectral behaviour. Recently published spectra of inactive MDH's, isolated from *Pseudomonas* AM1 mutants unable to oxidize methanol [14], show unusual maxima, pointing to a relationship between enzyme activity and interaction of PQQ with an amino acid residue of the protein chain. The interaction could be responsible for the characteristic absorption spectrum of MDH.

Reconstitution *in vitro* of apoenzyme and PQQ has been claimed. Unfortunately, details about the paramagnetic properties and biological activity of the reconstituted enzyme were not given [5]. It is expectable that further characterization of inactive MDH's from mutated organisms and the 3-dimensional structure elucidation of the enzyme, which is underway, will provide rapid progress in knowledge about structure and functioning of the enzyme.

2.4. Mechanism of action

The mechanism of action of the enzyme *in vitro* is known in broad outline [15,16]. Although Abeles and coworkers have postulated different schemes [17,18], there are good reasons to reject them [15,16]. Recent results [19] with stopped-flow spectrometry are in accordance with our original scheme. The preliminary data reveal that not only MDH_{oxI} oxidation but in particular alcohol oxidation (step b and d in the scheme, respectively) are activated by NH_3, that the rate-determining step concerns one of the reactions taking place during alcohol-conversion (step d), and that the high pH is most probably related to substrate addition (step c).

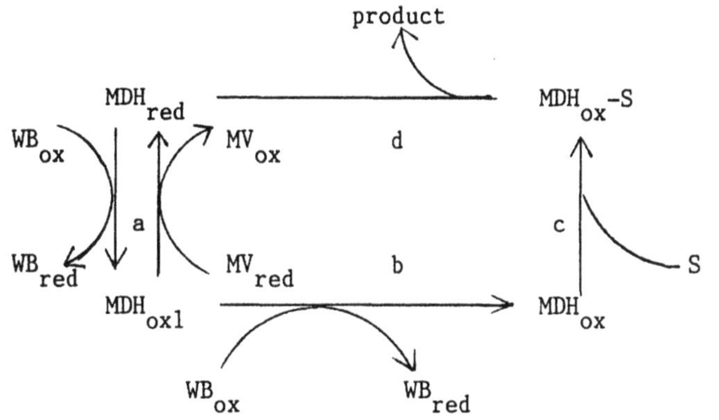

WB, Wurster's Blue; MV, methyl viologen.

2.5. Coupling to the respiratory chain

Research on whole cells, from different directions and laboratories, has provided evidence that MDH is coupled to the respiratory chain at the level of cytochrome *c*. Proof for this could also be delivered by experiments *in vitro* since reduction of oxidized cytochrome *c* via MDH was achieved in anaerobically isolated preparations [20]. In the same publication it was, however, mentioned that O_2 destroyed this activity, turning the enzyme from an NH_3-independent into an NH_3-dependent form. Subsequently, Anthony *c.s.* [21] showed that methanol could reduce oxidized cytochrome *c* via MDH, using aerobically isolated preparations. Depending on the origin of the preparations NH_3 was a stimulator or not. The reduction rate was, however, much lower than that observed in the common dye-assay, which could indicate that the low rate was either due to residual activity from components which had partly lost the ability to transfer the electrons to each other or even to an artificial activity, different from that operating *in vivo*.

Further work in our laboratory on the anaerobically isolated preparations has partially provided an explanation for the anomalies. The preparations appear to contain a heat stable, low molecular weight factor which substantially enhances the reaction rate of cytochrome *c* reduction via MDH. Moreover, it is able to replace NH_3 in the dye-assay at pH 9. Although the nature of the coupling factor is still unknown, it is already clear that it is not NH_3 and that its activity is destroyed by O_2 in the presence of MDH and cytochrome *c* [22]. Since introduction of O_2 in the anaerobic preparations concomittantly changed the absorption spectrum of MDH [20], interesting avenues are open now to study the coupling of MDH to the respiratory chain and possible differences between MDH *in vitro* and *in vivo*.

3. (FORM)ALDEHYDE DEHYDROGENASES

3.1. Introduction

Formaldehyde is an intermediate in C_1 metabolism. Several cyclic conversion routes have been discovered but also a number of formaldehyde dehydrogenases for oxidation to formate. The occurrence of these enzymes is not restricted to autotrophs so that they may also have a more general function such as maintaining unharmful levels of formaldehyde in the cell or in the regulation of the redox balance. Aldehydes are toxic compounds and since they are ubiquitous in nature, organisms have to cope with this problem. This is expecially the case for methylotrophs: methane mono-oxygenase as well as MDH have a very broad substrate specificity, tools to provide extra aldehydes from alkanes and alcohols in the natural environment. Not surprising, therefore, (form)aldehyde dehydrogenases have been found with a very broad substrate specificity, pointing to a detoxyfying function.

There are some data which could be interpreted to suggest that quinoproteins are involved in formaldehyde conversion. MDH oxidizes formaldehyde very efficiently *in vitro.* It has been claimed that a quinoprotein aldehyde dehydrogenase exists in acetic acid bacteria [23] so that comparable enzymes in methylotrophs could have the same cofactor. NAD-dependent formaldehyde dehydrogenase has been found in *Rhodococcus erythropolis,* requiring an unknown, low molecular weight cofactor which could be PQQ [24]. In the following these data will be scrutinized.

3.2. NAD-dependent formaldehyde dehydrogenase, requiring an extra cofactor which is not glutathione

In studying the degradation of methoxylated aromatic compounds by *Rhodococcus erythropolis,* Eggeling & Sahm [24] discovered an enzyme catalyzing the following reaction: formaldehyde + NAD + X → NADH + product. The identity of the product is not known but on heating the solution, it decomposes into formate. X is a heat stable reducible compound which is neither GSH nor THF. Alcohols (C_2-C_8) are oxidized by the enzyme in the absence of X and although methanol is not, it can replace factor X. Therefore, the authors speculated that a hemiketal of formaldehyde is the genuine substrate for the enzyme.

Research in our laboratory has shown that factor X is neither PQQ, coenzyme A, nor the coupling factor mentioned in section 2.5. and that it is not present in *Methanosarcina barkeri.* A similar factor and enzyme appear to be present in methylamine-grown *Thiobacillus versutus* and, surprisingly, also in methanol-grown *Hyphomicrobium* X. Thus the long standing problem about the nature and presence of a formaldehyde oxidizing enzyme in the latter organism [25] seems now to be solved.

Since the dependecy on the factor develops during purification, one could imagine that other types of NAD-dependent formaldehyde dehydrogenase might contain factor X in a more tightly bound form. In this context, it is relevant to mention the so-called formaldehyde dismutase (EC 1.2.99.4) from *Pseudomonas putida* F61 (although the enzyme converses several alcohols and aldehydes, it should be noted that this is not the case for methanol and since a methanol dissimilating enzyme in this organism is unknown, the significance of formaldehyde dismutation is unclear and the name may be inappropriately chosen). The enzyme operates without added cofactors but analysis of denatured enzyme showed stoichiometric amounts of NAD/NADH [26]. Although indications for the presence of factor X are absent in the quoted paper, investigations on its occurrence might be useful. Reversibly, in view of the similarities in substrate specificity, the formaldehyde dehydrogenase should be examined on dismutase activity.

3.3. Dye-linked (form)aldehyde dehydrogenase (EC 1.2.99.3)

Such a type of enzyme has been observed in methylotrophs as well as in non-methylotrophs. That from *Hyphomicrobium* X was partly characterized but the nature of the cofactor not indicated [27]. Its significance for formaldehyde conversion was disputed [25], but other investigators, using other *Hyphomicrobium* strains as well, came to different conclusions [28]. Also the enzyme from *Methylosinus trichosporium* has been purified and since the authors could only detect haem c as chromophore, the enzyme has been classified as a haemoprotein [29]. Acetic acid bacteria contain an aldehyde dehydrogenase with similar broad substrate specificity and in this case it was considered a quinoprotein [23].

A quite different conclusion follows from our characterization of the aldehyde dehydrogenase from *Pseudomonas testosteroni*. The enzyme is a molybdopterido-flavoprotein with non-haem iron sulphur clusters (unpublished work). On the other hand, characterization of dye-linked aldehyde dehydrogenase from *Hyphomicrobium* X revealed that it is different from the former enzyme. In contrast to the results of Marison & Attwood [27], but in accordance with those of Patel *et al.* [29], the final enzyme preparation appears to contain haem c (unpublished work).

From the foregoing, it follows that the nature of the cofactor in this type of methylotrophic formaldehyde dehydrogenases is far from clear. To complicate matters even further, different growth conditions induce different types of enzymes [30] while monomeric and dimeric enzyme forms with different properties are observed during purification [31].

3.4. Methanol dehydrogenase

It has been suggested that this enzyme does not play a significant role in formaldehyde conversion *in vivo,* since for instance MDH$^-$-mutants show unimpaired formaldehyde conversion rates. Additional evidence was obtained by experiments in which whole cells of *Hyphomicrobium* X were treated with cyclopropanol. This led to complete blocking of MDH, as demonstrated by the absence of methanol oxidation. On the other hand, formaldehyde oxidation was not inhibited at all while acetaldehyde and propionaldehyde were for 75% [8]. Cell-free extracts showed indeed that the dye-linked aldehyde dehydrogenase was inactivated to a large extent. Thus, MDH as well as dye-linked formaldehyde dehydrogenase have certainly not a major role in formaldehyde conversion in this organism.

Interestingly, Page & Anthony have found that addition to MDH of the modifier protein, mentioned in section 2.2., blocked formaldehyde but scarcely methanol oxidation of MDH *in vitro* [32]. Since there are indications that alcohol and aldehyde oxidation by MDH occur at different sites [33], the selectivity of inhibition could be related to this dual substrate specificity. In order to indicate a role in prevention of formaldehyde oxidation *in vivo* by MDH, it would be interesting to know whether stoichiometric amounts of the modifier protein are synthesized under all growth conditions.

Summarizing the chapter, clear evidence for the participation of quinoproteins in formaldehyde conversion is not available.

4. METHYLAMINE OXIDOREDUCTASES

4.1. Methylamine dehydrogenase (MADH) (EC 1.4.99.3)

Several Gram-negative utilizers of methylated amines contain MADH. From the paramagnetic properties of the enzyme, its prosthetic group was concluded to be PQQ or a closely related derivative of it [34]. Though the structure of the prosthetic group still awaits elucidation, the primary structure of the small subunit of the enzyme from *Ps.* AM1 has been partly solved and a 3-dimensional structure determination of that from *Thiobacillus versutus* is under way [35]. Much attention has been paid to the natural electron acceptor amicyanin, for the first time isolated from *Ps.* AM1 [36]. The protein has a M_r of 14,000, contains one copper, and is active as electron carrier between MADH and cytochrome *c*. Copper proteins with the same function have now been isolated from a variety of bacteria [37,38,39]. The protein is only present if the organism is grown on methylated amines and it is different from azurin, a copper protein frequently encountered in the same organism. Redox potentials, absorption spectra and ESR spectra [37] indicate that amicyanin is a new type blue copper protein. The amino acid sequence of the protein from *Ps.* AM1 has been elucidated [40].

There exist large differences in i.e.p.'s among MADH's and amicyanins. However, comparison of the amino acid composition of some amicyanins shows that they are similar, suggesting that the difference is due to amidation of acidic groups in the basic proteins.

4.2. Methylamine oxidase (MeAO)

Also in the case of methylamine conversion it appears that Gram-positives have developed oxidoreductases different from that in Gram-negatives. This is not surprising since they do not have the periplasmic space of Gram-negatives in which MADH and amicyanin are located. The Gram-positive *Arthrobacter* P1 contains MeAO that belongs to the type of copper-containing amine oxidases (EC 1.4.3.6). Besides copper it has also covalently-bound PQQ, just like bovine serum amine oxidase [41], diamine oxidase from hog kidney, and human lysyl oxidase [42]. Most probably, yeasts using methylamine as a nitrogen source, contain a similar enzyme [43].

It is a striking fact that the amine-oxidizing systems MADH/amicyanin and MeAO, both contain copper as well as PQQ. In this respect it would be interesting to know the extent of sequence similarity of the amino acids in these systems so that a possible evolutionary relationship could be delineated.

5. FUTURE PROSPECTS

From table 1 we see that a particular C_1-dissimilation step can be catalyzed by oxidoreductases containing different cofactors.

Table 1. Different oxidoreductases catalyzing the same step in C_1-dissimilation.

Enzyme	Cofactors	EC number
Methanol dehydrogenase	PQQ	1.1.99.8
Methanol dehydrogenase	PQQ, NAD	-
Methanol oxidase	FAD	1.1.3.13
Methylamine dehydrogenase	PQQ	1.4.99.3
Methylamine oxidase	PQQ, copper	1.4.3.6
Formaldehyde dehydrogenase	NAD, GSH	1.2.1.1.
Formaldehyde dehydrogenase	NAD	1.2.1.46
Formaldehyde dehydrogenase	? (dye-linked)	1.2.99.3
Formaldehyde dismutase	NAD, NADH	1.2.99.4
Formaldehyde dehydrogenase	NAD, X	-

What is the reason for the diversity in cofactors? Assuming that the dissimilation rate, the affinity for substrate, and conversion efficiency (in terms of energy) are determinative for successful operating in the natural environment, the diversity should be a reflection of the variety in organisms. Comparative studies on the catalytic and bioenergetical properties of the enzymes could shed light on this. In this respect, the methylotrophic oxidoreductases are very attractive since they are synthesized in large quantities and it is expectable that the results of genetical studies will soon become a stimulating factor for achieving this goal.

REFERENCES

[1] Duine JA, Frank, Jzn J, Jongejan JA. (1986) FEMS Microbiol. Rev. 32, 165-178.
[2] Duine JA, Frank, Jzn J, Jongejan JA. (1986) Adv. Enzymol. 59, 169-212.
[3] Duine JA, Frank, Jzn J, Berkhout MPJ. (1984) FEBS Lett.168, 217-221.
[4] Duine JA, Frank, Jzn J, Jongejan JA, Dijkstra M. (1984) In: Microbial Growth on C1 compounds (Crawford RL, Hanson RS eds.) pp. 91-96, Washington, American Society for Microbiology.

[5] Davidson VL, Neher JW, Cecchini G. (1985) J. Biol. Chem. 260. 9642-9647.
[6] Mazodier P, de Biville F, Gasser F, van Kleef MAG, Duine JA. (1986) Abstracts of the fifth international symposium on microbial growth on C1 compounds (Duine JA, van Verseveld HW eds.), p. 152, Amsterdam, Free University Press.
[7] Dijkstra M, van den Tweel WJJ, de Bont JAM, Frank Jzn J, Duine JA. (1985) J. Gen. Microbiol. 131, 3163-3169.
[8] Groeneveld A, Dijkstra M, Duine JA. (1984) FEMS Microbiol. Lett. 25, 311-314.
[9] Loffhagen N, Babel W. (1984) Ztschr. Allg. Mikrobiol. 24, 143-149.
[10] Groen BW, Frank Jzn J, Duine JA. (1984) Bioch;em. J. 223, 921-924.
[11] Groen BW, van Kleef MAG, Duine JA. (1986) Biochem. J. 234, 611-615.
[12] Ford S, Page D, Anthony C. (1985) J. Gen. Microbiol. 131, 2173-2182.
[13] Duine JA, Frank Jzn J, Westerling J. (1978) Biochim. Biophys. Acta 524, 277-287.
[14] Nunn DN, Lidstrom ME. (1986) J. Bacteriol. 166, 581-590.
[15] de Beer R, Duine JA, Frank Jzn J, Westerling J. (1983) Eur. J. Biochem. 130, 105-109.
[16] Dijkstra M, Frank Jzn J, Jongejan JA, Duine JA. (1984) Eur. J. Biochem. 140, 369-373.
[17] Mincey T, Bell JA, Mildvan AS, Abeles RH. (1981) Biochemistry 20, 7502-7509.
[18] Parkes C, Abeles RH. (1984) Biochemistry 23, 6355-6363.
[19] Frank Jzn J, Balny C, Duine JA. (1986) In: Abstracts of the fifth international symposium on microbial growth on C1 compounds (Duine JA, van Verseveld HW eds.), p. 62, Amsterdam, Free University Press.
[20] Duine JA, Frank Jzn J, de Ruiter LG. (1979) J. Gen Microbiol. 115, 523-526.
[21] Beardmore-Gray M, O'Keeffe DT, Anthony C. (1983) J. Gen. Microbiol. 129, 923-933.
[22] Dijkstra M, Frank Jzn J, Duine JA. (1986) In: Abstracts of the fifth international symposium on microbial growth on C1 compounds (Duine JA, van Verseveld HW eds.), p. 63, Amsterdam, Free University Press.
[23] Ameyama M, Adachi O. (1982) Meth. Enzymol. 89, 491-497.
[24] Eggeling L, Sahm H. (1985) Eur. J. Biochem. 150, 129-134.
[25] Attwood MM, Quayle JR. 1984. In: Microbial Growth on C1 Compounds (Crawford RL, Hanson RS eds.), pp. 315-323, Washington, American Society for Microbiology.
[26] Kato N, Yamagami T, Shimao M, Sakazawa C. (1986) Eur. J. Biochem. 156;, 59-64.
[27] Marison IW, Attwood MM. (1980) J. Gen. Microbiol. 117, 305-313.
[28] Kohler J, Schwartz AC. (1982) Can. J. Microbiol. 28, 65-72.
[29] Patel RN, Hou CT, Derelanko P, Felix A. (1980) Arch. Biochem. Biophys. 203, 654-662.
[30] Attwood MM, Levering PR, Dijkhuizen L. (1986) In: Abstracts of the fifth international symposium on microbial growth on C1 compounds (Duine JA, van Verseveld HW eds.), p. 70, Amsterdam, Free University Press.
[31] Kesseler FP, Schwartz AC. (1986) In: Abstracts of the fifth international symposium on microbial growth on C1 compounds (Duine JA, van Verseveld HW eds.), p. 52, Amsterdam, Free University Press.
[32] Page D, Anthony C. (1986) J. Gen. Microbiol. 132, 1553-1563.
[33] Duine JA, Frank Jzn J. (1981) J. Gen. Microbiol. 122, 201-209.
[34] de Beer R, Duine JA, Frank Jzn J, Large PJ. (1980) Biochim. Biophys. Acta 622, 370-374.
[35] Vellieux FMD, Frank Jzn J, Swarte MBA, Groendijk H, Duine JA, Drenth J, Hol WGJ. (1986) Eur. J. Biochem. 154, 383-386.
[36] Tobari J, Harada Y. (1981). Biochem. Biophys. Res. Commun. 101, 502-508.
[37] van Houwelingen T, Canters GW, Stobbelaar G, Duine JA, Frank Jzn J, Tsugita A. 1985. Eur. J. Biochem. 153, 75-80.
[38] Husain M, Davidson VL. (1985) J. Biol. Chem. 260, 1426-1429.
[39] Lawton SA, Anthony C. (1985) Biochem. J. 228, 719-726.
[40] Ambler RP, Tobari J. (1985) Biochem. J. 232, 451-457.
[41] Lobenstein-Verbeek CL, Jongejan JA Frank, Jzn J, Duine JA. (1984) FEBS Lett. 170, 305-309.
[42] van der Meer RA, Jongejan JA, Frank Jzn J, Duine JA. (1986) FEBS Lett., in press.
[43] Large PJ, Green J. (1984) In: Microbial Growth on Cy1 Compounds (Crawford RL, Hanson RS eds.), pp. 155-164, Washington, American Society for Microbiology.

NMR SPECTROSCOPY IN THE STUDY OF C_1-METABOLISM

J.K.M. SANDERS

Chemical Laboratory, University of Cambridge, Cambridge CB2 1EW, United Kingdom

1. INTRODUCTION

Nuclear Magnetic Resonance (NMR) spectroscopy is a non-invasive technique that can be applied to whole cells, tissues, or even live animals. It tells us about the chemical environment that magnetic nuclei such as ^1H, ^2H, ^{13}C, ^{15}N and ^{31}P find themselves in. This paper will be concerned only with the use of the first three of these nuclei to probe the following aspects of C_1-biochemistry in whole cells:

(a) The discovery of metabolic pathways, and detection of intermediates on those pathways.

(b) The investigation of enzyme mechanism by kinetic and isotope labelling studies.

(c) The assay of important species within cells.

References [1-3] give complete descriptions of the theory and application of NMR in Biology and Chemistry.

2. THEORY

The NMR experiment may be compared to the plucking of stretched violin strings or the striking of bells. We apply a strong, short-lived pulse, or perturbation, and then detect the response as the system relaxes back towards equilibrium. We can measure the intensity of the response, the frequency of the response, and the rate at which the response decays. Figure 1(a) shows the simplest response, from a set of equivalent nuclei all having the same characteristic frequency. Protons, or carbons, in different environments will, in general, have different frequencies. Furthermore, if magnetic nuclei are in different environments but are within about three bonds of each other, they split each other's frequencies into extra lines. These splittings, known as spin-spin couplings, *J*, can tell us which nucleus is attached to which. So, Figure 1(b) shows the response of ^{13}C-labelled acetate, made more complex by ^1H-^{13}C coupling. Figure 1(c), which is the Fourier transform of Figure 1(b), shows the frequency response for the labelled acetate: the two major lines derive from the 90% of protons that are coupled to ^{13}C, while the small central line is from ^{12}C-acetate. ^{12}C and ^{16}O are magnetically inactive, so they give no NMR signals and cause no splittings. Provided that suitable experimental conditions are used, the size of NMR signals is proportional to the number of spins giving the response; it is, therefore, a powerful analytical technique.If we put live cells into a spectrometer and record a spectrum, what will we see? A ^1H spectrum will consist of a massive signal for water, which contais 110 molar protons, plus weak signals for all the cellular components. The ^{13}C spectrum will also be vey complicated. The ^{31}P spectrum will show inorganic phosphate and nucleoside phosphates; it will be a useful guide to the energetic state of the

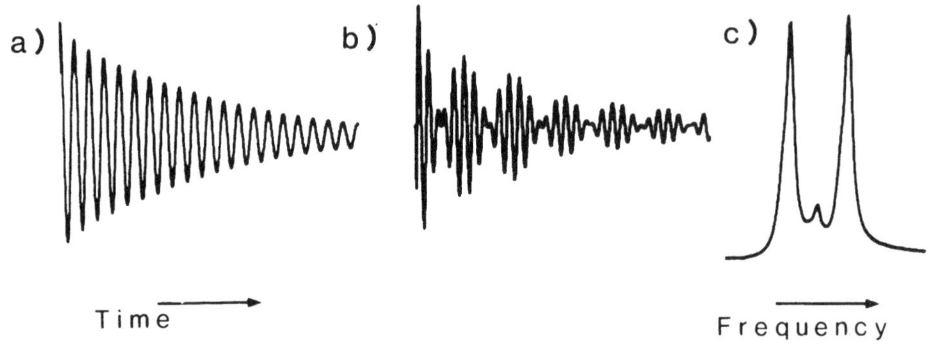

Figure 1. a) The time response of the methyl protons in acetate, $MeCOO^-$ to NMR excitation. b) The time response of the methyl protons in 90% labelled 1-[^{13}C]-acetate, $Me^{13}COO^-$. c) The frequency spectrum corresponding to (b).

cell [2], but can tell us little about metabolic pathways.

The great power of NMR is that isotopic labels allow us to see selectively what we want to see. So if we feed bacteria with [D_2]-formaldehyde we can follow the fate of the deuterium atoms, while if we feed [^{13}C]-formaldehyde we can follow the fate of the carbon, which may be different.

The great drawback is poor sensitivity. We need the compound of interest to be about 1 mM if we are to detect it in a short time. At present, this limits the molecules and pathways that we can study, but as NMR sensitivity improves, so will the range of processes studied.

3. DISCOVERING PATHWAYS

We have used NMR to follow the metabolic fate of labelled formaldehyde in a variety of bacterial and other cells [4-7]. Figure 2(a) shows a ^{13}C spectrum of a culture of live *E. coli* just after challenge with 10 mM [^{13}C]-formaldehyde, while Figure 2(b) shows the same culture some hours later [4]. The formaldehyde has been oxidised and reduced to formate and methanol respectively [4], and also incorporated into the terminal position of a variety of C_3-alcohols. The structure determinations of these alcohols were carried out by a variety of NMR methods without isolation [5]. To the best of our knowledge, these are novel biochemical pathways for formaldehyde. The power of NMR here is that it required no previous guesswork or prejudice about the likely nature of the metabolic products.

All the products shown in Figure 2 are end-products that accumulate in the medium. Varying the bacterial strain and culture conditions gives different quantitative results, but essentially the same metabolic pathways are observed. If we use very short spectrum accumulation times (~ 5 minutes) we can detect *S*-hydroxymethyl glutathione as a transient *intracellular* intermediate [6]. This is formed in a chemical equilibrium,

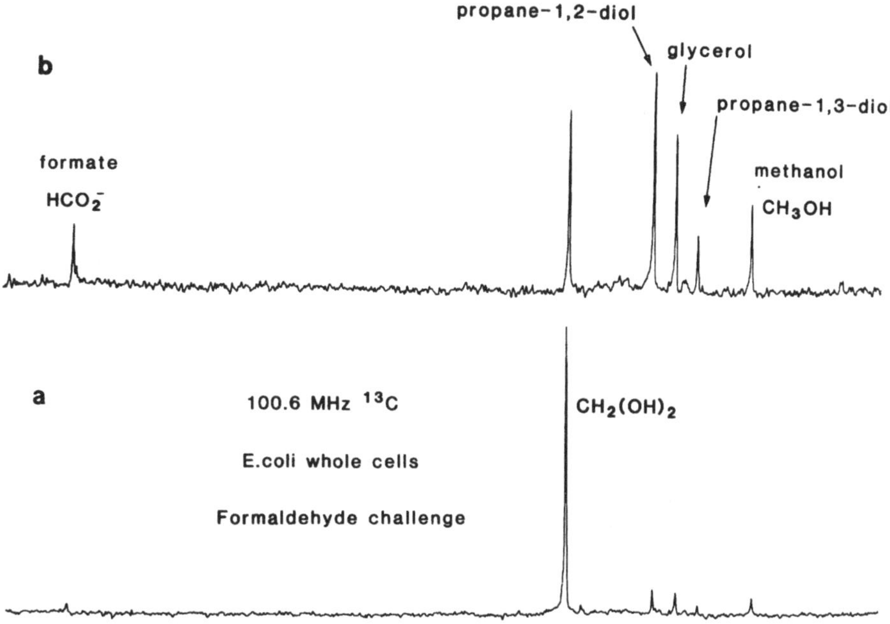

Figure 2. Carbon-NMR spectra of a suspension of *E coli* cells following challenge with labelled formaldehyde. (a) Immediately after challenge, (b) some hours later. Each spectrum took about 20 minutes to acquire.

$$GSH + CH_2(OH)_2 \leftrightarrow GSCH_2OH + H_2O)$$

but consumed enzymically. We presume that $GSCH_2OH$ is the true substrate for formaldehyde dehydrogenase; this is the first reliable evidence for its formation *in vivo.*

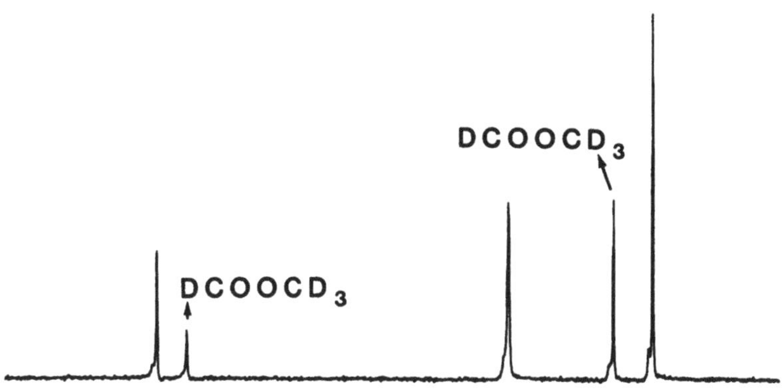

Figure 3. Deuterium NMR spectrum of *S. aureus* culture following challenge with deuterium labelled formaldehyde [7]. Centre peak is water; left and right-most peaks are formate and methanol respectively.

Other organisms behave differently towards formaldehyde. Pseudomonas putida produces essentially only formate and methanol (see 4.2. below), while Staph. aureus also produces methyl formate (Figure 3) [7]. Exogenously-added methanol is incorporated only into the ester methyl group; added formate ion is not incorporated at all. Again, these are NMR results from whole cells. Thus the apparent biosynthetic pathway is via the hemiacetal:

$$CH_2(OH)_2 \xrightarrow{\text{reduction}} CH_3OH \xrightarrow{CH_2(OH)_2} HOCH_2OMe \xrightarrow{\text{oxidation}} HCOOMe$$

4. KINETICS AND MECHANISM

4.1. Methanol oxidation

At the simplest level, NMR can be used to measure concentrations of labelled compounds, and hence to observe kinetics in cells. Figure 4 shows the successive oxidation of [^{13}C]-methanol to formaldehyde and formate by *Methylosinus trichosporium* OB3b [8]. Under different culture conditions, different relative activities of the various enzymes are observed: in some cases significant formaldehyde buildup is observed, while in others formaldehyde oxidation or incorporation into biomass is too fast. The biological significance of this is not yet clear.

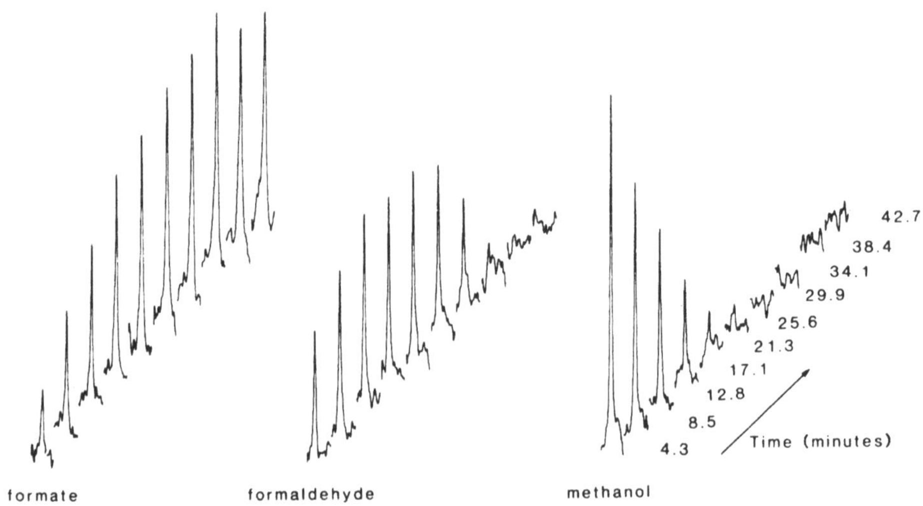

Figure 4. Carbon spectra demonstrating methanol oxidation by *M. trichosporium* OB3b (low biomass culture).

If we feed a bacterium a 1:1 mixture of $^{13}CH_3OH$ and $^{13}CD_3OH$ and acquire ^{13}C spectra under conditions where coupling to H and D is not visible, then initially we see *two* methanol signals, corresponding to the two isotopically different species. As oxidation proceeds, $^{13}CH_3OH$ is consumed more rapidly [7] (Figure 5). The enzymic kinetic isotope effect, k_H/k_D is 1.7 (±0.2); the small value indicates that C-H bond cleavage is not the rate limiting step in this oxidation.

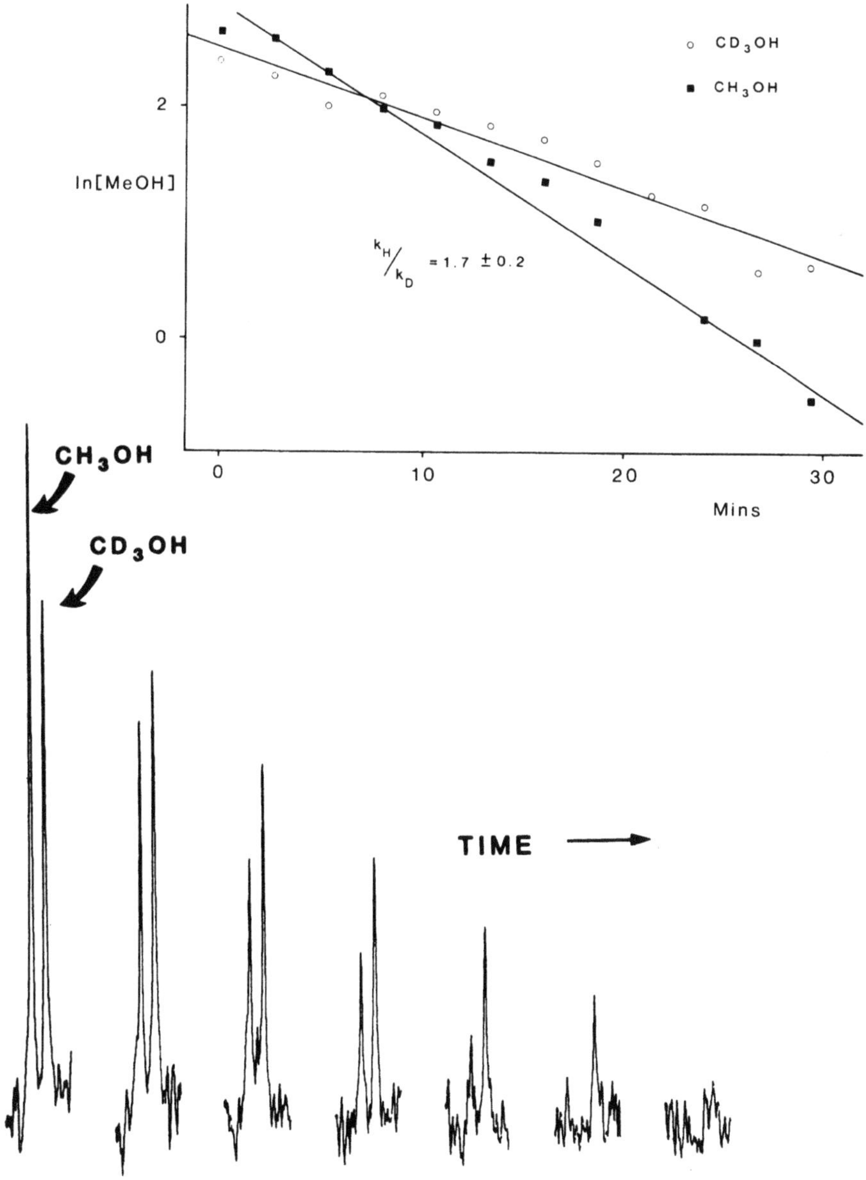

Figure 5. Demonstration of an enzymic isotope effect in methanol oxidation by *M. trichosporium* OB3b [7].

4.2. Formaldehyde dismutation

Is the production of roughly equal amounts of methanol and formate from formaldehyde by *E. coli, S. aureus* and *P. putida* the fortuitous result of similar levels of oxidation and reduction enzymes, or is there a formaldehyde dismutase (Cannizzarase) operating? Deuterium NMR may allow us to make a distinction:

$$2CD_2(OH)_2 \xrightarrow{\text{dismutase}} DCO_2H + CD_3OH$$

but

$$CD_2(OH)_2 \xrightarrow{\text{NADH}} CD_2HOH$$

The deuterium NMR spectrum of CD_3OH will be a singlet, while CD_2HOH will show a doublet due to H-D coupling. *E. coli* produces CD_2HOH, and possesses little dismutase (Figure 6), as would be expected from the glutathione results presented above. *S. aureus* (Figure 3) and *P. putida* (Figure 6) produce essentially pure CD_3OH, indicating dismutase activity. For *P. putida,* at least, this is no surprise since Kato has described a formaldehyde dismutase from this organism [9].

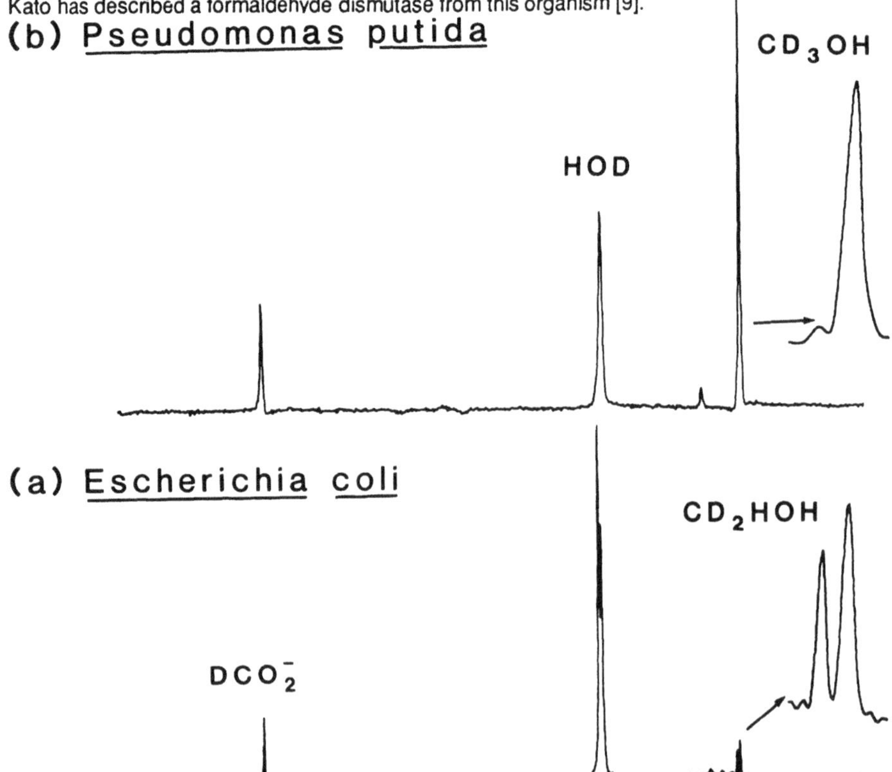

Figure 6. Deuterium NMR spectra of *E. coli* and *P. putida* following callenge with deuterium-labelled formaldehyde [7].

Again we can probe kinetics and mechanism by isotopic labelling: crossover experiments (Figure 7), followed by NMR spectroscopic assessment of the product ratios, give a value of $k_H/k_D \sim 3.7$ for the H transfer step

5. ASSAY OF CELL COMPONENTS

In some cells or tissues, dominant cell compounds can be assayed directly by ^{1}H, ^{13}C, or ^{31}P NMR as appropriate. In general, however, the spectra are too complex for this approach. An alternative approach is summarised in the following equation:

$$\text{cell component + labelled marker} \overset{k}{\leftrightarrow} \text{labelled adduct}$$

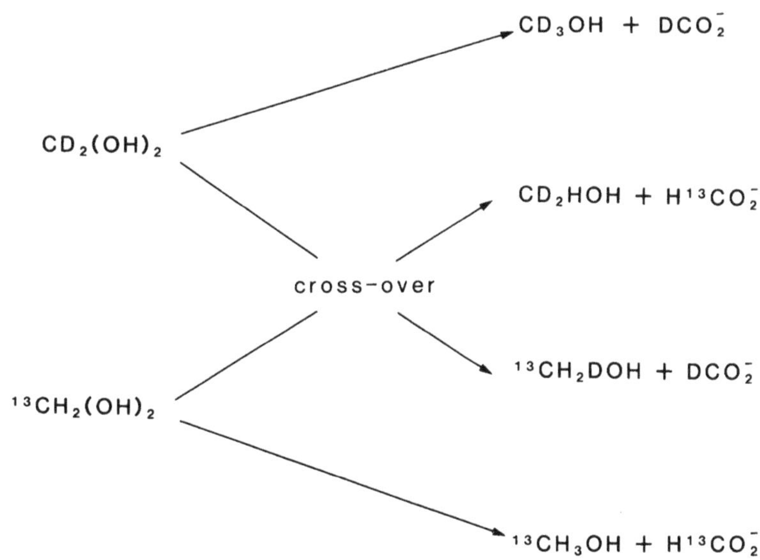

Figure 7. Crossover experiment for estimating isotope effects in formaldehyde dismutation.`

If we know the equilibrium constant *k*, can get the marker into the cell, and can recognise the adduct in an NMR spectrum, then we have a way of assaying the component of interest. Thus, NMR indicators of pH and metal ion concentration have been developed by others in Cambridge [10]. The *S*-hydroxymethyl glutathione results [6] mentioned above do not only elucidate a pathway; they also allow us to estimate the glutathione concentration to be a few millimolar in *E. coli* cells. Similarly, we can also estimate the free asparagine concentration in potato cells [11] by adduct formation with formaldehyde (see figure 8). Interestingly, the concentration of free glutamine is similar, but it does not interfere with the assay because the corresponding seven membered ring adduct is less stable and is not formed to any great extent.

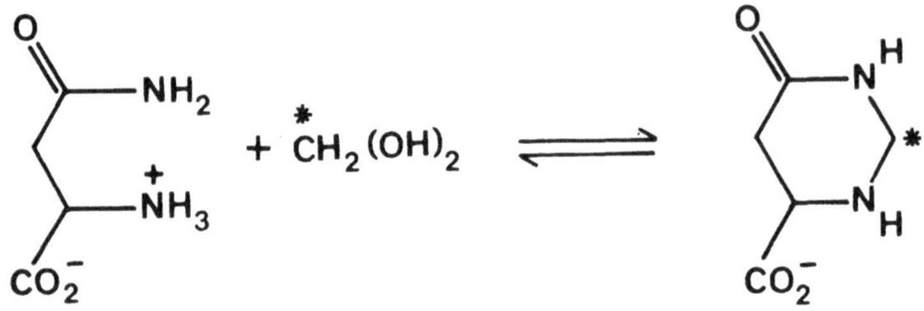

Figure 8. Adduct formation with formaldehyde

6. CONCLUSION

Whole cell or *in vivo* NMR can be used to follow metabolic processes and investigate the kinetics and machanism of biological processes *provided*

(1) The compounds of interest can be obtained with suitable isotopic labelling.

(2) The organism can tolerate, take up and metabolise roughly millimolar concentrations of the compounds in question.

(3) The organism can be squeezed into the NMR spectrometer!

ACKNOWLEDGEMENTS

I thank Professor N. Kato for supplying the *P. putida* culture. I am indebted to all those co-workers whose names appear in the references for their hard work, inspiration and company, with particular thanks to Dr B.K. Hunter (Queen's University, Ontario, Canada); I am also grateful for financial support from Pfizer, Shell, Unilever, NATO, the RSC, and SERC.

REFERENCES

[1] Sanders JKM and Hunter BK (1986) Modern NMR Spectroscopy - A guide for chemists, Oxford.
[2] Gadian DG (1982) NMR and its application to living systems, Clarendon, Oxford.
[3] Morris PG (1985) NMR imaging in medicine and biology, Clarendon, Oxford.
[4] Hunter BK, Nicholls KM, and Sanders JKM (1984) Biochemistry, 23, 508.
[5] Hunter BK, Nicholls KM, and Sanders JKM (1985) Biochemistry, 24, 4148.
[6] Mason RP, Sanders JKM, Crawford A, and Hunter BK (1986) Biochemistry 25, in press.
[7] Mason RP and Sanders JKM, unpublished.
[8] Cornish A, Nicholls KM, Scott D, Hunter BK, Aston WJ, Higgins IJ, and Sanders JMK (1984) J. Gen. Microbiol, 130, 2565.
[9] Kato N, Kobayashi H, Shimao M, and Sakazawa C (1984) Agric. Biol. Chem, 48, 2017.
[10] Smith GA, Hesketh RT, Metcalfe JC, Feeney J, and Morris PG (1983) Proc. Natl. Acad. Sci USA, 80, 7178.
[11] Mason RP, Sanders JKM and Gidley MJ, (1986) Phytochemistry, 25, 1567.

COMPARATIVE BIOCHEMISTRY OF C_1-CARRIERS

Lars G. LJUNGDAHL
Center for Biological Resource Recovery and Department of Biochemistry, University of Georgia, Athens, Georgia, 30602, U.S.A.

1. INTRODUCTION

C_1-compounds of oxidation levels from CO_2 to CH_4 are important in the metabolism of all cells. They are precursors or products of the biosynthesis and biodegradation of purines, pyrimidines, several amino acids and methylated compounds. In addition modification methylases alters DNA, methylation processes are involved in regulation of chemotaxis, and protein synthesis in prokaryotes and eukaryotic organelles such as the mitochondrian and the chloroplast is initiated by formyl-methionyl-tRNA. Many reactions of C_1-compounds involve C_1-carriers; the most commonly known are biotin, tetrahydrofolate, corrinoids and S-adenosyl methionine. Newly discovered C_1-carriers found in methanogenic bacteria are coenzyme M, methanofuran, and tetrahydromethanopterin. C_1-carriers are central in the newly ascertained pathways of autotrophic CO_2 fixation in several types of anaerobic bacteria and of the syntheses of acetate and methane from CO_2 in acetogenic and methanogenic bacteria. Recent reviews cover the acetate synthesis from CO_2 in acetogens [1,2] and the formation of methane from CO_2 [3,4]. In this presentation I wish to discuss the roles of C_1-carriers in the acetogens and the methanogens.

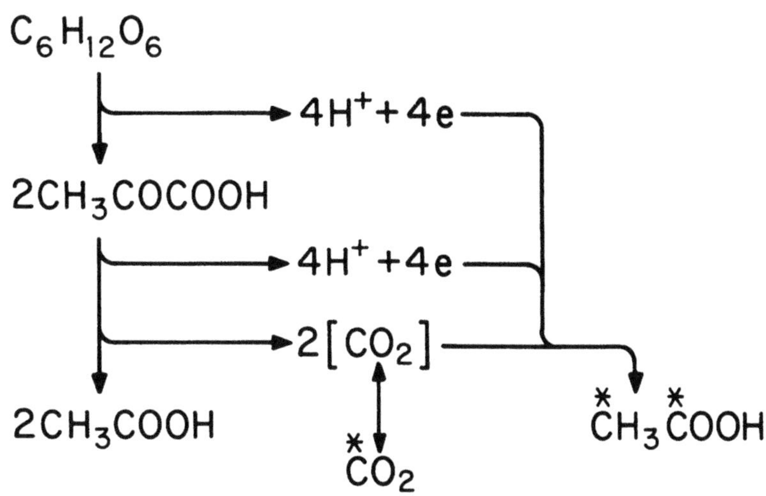

Figure 1. Fermentation of glucose to 3 moles of acetate showing CO_2 as electron acceptor and the synthesis of one third of the acetate from CO_2 [2].

2. Pathways of Acetate and Methane Formation via C_1-Carriers

The most detailed studies of the synthesis of acetate from CO_2 have been carried out with *Clostridium thermoaceticum.* This bacterium discovered in 1942 [5] converts fructose, glucose and xylose quantitatively to acetate. The convertion, that involves fermentation as well as the synthesis of acetate from CO_2, is outlined in Figure 1. As shown, CO_2 formed in the fermentation of the sugars, serves as an electron acceptor and is reduced to acetate. According to the stoichiometry of the glucose fermentation, two thirds of the acetate are formed directly from the sugar, and one third is synthesized from CO_2 [6].

Rather recently hydrogenase was demonstrated in *C. thermoaceticum* [7]. This was the key to the discovery that this bacterium can grow autotrophically on a gas mixture of H_2/CO_2 [8]. It is now realized that other one-carbon compounds such as CO, formate, and methanol serve as substrates. The pathway of autotrophic CO_2 fixation leads to the formation of acetyl-CoA, which can be regarded as the first stable intermediate. Acetyl-CoA is then either converted to acetate or utilized for synthesis of cell material.

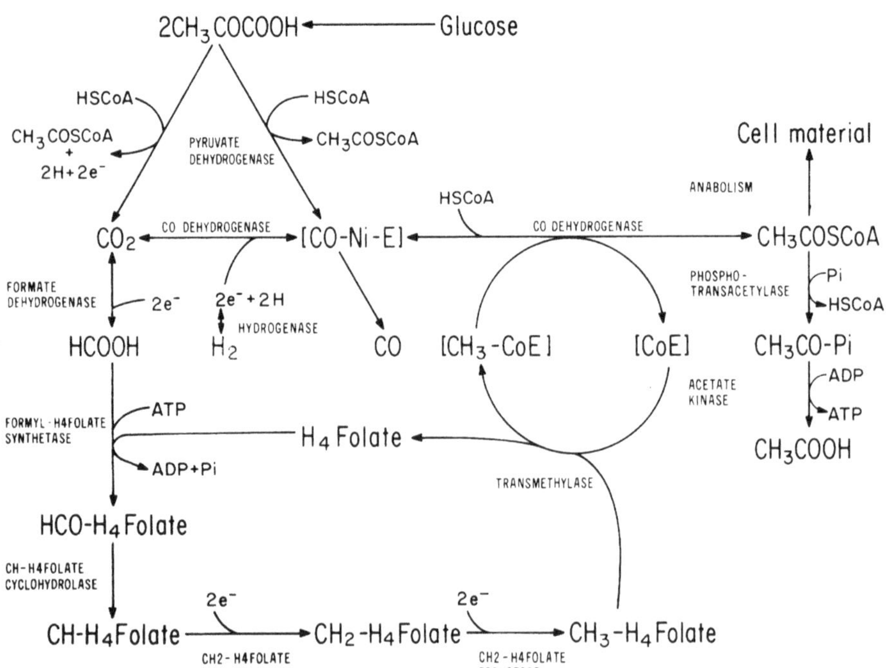

Figure 2. Detailed outline of the fermentation of glucose by *Clostridium thermoaceticum* and the pathway of autotrophic fixation of CO_2 with acetyl-CoA as product [2].

The pathway of acetyl-CoA synthesis in *Clostridium thermoaceticum* is detailed in Figure 2. It involves the reduction of CO_2 to formate, the synthesis of 10-formyltetrahydrofolate and its reduction to 5-methyltetrahydrofolate, and the transfer of the 5-methyl group to a protein bound corrinoid likely 5-methoxybenzimidazolyl-Co-methylcobamide (Co-methyl Factor IIIm). It also involves the reduction of CO_2 to a "CO" group. This reduction is catalyzed by carbon monoxide dehydrogenase, a nickel, FeS, Zn enzyme. The "CO" group is apparently bound to the Ni in the enzyme. The final step in acetyl-CoA synthesis is the condensation catalyzed by CO dehydrogenase of the Co-methyl group with the "CO"-Ni group and coenzyme A to form acetyl-CoA. The C_1-carriers in acetogens are obviously tetrahydrofolate and Factor IIIm, but CO dehydrogenase may also be included since the Ni in it binds the "CO" group. In addition, the formate dehydrogenase, a tungsten, selenium, FeS enzyme contains a tungsten-pterin cofactor, that may carry CO_2 when it is reduced to formate.

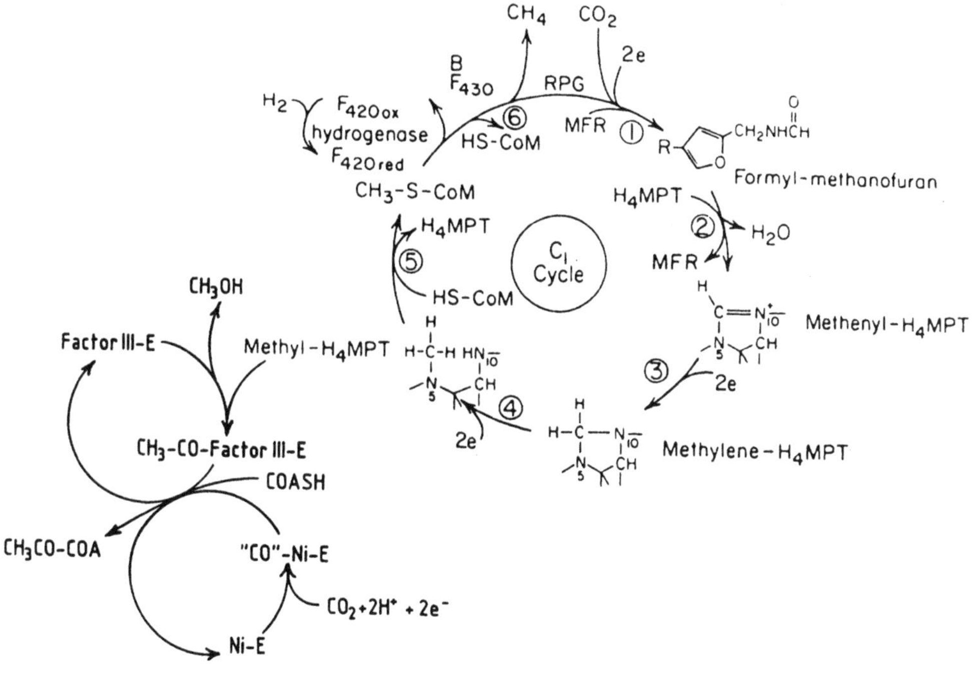

Figure 3. The C_1 cycle for reduction of CO_2 to methane [from ref.4]. MFR= methanofuran; H_4MPT= tetrahydromethanopterin; HS-CoM= coenzyme M; F_{420}= deazaflavin; F_{430}= nickel tetrapyrrole; B= 7-mercaptoheptanoylthreonine; RPG= effect of coupling methane formation with CO_2 activation and reduction. Autotrophic acetyl-COA synthesis via 5-hydroxybenzimidazolylcobamide-enzyme (factor III-E) and carbonmonoxide dehydrogenase (Ni-E) starts with methyl-H_4MPT. Methanol enters the pathways apperently via factor III-E.

The pathway of methane synthesis from CO_2 established in *Methanobacterium thermoautotrophicum* and other methanogens is outlined in Figure 3 [4]. In many ways the methane synthesis is similar to that of acetyl-CoA synthesis in acetogens. However, the C_1-carriers are different and there is no evidence of free formate as an intermediate in the reduction of CO_2 to form methane. The first step involves a reductive fixation of CO_2 to form formyl-methanofuran. Methanofuran is a unique C_1-carrier. The reaction requires coupling with the reduction of methyl-coenzyme M to methane and is complicated and not yet completely understood. The formyl group of formyl-methanofuran is then transferred to tetrahydromethanopterin, also a unique C_1-carrier in methanogens, to yield methenyltetrahydromethanopterin. The methenyl moiety is thereafter reduced with the formation of methyltetrahydromethanopterin. These reactions are analogous to the reduction of methenyltetrahydrofolate to methyltetrahydrofolate in the pathway of acetyl-CoA synthesis in acetogens. The methyl of methyltetrahydromethanopterin is transferred to coenzyme M before being converted to methane. The final reaction, the formation of methane from methyl-coenzyme A, requires F_{420}, a 5-deazariboflavin derivative [9]. F_{430}, a nickel tetrapyrrole [10] and a component B recently identified as 7-mercaptoheptanoylthreonine phosphate [11]. The reaction is stimulated by CO_2 [The RPG effect, 4].

The synthesis of acetyl-CoA from CO_2 for cell carbon by autotrophic methanogens appears to proceed along the pathway of methanogenesis to the point of formation of methyltetrahydromethanopterin. The methyl group instead of being transferred to coenzyme M is apparently accepted by a corrinoid-protein (perhaps with the formation of 5-hydroxybenzimidazolyl-Co-methylcobamide, Co-methyl Factor III). The final step is the condensation of the Co-methyl group with CO and coenzyme A to form acetyl-CoA in a manner exactly as occurs in acetogenic bacteria [12,13]. Like the acetyl-CoA synthesis in acetogens the final step of this synthesis is in methanogens catalyzed by a nickel-containing carbon monoxide dehydrogenase [14]. Supporting this is the fact that carbon monoxide dehydrogenase is lacking in methanogenic bacteria that require acetate for growth [15].

The methanogens that have the ability to cleave acetate to form methane and carbon dioxide may carry out this reaction by a reversal of the reactions leading to the synthesis of acetyl-CoA. The involvements of carbon monoxide dehydrogenase [16,17,18] and a corrinoid enzyme [19] are clearly indicated as is the coenzyme M methylreductase system [20].

The reduction of methanol to methane depends also on a corrinoid protein. Already in 1966 Blaylock and Stadtman [21] obtained a soluble enzyme system from *Methanosarcina barkeri*, that converts methanol to methane and also forms methyl-B_{12} with the methyl group derived from methanol [22]. The formation of methyl-coenzyme M from methanol requires two enzymes; methanol: 5-hydroxybenzimidazolylcobamide (methanol:Factor III) methyltransferase and 5-hydroxybenzimidazolyl-Co-methylcobamide:2-mercaptoethanesulfonate (Co-methyl Factor III:coenzyme M) methyltransferase [23]. The methanol: Factor III methyltransferase has been purified [24]. It has two different subunits, α=34,000 daltons and β=53,000 daltons, with an $\alpha_2\beta$ structure and contains three to four molecules of Factor III.

A third pathway dependent on tetrahydrofolate for acetate synthesis from CO_2 has been found in purine fermenting clostridia [25,26] and *Eubacterium angustum* [27]. (Figure 4). In the pathway CO_2 is reduced to formate that after condensation with tetrahydrofolate is reduced to methylenetetrahydrofolate. It reacts with a second CO_2 and NH_3 to form glycine; a reaction that is catalyzed by glycine

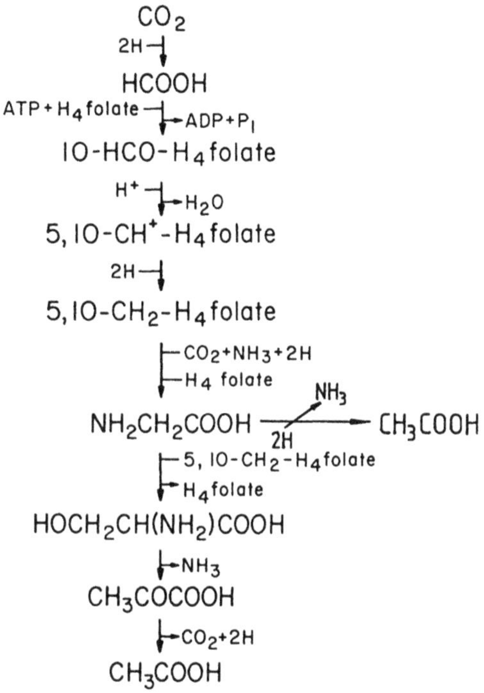

Figure 4. Acetate synthesis from CO_2 in purinolytic anaerobic bacteria. The key enzyme is glycine synthase forming glycine from methylene-H4 folate CO_2 and NH_3. Acetate is then formed either by a direct convertion to acetate in the glycine reductase reaction or via serine and pyruvate.

synthase. Glycine may then be directly converted to acetate by a selenium dependent glycine reductase system [28]. Alternatively, glycine may react with an additional methylenetetrahydrofolate fo form serine, which in turn via pyruvate yields acetyl-CoA and CO_2 [29]. This pathway is probably not used for acetyl-CoA synthesis in anaerobic bacteria.

3. Enzyme Reactions Involving C_1-compounds and C_1-carriers

The first step in the synthesis of acetate from CO_2 by acetogenic bacteria is catalyzed by formate dehydrogenase. The enzyme from various sources use diverse electron acceptors, differ in sizes and contain different cofactors and trace metals [30]. The *C. thermoaceticum* enzyme is NADP-dependent and contains pterin, tungsten, selenium and iron-sulfur centers [31]. The pterin and tungsten form a tungsten-pterin cofactor, which apparently is very similar to the molybdenum-pterin cofactor of several molybdenum dependent enzymes [32]. The side chain of the pterin has two sulfurs which are liganded to molybdenum. EXAFS-studies of the formate dehydrogenase form *C. thermoaceticum* indicate that the tungsten is also liganded to two sulfurs presumably those of the pterin [33]. Other ligands are either to N or O atoms. There is no evidence for the tungsten cofactor to be a carrier or to form a complex with CO_2 or formate. However it is well known that tungsten forms ligands with CO_2 or CO and the possibility that it can serve as C_1-carrier has not been excluded.

The physiological role of formate dehydrogenases in methanogens is apparently to oxidize formate to CO_2 and to produce reducing equivalents for the reduction of CO_2 to methane. Formate is not a free intermediate in the formation of methane from CO_2 [3,4]. The formate dehydrogenase of *Methanobacterium formicicum* contains the molybdenum-pterin cofactor, FAD and iron-sulfur centers [34]. It couples with coenzyme Factor $_{420}$.

The first step in the reduction of CO_2 for methane production involves a reductive fixation of CO_2 with the formation of formyl-methanofuran. This reaction is not completely understood, but as has already been discussed it is coupled to the formation of methane from methyl-coenzyme M and involves F_{420}, F_{430}, component B, and the RPG-effect [3,4]. An analogous mechanism of forming formyltetrahydrofolate in acetogenic bacteria has not been observed. The structures of methanofuran [36] and formyl-methanofuran [37] have been determined.

In the acetogenic bacteria the reduction of formate to the methyl group occurs with tetrahydrofolate (H_4folate) as C_1-carrier. The first reaction is the formation of formyl-H_4folate at the expense of ATP. The reaction is catalyzed by formyl-H_4folate synthetase. Three additional enzymes are involved, methenyl-H_4folate cyclohydrolase, methylene-H_4folate dehydrogenase, and methylene-H_4folate reductase. The four H_4folate enzymes have been purified and characterized from *C.thermoaceticum* and *C.formicoaceticum* [2]. Extracts of acetogenic bacteria have high activities of these enzymes and the purified enzymes have high specific activities. This is in support of their importance in acetogenic bacteria.

It is of interest that homogeneous H_4folate enzymes from eukaryotes are trifunctional and catalyze the synthetase, cyclohydrolase and dehydrogenase reactions [38]. In *C.thermoaceticum* the dehydrogenase and cyclohydrolase activities are catalyzed by the same enzyme [39] whereas in *C.formicoaceticum* all activities are catalyzed by separate enzymes [40]. The methylene-H_4folate dehydrogenase in *C.thermoaceticum* is NADP specific, whereas in *C.formicoaceticum* it is NAD-dependent. The methylene-H_4Folate reductase contains one Fe_4S_4 center and 2 FAD. The physiological reductant probably is ferredoxin [41].

The methanogenic bacteria have instead of H_4folate tetrahydromethanopterin as equivalent C_1-carrier [42]. The active sites of the two carriers are almost identical and it seems obvious that tetrahydromethanopterin functions in the same way as H_4folate. According to the scheme in figure 3 methenyltetrahydromethanopterin is formed by transfer of the formyl group of formylmethanofuran to tetrahydromethanopterin. Such a reaction would involve a cyclohydrolase as well as a transfer at the same time. Recently a methenyltetrahydromethanopterin cyclohydrolase was demonstrated in *M.thermoautotrophicum* [43]. Thus, it is possible that formyltetrahydromethanopterin is the product of the transfer and that cyclohydrolysis follows. Furthermore, formyltetrahydromethanopterin in similarity with formyl-H_4folate undergoes chemical cyclohydrolysis; a reaction which is strongly pH-dependent.

The reduction of methenyltetrahydromethanopterin to methylenetetrahydromethanopterin is dependent on coenzyme F_{420} and not on NADH or NADPH [44]. Chemically methylenetetrahydromethanopterin is also formed in a reaction between formaldehyde and tetrahydromethanopterin. A similar reaction occurs between formaldehyde and H_4folate. The reduction of methylenetetrahydromethanopterin to the methyl derivative has not been elucidated. An apparent redox potential of the methylene-H_4folate/methyl-H_4folate couple is -120mV. If the redox potential is similar for the corresponding tetrahydromethanopterin derivatives, coenzyme F_{420} may function as reductant also in this reaction. The formation of methyl-coenzyme M from methyltetrahydromethanopterin and subsequently methane, the unique reactions in methanogens have already been discussed.

Acetyl-CoA formation for production of acetate and for cell carbon in autotrophic acetogens and autotrophic methanogens clearly occurs in a similar manner. Enzymes involved are transmethylases, corrinoid proteins, and carbon monoxide dehydrogenases as has already been pointed out. Both acetogens and methanogens contain large quantities of corrinoids. From *C. thermoaceticum* not less than 20 corrinoids have been isolated [45]. Also methanogens contain many different corrinoids [46]. Interestingly, the apparently functional complete corrinoids in acetogens is 5-methoxybenzimidazolyl cobamide (Factor IIIm) [45] and in methanogens 5-hydroxy-benzimidazolyl cobamide [47]. The benzimidazole moiety of the corrinoids in acetogens and methanogens thus differ from the common vitamin B_{12} which has 5,6-dimethylbenzimidazole. The relation of the different benzimidazoles of any to function is not understood.

The final reaction in acetyl-CoA synthesis is catalyzed by carbon monoxide dehydrogenase [48]. The enzyme is present in autotrophic acetogens, methanogens and sulfur reducing bacteria. The most obvious property of the enzyme is that it contains nickel. EPR studies strongly indicate that the nickel is carrier of the "CO" group [49]. In addition the enzyme must have sites for a methyl group, that is transferred from the Co-methyl-corrinoid protein, and coenzyme A.

The elucidations of methane and acetate biosynthesis from CO_2 have followed parallel paths. Around 1965 many of us believed the two pathways were almost identical with a Co-methylcorrinoid and methyltetrahydrofolate as intermediates. With the discovery of coenzyme M in methanogens [50] the concept of common pathway was questioned. Now we see the diversity but also the similarity of the two pathways. The story not yet completed has been full of surprises and excitements.

REFERENCES

[1] Wood HG,Ragsdale SW, Pezacka E. (1986) Tr. Biochem. Sci. 11:14-18.
[2] Ljungdahl LG. (1986). Ann. Rev. Microbiol. 40:415-50.
[3] Whitman WB. (1985). In "The Bacteria, Vol VIII, Archaebacteria" (Eds. Woese CR, Wolfe RS) Academic Press Inc. New York pp. 3-84.
[4] Wolfe RS. (1985) Tr. Biochem. Sci. 10:396-399.
[5] Fontaine FE, Peterson WH, McCoy E, Johson MJ, Ritter G. (1942) J. Bacteriol. 43:701-15.
[6] Wood HG. (1952) J. Biol. Chem. 194:905-31.
[7] Drake HL. (1982) J. Bacteriol. 150:702-709.
[8] Kerby R, Zeikus JG. (1983) Curr. Microbiol. 8:27-30.
[9] Eirich LD, Vogels GO, Wolfe RS. (1978) Biochemistry. 17:4583-4593.
[10] Pfaltz A, Jaun M, Fassler A, Eschenmoser A, Jaenchen R, Gilles HH, Diekert G, Thauer RK. (1982) Helv. Chim. Acta. 65:828-865.
[11] Noll KM, Rinehart KL Jr, Tanner RS, Wolfe RS. (1986) Proc. Natl. Acad. Sci. U.S.A. 83:4238-4243.
[12] Stupperich E, Hammel KE, Fuchs G, Thauer RK. (1983) FEBS Lett. 152:21-23.
[13] Stupperich E, Fuchs G. (1984) Arch. Microbiol. 139:14-20.

[14] Hammel KE Cornwell KL, Diekert GR, Thauer RK. (1984) J. Bacteriol. 157:975-978.
[15] Bolt MH, Eikmanns B, Thauer RK. (1985) Arch. Microbiol. 143:266-269.
[16] Eikmanns B, Thauer RK. (1984) Arch. Microbiol. 138:365-370.
[17] Smith MR, Lequerica JL, Hart MR. (1985) J. Bacteriol. 162:67-71.
[18] Krzycko JA, Zeikus JG. (1984) J. Bacteriol. 158:231-237.
[19] Eikmanns B, Thauer RK. (1985) Arch. Microbiol..142:175-179.
[20] Nelson MJK, Ferry JG. (1984) J. Bacteriol 160:526-532.
[21] Blaylock BA, Stadtman TC. (1966) Arch. Biochem. Biophys. 116:138-158.
[22] Blaylock BA. (1968) Arch. Biochem. Biophys. 124:314-324.
[23] van der Meijden P, van der Lest C, van der Drift C, Vogels GD. (1984) Biochem. Biophys. Res. Commun. 118:760-766.
[24] van der Meijden P, te Brommelstroet BW, Poirot CM, van der Drift C, Vogels GD. (1984) J. Bacteriol. 160:629-635.
[25] Schiefer-Ullrich H, Wagner R, Dürre P, Andreesen JR. (1984) Arch. Microbiol. 138:345-53.
[26] Ljungdahl LG. (1984) In "Folates and Pterins" (Eds. Blakley RL. Bencovic SJ) Vol 1. Wiley, NY. pp 555-579.
[27] Beuscher HV, Andreesen JR. (1984) Arch. Microbiol. 140:2-8.
[28] Stadtman TC. (1980) Ann. Rev. Biochem. 49:93-110.
[29] Waber LJ, Wood HG. (1979) J. Bacteriol. 140:468-478.
[30] Ljungdahl LG. (1980) In "Molybdenum and molybdenum-containing enzymes" (Ed. Coughlan MP). Pergamon Press. NY. pp 347-383.
[31] Yamamoto I, Saiki T, Liu SM, Ljungdahl LG. (1983) J. Biol. Chem. 258:1826-1832.
[32] Hageman RV, Rajagopalan KV. (1986) Methods Enzymol. 122:399-412.
[33] Cramer SP, Liu SL, Mortenson LE, Spence JT, Liu SM, Yamamoto I, Ljungdahl LG. (1985) J. Inorg. Biochem. 23:119-124.
[34] Schauer NL, Ferry JG. (1986) J. Bacteriol. 165:405-411.
[35] Jones JB, Stadtman TC. (1981) J. Biol. Chem. 256:656-663.
[36] Leigh JA, Rinehart KL Jr, Wolfe RS. (1984) J. Amer. Chem. Soc. 106:3636-3640.
[37] Leigh JA, Rinehart KL Jr, Wolfe RS. (1985) Biochemistry. 24:995-999.
[38] MacKenzie RE. (1984) In "Folates and Pterins" (Eds. Blakley RL, Bencovic SJ) Vol. 1. Wiley. NY. pp 255-306.
[39] Ljungdahl LG, O'Brien WE, Moore MR, Liu MT. (1980) Methods Enzymol. 66:599-609.
[40] Clark JE, Ljungdahl LG. (1982) J. Biol. Chem. 257:3833-3836.
[41] Clark JE, Ljungdahl LG. (1984) J. Biol. Chem. 259:10845-10849.
[42] Escalante-Semerena JC, Rinehart KL Jr, Wolfe RS. (1984) J. Biol. Chem. 259:9447-9455.
[43] Donnelly MI, Escalante-Semerena JC, Rinehart KL Jr, Wolfe RS. (1985) Arch. Biochem. Biophys. 242:430-439.
[44] Hartzell PL, Zvilius G, Escalante-Semerena JC Donnelly MI. (1985) Biochem. Biophys. Res. Commun. 133:884-890.
[45] Ljungdahl LG, Irion E, Wood HG. (1966) Fed.Proc.25:1642-1648
[46] Whitman WB, Wolfe RS. (1984) Anal.Biochem.137: 261-265
[47] Pol A, van der Drift C, Vogels GD. (1982) Biochem.Biophys.Res.Commun.108:731-737
[48] Ragsdale SW, Wood HG. (1985) J.Biol.Chem.260:3970-3977
[49] Ragsdale SW, Ljungdahl LG, DerVartanian DV. (1983) Biochem.Biophys.Res.Commun.115:658-665
[50] Taylor CD, Wolfe RS. (1974) J.Biol.Chem.249:4879-4885

PHYSIOLOGY, BIOCHEMISTRY AND GENETICS OF METHYLOTROPHIC YEASTS

METHYLOTROPHIC YEASTS -- 1986

Gene H. WEGNER and Wim HARDER

Provesta corporation, Phillips Petroleum Company
Bartlesville, OK, U.S.A. , and Department of Microbiology , University of Groningen , The Netherlands

1. INTRODUCTION

The purpose of this paper is to set the scene for the session by highlighting historical and recent developments in the physiology, biochemistry and genetics of the methylotrophic yeasts from both a basic scientific, as well as an applied point of view.

Of the one-carbon compounds, methanol is the substrate that predominantly is used as a carbon and energy source by yeasts and can support growth of methylotrophic yeasts at respectable growth rates and to high cell densities . Although methylated amines (as a N-source) and even methane have been reported to be metabolized by yeasts, this session will focus primarily on the utilization of methanol by these organisms.

2. HISTORICAL OVERVIEW

From a historical standpoint, our understanding of methylotrophic yeasts is relatively recent. Table 1 lists the published historical milestones in the development of our knowledge of these organisms.

Table 1. Methylothrophic yeasts - Historical Overview

1969	Methanol utilization first reported
1970s	Isolation technique worked out
	Single - cell - protein processes developed
1975	Biochemistry of energy generation established
1978	Carbon assimilation pathway worked out
1975 - 1985	Compartmentation of key enzymes established
1983 - Future	Genetics of methylotrophic yeasts
	Use as hosts for heterologous gene expression
	Topogenesis of peroxisomal enzymes

Bacterial utilization of one-carbon compounds has been known from the beginning of this century. However, growth of yeasts on one-carbon compounds was not established until the late 1960s when Ogata reported the isolation of a methanol-utilizing yeast [1]. Similar independent results by several research groups soon followed this first report.

The isolation of methanol-utilizing yeasts from nature is not as readily accomplished as it is with bacteria. During the early 1970s, enrichment and isolation techniques were worked out. These usually employed batch-type enrichment cultures containing anti-bacterial compounds such as penicillin or cycloserine at pH 4.5-5.0 [2] or continuous flow enrichments at low pH [3]. The yeast isolated from natural sources belong to a limited number of genera. Similar results have been obtained from screening culture collections [4]. All the strains identified so far belong to only four genera which are shown in table 2.

Table 2. Genera of methylotrophic yeasts

Ascomycetous	Asporogenous
Hansenula	*Candida*
Pichia	*Torulopsis*

Recently it has been argued that these yeast genera are closely related, the ascomycetous yeasts, *Hansenula* and *Pichia* , being distinguished mainly on the basis of their ability to use nitrate as nitrogen source. Furthermore, it has been suggested that the methanol-utilizing strains within the genera *Candida* and *Torulopsis* may be regarded as imperfect forms of *Hansenula* and *Pichia* [5, 6].

.During the early 1970s, there was considerable interest in producing single-cell-protein from methanol using methylotrophic yeasts and several companies developed methanol-yeast processes. This potential commercial application acted as a stimulant to much of the subsequent research into the biochemistry and physiology of the methylotrophic yeasts.

By 1975 the biochemistry of energy generation in these yeasts was firmly established and found to differ fundamentally from that in bacteria in a number of respects, including the nature of the enzymes involved, their compartmentation in peroxisomes and the involvement of mitochondria in energy transduction.

After considerable initial confusion about the C_1 assimilation pathway utilized by these yeasts, by 1978 it became clear that the route was different from those encountered in bacteria [7] and involved the key enzyme dihydroxyacetone synthase [8].

The compartmentation of key enzymes involved in methanol metabolism in peroxisomes with a crystalline matrix is a striking feature of all yeasts grown on methanol . Although alcohol oxidase and catalase had been recognized as being peroxisomal enzymes since about 1975, it was not until 1985 that it was established that dihydroxyacetone synthase is also contained in the peroxisomes of methanol grown yeasts [9, 10].

Since 1983 there has been a rapidly expanding interest in the classical and molecular genetics of methylotrophic yeast. This has largely been driven by the realization that these yeasts contain enzymes that are well regulated and highly expressed during growth of methanol (in particular alcohol oxidase and dihydroxyacetone synthase) and may have key genetic features to make excellent hosts for the production of recombinant DNA proteins. Also, in the last few years, research has begun in several laboratories which will hopefully lead to an understanding of the molecular mechanisms involved in the topogenesis or targeting of alcohol oxidase, catalase and dihydroxyacetone synthase to the peroxisomes so abundantly present in these yeasts.

3. METHANOL METABOLISM IN YEASTS

Yeasts do not contain the typical bacterial methanol dehydrogenase. Instead, in these organisms peroxisomal alcohol oxidase, using oxygen as an electron acceptor, catalyses the oxidation of methanol

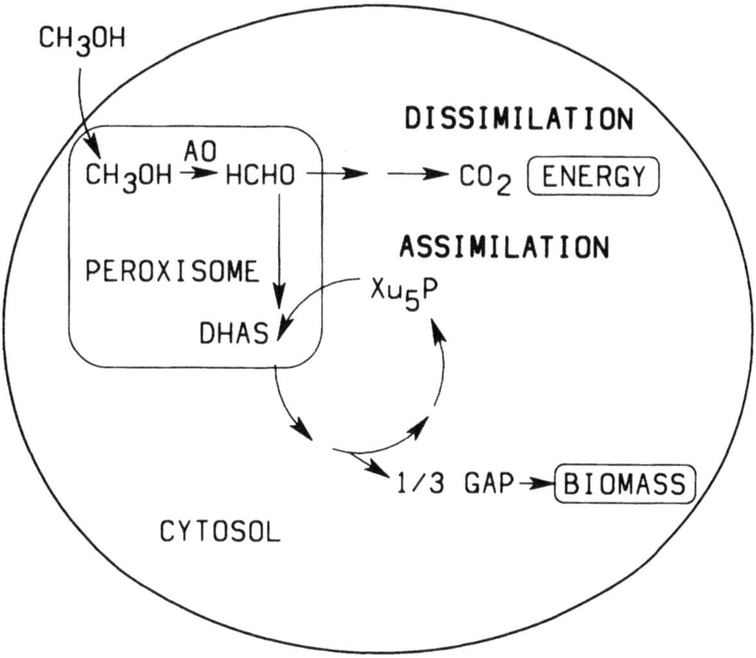

Figure 1. Overview of methanol metabolism in yeasts.
AO, alcohol oxidase; DHAS, dihydroxyacetone synthase; Xu_5P, xylulose 5-phosphate; GAP, glyceraldehyde 3-phosphate.

to yield formaldehyde and hydrogen peroxide. The hydrogen peroxide is subsequently broken down by peroxisomal catalase. Formaldehyde, generated in the peroxisome may either be dissimilated by cytosolic enzymes to yield CO_2 and energy or assimilated into cellular biomass (figure 1).

Alcohol oxidase is the first enzyme in the oxidation of methanol and under methanol limiting conditions can be present in large amounts, in crystalline form, neatly packaged in peroxisomes. This enzyme has been purified from various yeasts and partially characterized. Our current knowledge of alcohol oxidase and its biochemistry will be discussed in depth in another paper in this session [11].

The compartmentation in peroxisomes of three key enzymes in the metabolism of methanol (alcohol oxidase, catalase and dihydroxyacetone synthase) has important implications on the regulation of metabolism and control of the fluxes of formaldehyde over dissimilatory and assimilatory pathways. Metabolic regulation of the methylotrophic yeasts is discussed in details in the paper that follows [12].

4. GENETICS OF METHYLOTROPHIC YEAST

In recent years, research efforts in both classical and molecular genetics has increased significantly. Classical genetic manipulation has been limited to the two ascomycetous genera, *Hansenula* and *Pichia* . Within these genera, genetic manipulation has been reported for *P. pinus* [13, 14] and *H. polymorpha* [15]. Mutants defective in methanol utilization, as well as other markers, have been isolated and placed in complementation groups.

The molecular aspect of genes involved in methanol utilization have been investigated recently in *H. polymorpha* and *P. pastoris.* The genes for alcohol oxidase and dihydroxyacetone synthase have been cloned from these two yeasts [16-18] and transformation systems have also been developed [19, 20]. These important new developments and their impact on our understanding of the regulation of gene expression and the targeting of proteins to peroxisomes are discussed in greater detail in other papers in this session [21,22].

High level expression systems for heterologous genes have very recently been developed employing *P. pastoris* using the promotors and regulatory regions of the alcohol oxidase and dihydroxyacetone synthase genes [23]. These new developments will be discussed in the applied session of this symposium.

5. APPLIED ASPECTS

5.1 Single-Cell-Protein

The application of methylotrophic yeasts to produce single-cell-protein (SCP) from methanol was the subject of considerable research and development efforts by several groups during the last decade. Perhaps the most extensive work has been done at Phillips Petroleum Company. This work started in the late 1960s with the isolation of some of the first methanol utilizing yeasts. After extensive screening of numerous yeast cultures isolated from natural sources and obtained from culture collections, a strain of *Pichia pastoris* was chosen because of its greater cell mass yield from methanol, higher protein content and stable fermentation characteristics.

This yeast grows relatively slowly on methanol (doubling time 5.5-6.0 hours). Therefore, to increase productivity of the fermentation system, it was necessary to increase the cell density, so research was directed towards this goal. It was found that this culture of *P. pastoris* was capable of growth at high cell densities if provided with the proper balance of mineral nutrients. Upon scale-up, oxygen transfer and heat removal soon became the limiting factors. Therefore, extensive fermentor design and modification was undertaken to maximize these limiting factors. From this research was developed what is known as the Phillips/Provesta continuous high cell density, direct-dry process [24].

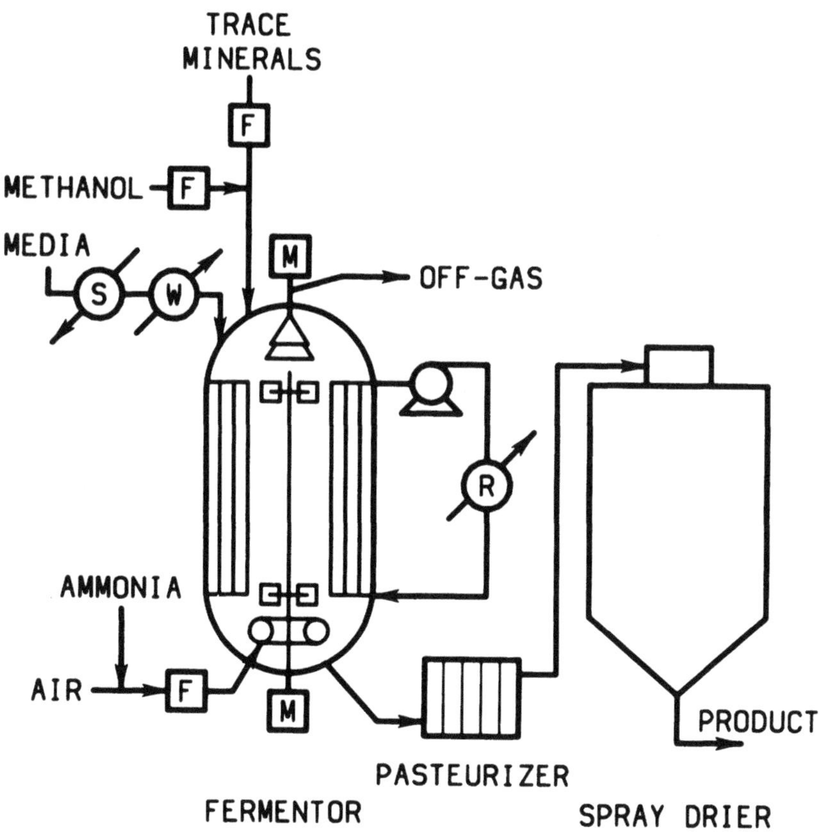

Figure 2. The Phillips/Provesta continuous, high-cell-density, direct-dry process for single-cell-protein.

In this process (see Fig. 2), yeast cell concentrations of over 125 g/l (dry wt.) are routinely obtained in the fermentor effluent. This yeast slurry is given a heat treatment and dried directly without intervening centrifugation steps. Consequently this process is simple to operate and has the advantages of eliminating cell concentration steps and the resulting waste streams and allows 100% recovery of the yeast and soluble products.

This process, like other SCP processes, has serious problems competing in a free market situation with other sources of protein and would require very selective situations in order to be economically attractive. However, with the greatly increased methanol production coming on-stream in the Middle East, methanol prices are likely to drop significantly and attractive situations may develop.

5.2 Fermentation Technology

A useful by-product of the research with methylotrophic yeasts is that of the advanced fermentation technology. Significantly, it has been found that the high productivity fermentor technology developed for the methanol-yeast SCP process has applications in other fermentations which can benefit from oxygen transfer and heat transfer rates 2-4 times that achieved with conventional fermentors.

5.3 Alcohol Oxidase

A special application of methylotrophic yeasts is their use as a source of alcohol oxidase. This enzyme may constitute up to 35% of the soluble protein and due to breakthroughs in purification and isolation techniques, it is now commercially available in large quantities from Provesta Corporation. This enzyme, isolated from *P. pastoris* , is finding application as a diagnostic tool for determining alcohol concentrations and has other potential large volume uses.

5.4 rDNA Expression Systems

Recent research has shown that methylotrophic yeasts are excellent hosts for the expression of heterologous genes. The details of these developments are discussed in other papers in this symposium [21, 23]. This development may prove to be the most important application of these organisms yet uncovered.

5.5 Areas of Current Interest

In the papers to follow in this session, the contributors will discuss the main areas of interest involving methylotrophic yeasts (table 3).

The first two areas, metabolic regulation and biochemistry of alcohol oxidase, have been the subject of considerable research in the past decade and a significant body of knowledge has been built up on these subjects. However, as the other contributors to this session will discuss, there are still key areas which need explanation, for instance, in which way is the flux of formaldehyde regulated between dissimilation and assimilation, and why alcohol oxidase accumulates to such high levels.

Table 3. Current areas of interest

Metabolic Regulation
Biochemistry of Alcohol Oxidase
Classical and Molecular Genetics
Topogenesis of Peroxisomal Enzymes

Although important contributions have been made on the genetics and targeting of peroxisomal proteins in the methylotrophic yeast, much remains to be discovered. These areas of active interest and ones we look forward to understand more clearly in the future. It is in these areas where the future direction will be focussed.

REFERENCES

[1] Ogata K, Nishikawa H and Ohsugi M. (1969) Agricultural and Biological Chemistry 33, 1519.
[2] Dijken JP van and Harder W. (1974) Journal of General Microbiology 84, 409.
[3] Levine DW and Cooney CL. (1973) Applied Microbiology 26, 982.
[4] Hazeu W, Bruijn JC de and Bos P. (1972). Archiv für Mikrobiologie 87, 185.
[5] Lee JD and Komagata K. (1980) Journal of General and Applied Microbiology 26, 133.
[6] Komagata K. (1981) In "Microbial Growth on C_1 Compounds" (Dalton CH ed.) pp.301-311. Heyden and Sons, Ltd., London.
[7] Dijken JP van, Harder W, Beardsmore AJ and Quayle JR. (1978) FEMS Microbiology Letters 4, 97.
[8] O'Connor M and Quayle JR. (1980) Journal of General Microbiology 120, 219.
[9] Goodman JM. (1985). The Journal of Biological Chemistry 260, 7108.
[10] Douma AC, Veenhuis M, Koning, W de, Evers, M and Harder, W. (1985) Archives of Microbiology 143, 237.
[11] Hopkins TR. and Müller F. (1986). This Symposium.
[12] Trotsenko YA, Egli T and Harder W. (1986) This Symposium
[13] Popova A and Pugovkina NA. (1982) Genetika 18, 322
[14] Tolstroyukov II, Efremov BD and Bliznik KM. (1983) Genetika 19, 897.
[15] Gleeson MA, Waites MJ and Sudbery PE. (1984) In "Microbial Growth in C_1 Compounds," (Crawford, RL and Hanson,RS ed.) pp. 228-235. American Society for Microbiology, Washington, D.C.
[16] Janowicz ZA, Eckart MR, Drewke C, Roggenkamp RO and Hollenberg CP. (1985) Nucleic Acids Research 13, 3043.
[17] Ledeboer AM, Edens L, Maat J, Visser C, Bos JW and Verrips CT. (1985) Nucleic Acids Research 13, 3063.
[18] Ellis SB, Brust PF, Koutz PJ., Waters AF, Harpold MM and Gingeras TR. (1985) Molecular and Cellular Biology 5, 1111.
[19] Cregg JM, Barringer KJ, Hessler AY and Madden KR. (1985) Molecular and Cellular Biology 5, 3376.
[20] Roggenkamp R, Hansen H, Eckart M, Janowicz Z, Hollenberg CP. (1986) Molecular and General Genetics 202, 302.

[21] Cregg, JM. (1986) This Symposium.
[22] Tabak HF and AB G. (1986) This Symposium.
[23] Stroman DW and Thill G. (1986) This Symposium.
[24] Wegner EH. (1983) U.S.Patent 4, 414, 329. Nov.8, 1983

METABOLIC REGULATION IN METHYLOTROPHIC YEASTS

W. HARDER, Y. A. TROTSENKO, L.V. BYSTRYKH and T. EGLI
Dept. Microbiology, University of Groningen, Haren, The Netherlands; USSR Academy of Sciences, Pushchino, USSR and EAWAG/ETH, Dübendorf-Zürich, Switzerland.

1. INTRODUCTION

The ability to grow in media containing methanol (the C_1 substrate predominantly used as a carbon- and energy source) is not nearly as widespread among yeasts as it is in bacteria. Whereas bacterial utilization of methanol has been reported for many widely different genera and species, the relatively small numbers of methylotrophic yeasts strains identified so far appear to be taxonomically closely related [1]. This kinship is reflected in a remarkable uniformity in the pathways used for methanol metabolism. Although the properties of the various enzymes may slightly differ, all yeasts capable of growth on methanol possess the alcohol oxidase/catalase system for the initial oxidation of the substrate, they oxidise formaldehyde to carbon dioxide via NAD-dependent dehydrogenases and employ the xylulose mono-phosphate (XuMP) pathway for carbon assimilation. A further characteristic that is shared by all species examined is the specific involvement of subcellular organelles and as a result methanol metabolism in yeasts is strictly compartmentalized. This metabolic uniformity is probably also reflected in the mode of regulation of methanol metabolism and although this aspect has been studied in only a limited number of species, the results obtained may be considered representative for the group as a whole.

Yeasts that are capable of growth in media containing methanol as a sole source of carbon and energy are without exception facultative methylotrophs. They grow equally well or better on multi-carbon substrates and under those conditions several of the enzymes required for methanol metabolism are repressed. Metabolic regulation in these organisms therefore involves both control over the synthesis and turn over of C_1-specific enzymes ("long-term" regulation) and control of the activity of these enzymes ("short-term" regulation). Most of our current understanding of metabolic regulation in methylotrophic yeasts is of relatively recent date and stems from studies of various strains of *Hansenula polymorpha* and *Candida boidinii* (note that *Kloeckera* sp.2201, the first methanol-utilizing yeast that was described [2], has been re-identified as a strain of *C. boidinii* [1]; reference to this particular strain will be made as in the original publication). Two kinds of approaches have been used in these studies, namely a) analysis of batch and chemostat cultures relating to the regulation of the synthesis and degradation of various enzymes in response to growth conditions and b) studies of the kinetic properties of relevant enzymes and assessment of metabolic fluxes through regulation of the activity of these enzymes, taking into consideration the role of low molecular weight effector molecules and the involvement of subcellular compartments. The intention of this contribution is to review the results of these physiological and biochemical studies. But before we embark on this, a description of the current views on the enzymology of methanol metabolism in yeasts and the subcellular localization of the various enzymes is given.

2. ENZYMOLOGY AND SUBCELLULAR LOCALIZATION OF METHANOL METABOLISM IN YEASTS

A schematic representation of the pathways involved in the metabolism of methanol in yeasts and their compartmentation is shown in Fig.1. The initial reaction is an oxidation of methanol to formaldehyde which is catalyzed by alcohol oxidase (AO). The enzyme has been purified from various yeasts and studied extensively. It is a homo-octameric protein with a molecular mass of around 600 kDa, contains FAD as a prosthetic group and is able to oxidize a variety of primary aliphatic alcohols, their halogen derivatives and formaldehyde. AO possesses a variable affinity for methanol (K_m 0.2 - 3.1 mM) and O_2 (K_m 0.2 - 1.0 mM) (3,4,5) and it is possible that this is not only strain-dependent but is also related to growth conditions since there is evidence (6) that the FAD content of AO is not constant but

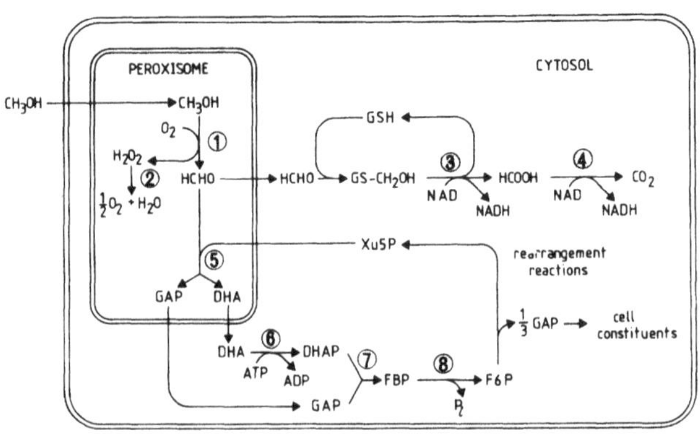

Figure 1. Compartmentation of methanol metabolism in *H. polymorpha*. 1, alcohol oxidase; 2, catalase; 3, formaldehyde dehydrogenase; 4, formate dehydrogenase; 5, dihydroxyacetone synthase; 6, dihydroxyacetone kinase; 7, fructose-1,6-bisphosphate aldolase; 8, fructose-1,6-bisphosphate phosphatase. Reproduced from [21].

may vary with growth rate. For further details on AO see [7]. Hydrogen peroxide, which is produced in the AO catalyzed reaction is decomposed by catalase. It has been postulated [8, but see also 16] that this enzyme may also play an important role in the oxidation of methanol by way of its peroxidatic activity using the H_2O_2 generated by AO, but recent kinetic studies performed with catalase purified to homogeneity from *Candida boidinii* 2201 [9] have made this unlikely. Both AO and catalase are localized in peroxisomes that are so conspicuous in yeasts growing on methanol [10].

Formaldehyde generated inside the peroxisomes can either be used for energy generation or be assimilated. The enzymes necessary for the oxidation of formaldehyde to CO_2 are located in the cytosol and in order to react with these enzymes, formaldehyde must pass the peroxisomal membrane. After having arrived in the cytosol, it first condenses non-enzymically with glutathione (GSH) by forming S-hydroxymethylgluthathione ($GS\text{-}CH_2OH$) [11] which is the true substrate for formaldehyde dehydrogenase [12, 13]. The product S-formylglutathione is either hydrolysed into formate and oxidized to carbon dioxide by formate dehydrogenase [12, 14] or oxidized directly [15].

Table 1. Some properties of formaldehyde- and formate-dehydrogenases purified from *Candida boidinii* (12; for comparison see also 13 and 14).

Property	Formaldehyde dehydrogenase	Formate dehydrogenase
Molecular mass (kDa) 81	74	
Subunits	2 (identical)	2 (identical)
pH optimum	7.5 - 7.9	7.5 - 8.5
apparent K_m (mM)	0.25 (HCHO)	13 (HCOOH)
	0.13 (GSH)	
	0.09 (NAD)	0.09 (NAD)
Inhibitors	NADH, ATP	NADH, ATP

Representative properties of the dehydrogenases involved in the oxidation of formaldehyde are shown in Table 1. A further enzyme, formaldehyde reductase, which reduces formaldehyde to methanol has been detected in methylotrophic yeasts. This enzyme is probably only important when the cytosolic formaldehyde concentration is significantly increased over the levels normally found in growing cells [16]. Since both dehydrogenases involved in the oxidation of formaldehyde are located in the cytosol, bulk of the NADH required for the generation of ATP is generated outside the mitochondria and the TCA cycle is mainly used for biosynthetic purposes [15, 17]. Because no evidence for the involvement of an NADH shuttle system has been found [17], it is thought that cytosolic NADH is oxidized by a mitochondrial NADH dehydrogenase which is localized at the outer face of the inner membrane [18].

Assimilation of the formaldehyde that is generated inside the peroxisomes is by way of the XuMP cycle which was postulated [19] after considerable initial confusion, but is now firnly established. Recent studies [20, 21] have demonstrated that the first enzyme involved in the assimilatory sequence, dihydroxyacetone synthase (DHAS), is located inside the peroxisomes. The enzyme is a special transketolase which catalyses the glycolaldehyde transfer from xylulose-5-phosphate as a donor to the acceptor molecule formaldehyde by way of a ping-pong mechanism [22]. Representative properties of the enzyme are summarized in Table 2.

Table 2. Some properties of key enzymes of the XuMP cycle purified from *Candida boidinii* KD1 [24, 25, 26, 27; but see also 22].

Property	DHAS	DHA
Molecular mass (kDa)	155	140
Subuntis	2 (identical)	2 (identical)
pH optimum	6.8 - 7.1	7.5 - 8.2
Cofactors	TPP, Mg^{2+}	Mg^{2+}
Apparent K_m (mM)	0.5 (HCHO)	0.006 (DHA)
	0.35 (Xylulose-5P)	
	0.4 (glycolaldehyde)	0.006 (glyceraldehyde)
	0.75 (Ribose-5P)	0.35 (MgyATP)
	0.75 (hydroxypyruvate)	also active with MgITP, MgCTP
	0.90 (Fructose-6P)	MgUTP, MgGTP
	1.0 (Sedoheptulose-7P)	
Inhibitors	GHS, HCHO	ADP, excess ATP
Localization	peroxisome	cytosol

DHAS differs from the classical transketolase in several respects and recent studies with DHAS minus mutants of *H. polymorpha* [23] have shown that the latter enzyme cannot substitute for the *in vivo* function of the former. *In vitro* DHAS is an unstable enzyme [24], but purified preparations can be stabilized with GSH or dodecanthiol. This and its sensitivity to SH- reagents has been taken as an indication that -SH groups are essential for the proper functioning of the enzyme. The recent finding [23] that low levels of GSH occur in peroxisomes may indicate that it is also stabilized *in vivo*. Inhibition of DHAS by GSH under *in vivo* conditions, as was reported for the purified enzyme [24], is not likely in view of the properties of DHAS (Table 2), the low concentration of GSH in the peroxisomes and the K_d of S-hydroxymethylGSH [11]. DHAS is inhibited by higher concentration of CH_2O (K_i 8.2 mM). The fact that this enzyme is localized in peroxisomes requires that its substrate xylulose-5P and its two products dihydroxyacetone and glyceraldehyde-3P must be translocated across the peroxisomal membrane. Little is known about the transport properties of these organelles. Recent work [28] using ^{31}P nuclear magnetic resonance techniques has shown that *in vivo* these organelles are acidic inside (pH 5.8 - 6.0) and that a ΔpH of 1.2 - 1.4 units exist over their membranes. Subsequent biochemical studies revealed that this ΔpH is generated and maintained by a membrane-associated proton-translocating

ATPase [29]. The significance of this electrochemical proton gradient in the translocation across the organellar membrane of various low molecular weight (xylulose-5P, GSH, FAD, TPP, heme, dihydroxyacetone, glyceraldehyde-3P, formaldehyde) and/or high molecular weight (subunits of AO, DHAS, catalase) compounds required for sustained growth on methanol remains to be established.

A second enzyme specifically involved in the XuMP cycle is dihydroxyacetone kinase (DHAK). This enzyme is localized in the cytosol [21] and mutant studies [see 30] have indicated that it is essential for growth on methanol. The enzyme has a high affinity for both dihydroxyacetone and glyceraldehyde and requires MgATP as a second substrate (Table 2). Significantly excess ATP inhibited its activity, but it is doubtful whether this inhibition plays any role *in vivo* since the cellular Mg^{2+} pool measured in *C. boidinii* KD1 exceeds that of ATP by almost an order of magnitude [23].

Frucose-1, 6-bisphosphatase (FBPase) is also induced during growth of methylotrophic yeasts on methanol [15] and plays an indispensable role in the regeneration of the xylulose-5P required for formaldehyde assimilation. Attwood and van Dijken [see 16] demonstrated that the properties of the FBPases purified from methanol- or ethanol-grown cells of *H. polymorpha* were identical. Comparable results have been obtained in Trotsenko's laboratory [23] who found that the enzyme purified from similary-grown *C. boidinii* KD1 was composed of 4 identical subunits of 46 kDa molecular mass and that it was inhibited by AMP. In contrast FBPase purified from glucose-grown cells had a 30-fold lower specific activity and a reduced subunit molecular mass (40 kDa). The significance of these findings is currently unclear.

3. REGULATION OF THE SYNTHESIS AND TURNOVER OF C_1- SPECIFIC ENZYMES

During exponential growth of a variety of methylotrophic yeasts on methanol in batch culture, the activities of all the C_1- specific enzymes described in section 2 (see also Fig.1) are significantly increased over the levels observed during growth on glucose or ethanol [15, 31 - 35]. Detailed studies on the regulation of the synthesis of these enzymes, conducted mainly in strains of *C. boidinii* and *H. polymorpha*, have shown that although the fine tuning of the metabolic controls involved may be different, synthesis of the C_1- specific enzymes in none of the strains is completely dependent upon the presence of methanol (induction). For example, in *H. polymorpha* AO and formaldehyde- and formate- dehydrogenases were all virtually absent during exponential growth on glucose, but their synthesis ocurred after depletion of the substrate in the stationary phase of growth [32]. In this organism expression of these enzymes was strongly influenced by the growth substrate; in cells growing with xylose, glycerol, ribose or sorbitol specific activities as high as 60% of the values detected in methanol-grown cells were reported. Addition of methanol to a culture growing on ribose (but not glucose or ethanol) enhanced the activity of AO to a value close to that found in methanol-grown cells [33]. Although the reported evidence for the assimilatory enzymes is not as detailed, there are several indications that their synthesis is similarly controlled [34, see below].

This (partial) catabolite repression modified by methanol induction type of regulation, although encountered in several methylotrophic yeasts, does not always mediate expression of C_1- specific enzymes to the same extent in different yeasts when they are placed under identical growth conditions. Generally glucose and ethanol are most effective as a souce of catabolite repression; for other compounds control by catabolite repression in some strains is more tight than in others, while methanol may have a small or a significant inducing effect.For instance in *Kloeckera* sp. 2201 derepression of AO synthesis following glucose exhaustion does not occur as readily as in *H. polymorpha* [35].

Also in *C. boidinii* [33] as in *P. pastoris* [30] AO is not detected in glycerol-grown cells and these organisms are more dependent on methanol for a high level of expression. This means that the sensitivity of the various yeasts to catabolite repression by different carbon sources and the effectiveness of methanol as an inducer varies among different strains. Methanol also functions as an inducer of catalase and formaldehyde- and formate- dehydrogenases in an AO minus mutant of *H. polymorpha* [33], indicating that the compound itself rather than a metabolite derived from it exerts this effect. There is evidence [36] that also formaldehyde may act as an inducer for all the methanol-dissimilatory enzymes in *H. polymorpha* , whereas formate induces at least the synthesis of formate dehydrogenase in the same organism [37]. The molecular details of the regulation of the C_1-specific - enzymes in methylotrophic yeasts remain to be established. There is compelling evidence, at least for two enzymes, that their synthesis is primarily controlled at the level of transcription [30, but see also 6], but studies of the regulation of m-RNA synthesis at the promotor level have not yet been reported. In view of the recent advances made in the molecular genetics of these yeasts, it should now be possible to investigate the occurrence of promotor sequences specifically involved in catabolite repression and methanol induction as predicted by physiological studies.

Not unexpectedly, catabolite repression of the synthesis of the C_1-specific enzymes is generally relieved in carbon-limited chemostat cultures where, as a rule, increasing derepression is found with decreasing dilution rates [D; 38]. When methanol was added as a second substrate to the medium reservoir of carbon-limited cultures of *H. polymorpha* or *Kloeckera* sp. 2202 growing at constant D (submaximal for either of the substrates), an increased rate of synthesis of both the dissimilatory [39, 40] and assimilatory enzymes [41] was observed. In these experiments both subtrates were used to completion. The extent of the including effect exerted by methanol was dependent upon the ratio of methanol:glucose in the medium and increased approximately linearly with increasing parts of methanol in the mixture until a plateau was reached (Fig.2).

Interestingly, during growth of both organisms on these mixtures, the cells were able to utilize methanol at D values considerably in excess of the μ_{max} for methanol alone (0.16 - 0.20 h^{-1}). As expected, repression of C_1-specific enzymes eventually occurred as D was increased; the value D_T for transition at which this became apparent was dependent upon the mixture composition and increased with decreasing proportions of methanol [42]. For instance, during growth of *H. polymorpha* on a medium containing 61.2% methanol/ 38.8% glucose (w/w) D_T was 0.23 h^{-1}, whereas it increased to 0.30 h^{-1} when the mixture composition was changed to 19.3% methanol/ 80.7% glucose.

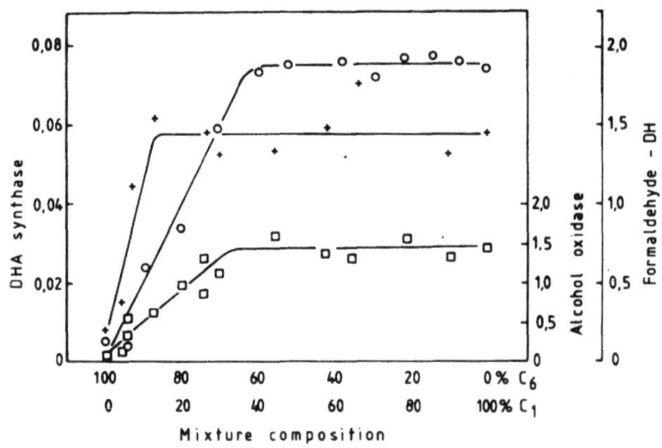

Figure 2. Specific enzyme activities (µmol.mg protein^{-1}. min^{-1}) measured in cell-free extracts of *Kloeckera* sp. 2201 grown in a carbon-limited chemostat at D = 0.14 h^{-1} on glucose/methanol mixtures (given in % w/w) of different composition. □-□AO, O-O DHAS and + + formaldehyde dehydrogenase. Data from [35] and [41].

In experiments using the former mixture, the inducing effect exerted by methanol on the synthesis of AO rapidly decreased with increasing values of D (probably caused by an increasingly overriding contribution of catabolite repression) and had disappeared at D = 0.30 h^{-1} where the specific activity of AO was the same as in glucose-limited cells [38, 43]. In contrast, the specific activities of catalase and formaldehyde- and formate dehydrogenases increased up to D = 0.29 h^{-1} and only became repressed above this dilution rate. This indicates that there are not only strain differences with respect to the sensitivity to catabolite repression and methanol induction of the C_1-specific enzymes, but that in addition the individual enzymes within one strain may display differences in their behaviour.

Since AO, DHAS and catalase are located in peroxisomes, synthesis of these enzymes in the cell is associated with proliferation of these organelles [10]. When cultures of *H. polymorpha* grown on methanol are transferred to media containing glucose or ethanol, the peroxisomes and the enzymes contained in them are actively destroyed. This active degradation, which has also been observed when glucose + methano-limited chemostat cultures were subjected to a condition of $NH4^+$-limitation [44] is due to fusion of peroxisomes with vacuolar vesicles followed by proteolysis and has been called degradative inactivation [see 10]. In contrast with these peroxisomal enzymes, the cytosolic C_1-specific dissimilatory enzymes, although their synthesis was completely repressed, were not degraded but were diluted out over newly formed cells. Compared to *H. polymorpha*, two of the peroxisomal enzymes in *Kloeckera* sp. 2201 are not as severely affected by degradative inactivation [44, 45]. This again suggests that different organisms may behave differently.

Because the inducible FAD-containing AO of methylotrophic yeasts may form a considerable portion (over 30% under certain conditions) of the total protein, an increased rate of synthesis of FAD is required during growth on methanol. Several studies have shown [31, 46] that this is brought about by increased levels of enzymes involved in FAD biosynthesis, whose synthesis is controlled by intracellular levels of free FAD [46].

4. REGULATION OF CARBON FLUX AND PARTITIONING OF FORMALDEHYDE OVER DISSIMILATORY AND ASSIMILATORY PATHWAYS

Before methanol carbon can enter the dissimilatory or assimilatory pathways in yeasts, it must first be converted into formaldehyde by AO. As we have seen in the previous section, the synthesis of this enzyme is carefully controlled, but there is little evidence for regulation of the activity of existing and properly packaged AO.

It must therefore be assumed that the flux of methanol through reaction 1 (Fig. 3) is controlled by the concentration of enzyme 1 in the cells and the concentration of its substrates and products. One condition is known in which AO is inactivated; this is generally seen when methanol-limited cells, which contain a high concentration of enzyme, are suddenly challenged with excess methanol. Then the rate of methanol oxidation far exceeds that of formaldehyde utilization and as a result formaldehyde rapidly accumulates in the cells. Concomitantly inactivation of AO occurs due to dissociation of FAD from the enzyme [see 10] and the cells eventually die.

In growing cells partitioning of formaldehyde over assimilatory and dissimilatory pathways (Fig. 3) must be carefully regulated because the yield of cells on methanol is remarkably constant under a wide range of conditions [3, 31, 42]. Under steady state conditions, when the appropriate enzymes are all present, this partitioning can be most readily accounted for on the basis of control by the energy status of the cells because it affects the rate at which xylulose-5P becomes available for the DHAS-catabolized reaction [21]. Regeneration of xylulose-5P, requires ATP (reaction 6, Figs. 1 and 3) and after its synthesis in the cytosol it must be translocated across the peroximal membrane. This process might involve the electrochemical proton gradient that exists over this membrane [28] and is maintained at the expenses of ATP [29]. Metabolic control may act in such a way that in case the energy status of the cells is high, the rate of supply of xylulose-5P to the formaldehyde fixing enzyme DHAS is stimulated which allows the flux of formaldehyde into the direction of assimilation to be increased. In the opposite case - during intracellular energy limitation - the rate of supply of xylulose-5P to the peroxisomes will be reduced which leads to accumulation of formaldehyde inside the peroxisomes. As a result of the chemical gradient over the peroxisomal membrane, formaldehyde that cannot be assimilated will diffuse out of the organelles where it is trapped by GSH that is present in significant concentrations in the cytosol [34, 47] and is dissimilated. Thus energy limitation increases the amount of formaldehyde available for dissimilation.

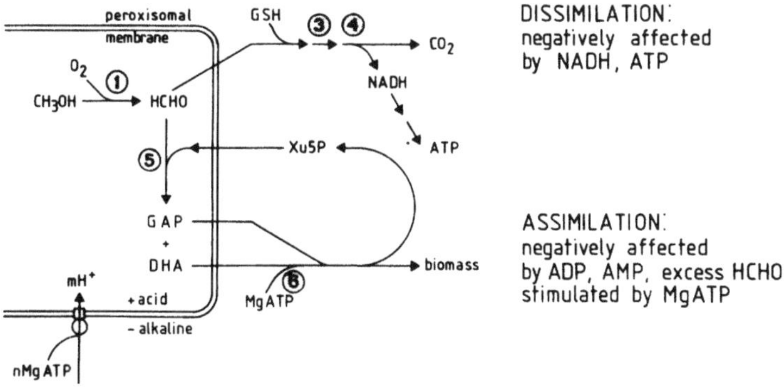

Figure 3. Partitioning of formaldehyde over dissimilatory and assimilatory pathways. See legend on Fig. 1 for details.

Thus energy limitation increases the amount of formaldehyde available for dissimilation. Some evidence in support of these ideas has been obtained in kinetic studies with permeabilized cells of *C. methylica* and DHAK partially purified from this organism [48, 49]. Also the finding [50] that addition of ^{14}C-formaldehyde to a methanol-limited culture of *H. polymorpha* was almost exclusively dissimilated is in accordance with the suggestion that formaldehyde once trapped in the cytosol by GSH is no longer available for assimilation.

Methylotrophic yeasts grow generally much faster on multi-carbon compounds than on methanol and the question must therefore be asked whether this may be due to a limited capacity of AO in supplying formaldehyde or to a limited capacity of either the assimilatory or dissimilatory pathways (or both). Several lines of evidence are available to suggest that the flux of formaldehyde over the assimilatory pathway may reach a maximal value above which synthesis of C_1-specific enzymes becomes repressed. In view of the fact that catabolite repression of these enzymes is not specifically exerted by one particular compound, this may indicate that the rate of supply of xylulose-5P to DHAS cannot exceed a maximal value (for instance because of a limited capacity for transport of this compound across the peroxisomal membrane) which in turn may lead to accumulation of intracellular metabolites involved in catabolite repression. But this is no more than a hypothesis; the evidence is as follows. During growth of *H. polymorpha* in a methanol-limited chemostat at different dilution rates, constant cell yields and a linear increase in the specific methanol consumption rates (q_{C1}) with increasing values of D were recorded up to D = 0.15 h^{-1}. At higher dilution rates where q_{C1} exceeded 0.42 g methanol.g dry wt $^{-1}$. h^{-1}, this linear relationship betwee q_{C1} and D no longer existed because a higher proportion of the methanol consumed was now oxidized to CO_2 [42]. The maximum rate of formaldehyde assimilation that can be calculated from these data (approx. 6 μmol.g dry wt^{-1}. h^{-1}) agrees remarkably well with the results of a computer simulation based on kinetic properties of the key assimilatory enzymes and realistic values for the concentrations of various metabolites [16]. This as well as another study [41] indicated that under these conditions DHAS, DHAK and FBPase operated close to their maximum capacity, whereas excess activities of the enzymes of the dissimilatory pathway were recorded. A maximum rate of methanol consumption of 0.42 g methanol.g dry wt^{-1}. h^{-1} was also found during growth of *H. poly,morpha* on glucose/methanol mixtures of various composition at D values in excess of 0.20 h^{-1} [42]. Apparently there is an upper limit to the flux of methanol carbon that a cell can handle and attempts to push the cells over this limit by increasing the methanol/glucose ratio in the mixture or D invariably lead to repression of the synthesis of C_1-specific enzymes. In the strain used by Egli et al. [41] there was little flexibility in the sense that a switch of metabolism towards the use of methanol as an energy source was very limited. In other organisms, however, and under other conditions, such a switch has been observed [51, 52], although methanol was never used for energy generation only as is the case in several prokaryotes [53].

5. CONCLUDING REMARKS

Physiological and biochemical studies conducted in the past decade have yielded a wealth of information on metabolic regulation in methylotrophic yeasts.

Control is exerted over the synthesis of C_1-specific enzymes, while the partitioning of C_1 units over the dissimilatory and assimilatory pathways is regulated by the energy status of the cells. Many of the details, particularly the precise role of the peroxisomes in the various transport processes and the molecular aspects of the control of expression of the various genes involved in methanol metabolism remain to be established. The fact that at least some of the results obtained in these fundamental studies have interesting practical applications (53, 54) may serve as a stimulus for further work.

6. ACKNOWLEDGEMENTS

We wish to thank Marry Pras for her excellent assistance in the preparation of this contribution.

REFERENCES

[1] Lee, J - D , Komagata K. 1980. J. Gen. Appl. Microbiol. 26: 133 - 158.
[2] Ogata K., Nishikawa H., Ohsugi M. 1969. Agr. Biol. Chem. 33: 1519 - 1520.
[3] Van Dijken JP , Otto R., Harder W. 1976. Arch. Microbiol. 111: 137 - 144.
[4] Kato N., Omori Y., Tani,Y., Ogata K. 1976. Eur. J. Biochem. 64: 341 - 350.
[5] Couderc R., Baratti J. 1980. Agric. Biol. Chem. 44: 2279 - 2289.
[6] Giuseppin MLF, van Eijk HML, Bes BCM, Hellendoorn M. 1986. Abstract, this symposium.
[7] Hopkins T., Müller F. 1986. This volume.
[8] Roggenkamp R., Sahm H., Wagner F. 1974. FEBS Lett. 41: 283 - 286.
[9] Mazaffar S., Ueda M., Kitatsuji K., Shumizu S., Osumi M., Tanaka A. 1986. Eur. J. Biochem. 155: 527 - 531.
[10] Veenhuis M., van Dijken JP, Harder W. 1983. Adv. Microb. Physiol. 24: 1 - 78.
[11] Uotila L., Koivusalo M. 1974. J. Biol. Chem. 249: 7653 - 7663.
[12] Schütte H., Flossdorf J., Sahm H., Kula MR. 1976. Eur. J. Biochem. 62: 151 - 160.
[13] Allais JJ, Louktibi A., Baratti J. 1983. Agric. Biol. Chem. 47: 1509 - 1516.
[14] Allais JJ, Louktibi A., Baratti J. 1983. Agric. Biol. Chem. 47, 2547 - 2554.
[15] Van Dijken JP, Harder W., Quayle JR. 1981. In: Microbial growth on C_1 compounds, pp. 191 - 201. London, Heyden.

[16] Trotsenko YA, Bystrykh LV, Ubiyvovk VM. 1984. In: Microbial growth on C_1 compounds, pp 118 - 122. Washington DC, ASM.
[17] Egli T, Lindley ND. 1984.J. Gen. Microbiol. 130: 3239-3249.
[18] Moore AL, Rich PR. 1980. Trends Biochem. Sci. 5: 284 - 287.
[19] Van Dijken JP, Harder W., Beardsmore AJ & Quayle JR. 1978. FEMS Microbiol. Lett. 4: 97 - 102.
[20] Goodman J. 1985. J. Biol. Chem. 260: 7108 - 7114.
[21] Douma AC, Veenhuis M., de Koning W., Evers M., Harder W. 1985. Arch. Microbiol. 143: 237 - 243.
[22] Kato N., Higuchi C., Sakazawa N., Nishizawa T., Tani Y., Yamada H. 1982. Biochim. Biophys. Acta 715: 143 - 150.
[23] Bystrykh LV., Trotsenko YA. Unpliblished observations.
[24] Bystrykh LV, Sokolov AP, Trorsenko YA. 1981. FEBS Lett. 132: 324 - 328.
[25] Waites MJ, Quayle JR. 1983. J. Gen. Microbiol. 129: 935 - 944.
[26] Bystrykh LV, Trotsenko YA. 1983. Biochemistry (USSR) 48: 1611 - 1616.
[27] Bystrykh LV. 1985. Biochemistry (USSR) 50: 1012 - 1017.
[28] Nicolay K., Veenhuis M., Douma, AC, Harder W. 1986. Arch. Microbiol., submitted for publication.
[29] Douma AC, Veenhuis M., Sulter GJ, Harder W. 1986. Arch. Microbiol., submitted for publication.
[30] Cregg JM, this volume.
[31] Sahm H. 1977. Adv. Biochem. Eng. 6: 77 -
[32] Eggeling L., Sahm H. 1978. Europ. J. Appl. Microbiol. Biotechnol.5: 197 - 202.
[33] Eggeling L., Sahm H. 1980. Arch. Microbiol. 127: 119 - 124.
[34] Koning W de, unpublished results.
[35] Egli T., Haltmaier T., Fiechter A. 1982. Arch. Microbiol. 131: 174 - 175.
[36] Zwart KB, Harder W. 1983. J. Gen. Microbiol. 129: 3157 - 3169.
[37] Babel W., Müller RH, Markuske D. 1983. Arch. Microbiol. 136: 203 - 208.
[38] Egli T., van Dijken JP, Veenhuis M., Harder W., Fiechter A. 1980. Arch. Microbiol.124: 115 - 121.
[39] Eggeling L., Sahm H. 1981. Arch. Microbiol. 130: 119 - 124.
[40] Egli T., Käppeli O., Fiechter A. 1982. Arch. microbiol. 131: 1 - 7.
[41] Egli T., Lindley ND, Quayle JR. 1983. J. Gen. Microbiol. 129: 1269 - 1281.
[42] Egli T., Bosshard C., Hamer G. 1986. Biotechnol. bioeng., in press.
[43] Egli T., Käppeli O., Fiechter A. 1982. Arch. microbiol. 131: 8 - 13.
[44] Egli T. 1982. Arch. Microbiol. 131: 95 - 101.
[45] Yasuhara S., Kawamoto S., Tanaka A., Osumi M., Fukui S. 1976. Agr. Biol. Chem. 40: 1771 - 1780.
[46] Brooke AG, Dijkhuizen L., Harder W. 1986. Arch. Microbiol. 145: 62 - 70.
[47] Ubiyvovk VM, Bystrykh LV, Trotsenko YA. 1982. Mikrobiologiya (USSR) 52: 383 - 387.
[48] Hofmann KH, Babel W. 1980. Z. allg. Mikrobiol. 20: 389 - 398.
[49] Hofmann KH, Babel W. 1984. Z. allg. Mikrobiol. 24: 713 - 717.
[50] Haltmaier T., Egli T., unpublished results.
[51] Eggeling L., Sahm H. 1981. Arch. Microbiol. 130: 362 - 365.
[52] Müller R., Uhlenhut GJ, Babel W. 1985. Arch. Microbiol. 143: 77 - 81.
[53] Egli T., Harder W. 1984. In: Microbial growth on C_1 compounds, pp 330 - 337. Washington DC, ASM.
[54] Stroman D., Thill G., this volume.

BIOCHEMISTRY OF ALCOHOL OXIDASE

T. R. HOPKINS and Franz MULLER

Biotechnology Division of Research and Development,
Phillips Petroleum Company, Bartlesville, OK 74004, U.S.A.;
and Department of Biochemistry, Agricultural University,
Wageningen, the Netherlands.

1. INTRODUCTION

A few yeasts are able to utilize methanol as a sole carbon and energy source. The metabolism of methanol has been studied in detail in *Hansenula polymorpha* and *Candida boidinii.* The enzyme involved in the metabolism of methanol is methanol or alcohol oxidase (EC 1.1.3.13). It converts methanol to formaldehyde in the presence of molecular oxygen. The co-product, hydrogen peroxide, can be dismutated to O_2 and H_2O in the presence of catalase which is intimately associated with alcohol oxidase in the cell. Formaldehyde in turn is dissimulated to CO_2 by formaldehyde dehydrogenase and formate dehydrogenase and, perhaps, by catalase.

Alcohol oxidase has been isolated and purified from different yeasts and molds. Most enzyme isolates have a molecular weight of about 600,000 and consist of eight subunits. Each subunit binds one FAD cofactor. The enzyme catalyzes the following ping-pong reaction using strict two electron redox stoichiometry.

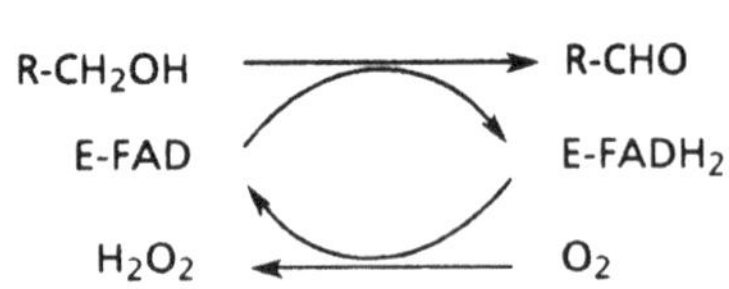

The reaction is considered to be irreversible.

The properties of alcohol oxidase and its regulation in the cell are unusual and complex. To give some examples, the specific activity of purified alcohol oxidase is approximately 10 EU/mg protein and it can be estimated that 0.5 to 1.0% of the total protein in the cell must be alcohol oxidase in order to maintain growth on methanol. Why is there some 20 times more enzyme found inside methanol utilizing yeast grown on methanol at low retention times? Alcohol oxidase is packaged in the cell as a crystalline material inside peroxisomes.What is the purpose of this compartmentation? Purified alcohol oxidase contains anionic flavosemiquinone. How are these semiquinones formed, kept air-stable and what role do they play in regulation of catalytic activity of alcohol oxidase? The aim of this paper is to discuss the properties of alcohol oxidase from *Pichia pastoris* (NRRL - Y - 4290) studied in our laboratory and to relate the relevance of these properties to the enzyme located in the peroxisomes of methanol-utilizing yeast.

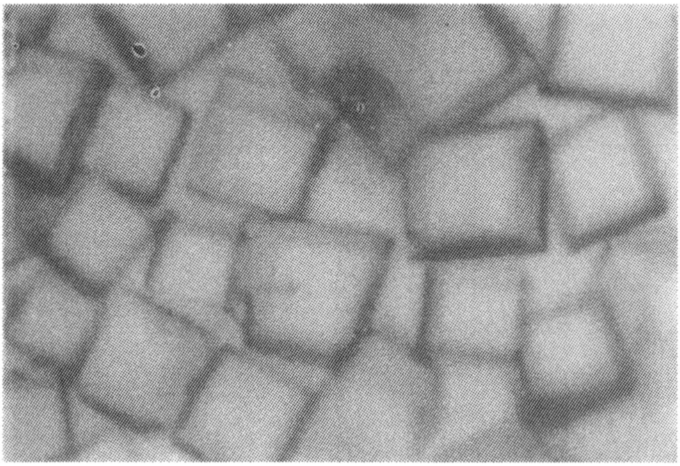

A

B

Figure 1. a) Light micrograph and b) scanning electron micrograph of crystals of alcohol oxidase.

2. PHYSICAL-CHEMICAL PROPERTIES

Table 1 tabulates fifteen properties of *Pichia pastoris* Y - 4290 alcohol oxidase. Most of these values are similar to other yeast alcohol oxidases reported in the literature (4, 6, 7, 9, 12, 13, 15, 17).

Table 1. Physical-Chemical properties of Alcohol Oxidase (*Pichia pastoris* , NRRL - Y - 4290).

Property	Value	Notes
Native Molecular Weight	630,000 ± 3%	Sedimentation velocity No carbohydrates
Monomer Molecular Weight	75,000 ± 5%	SDS - Polyacrylamide Electrophoresis
Subunits	8	
Cofactor	FAD	One per subunit, non-convalent bound
Temperature Optimum	40°C	
pH Optimum	6.0 8.5	Broad optimum range, pKa = 5.2 , pKb = 9.2
Specific Activity (Vmax)	12 - 30 IU/mg[1]	For both methanol and ethanol. Air-saturated buffer, 23°C.
Turnover Number	7,000 - 17,000 min^{-1}	
Km	0.7 mM (methanol) 9.0 mM (ethanol)	Air-saturated buffer, pH 7.5
Inhibitors	NaN_3 (iodoacetamide, HCHO, halides, CN^-)[2]	
$E^{1\%}_{1cm,\ 280\ mm}$	15.4 cm^2/mg	
Vis Absorption Maxima	383, 398, 456 nm	
Fluorescence Emission Maxima	340, 420, 550 nm	
Inhibition by H_2O_2[3]	>100 mM	
Isoelectric Point	5.7 ± 10%	

1. Preparations vary (see text for discussion).
2. Time-dependent inactivations (see text for discussion).
3. Concentrations giving 50% inhibition for a 20 - minute treatment at 30°C.

Substrate specificities for alcohol oxidase are given in Table 2. The best natural substrate was methanol followed by formaldehyde. Other good, but nonphysiological substrates, were ethanol, allyl alcohol and propargyl alcohol.

Dissolved oxygen concentrations in the assay reaction affected the enzymatic properties of alcohol oxidase. The Km calculated for methanol was about threefold higher when using oxygen saturated solutions over air-saturated solutions. Similar results were found by Couderc and Baratti (4) in the characterization of the alcohol oxidase from *Pichia pastoris* IFP206 and by van Dijken, et al (17) with alcohol oxidase from *Hansenula polymorpha*.

Table 2. Some Substrates of Alcohol Oxidase

Substrate	Relative Activity (%) [1]
Methanol	100
Ethanol	36
n - Propanol	20
Isopropanol	2
n - Butanol	10
Allyl Alcohol	30
Propargyl alcohol	50
2 - Butyne - 1, 4 - diol	15
2 - Chloroethanol	10
2 - Mercaptoethanol	7
Ethylmercaptan	16
Formaldehyde	33

1. Activities were assayed by following oxygen consumption. Alcohol oxidase was mounted on the tip of a model 98 Provesta Biosensor (Bartlesville, OK) (8). Final substrate concentrations were 0.0025% (w/v).

Other electron acceptors for flavoprotein oxidases besides molecular oxygen were examined but none were able to replace oxygen. Compounds surveyed included NAD^+, potassium ferricyanide, 2, 6-dichlorophenolindophenol, methyl viologen, mendola blue, cytochrone c and methylene blue.

3. RESISTANCE TO HYDROGEN PEROXIDE INACTIVATION

Pichia pastoris Y-4290 is remarkably resistant to inactivation by hydrogen peroxide. It remained active when exposed to H_2O_2 concentrations up to 1 M for 20 minutes at 30° C. Other yeast alcohol oxidases are reported to lose half their initial activity in 1 to 10 mM H_2O_2 (4, 9, 12).

4. SPECTRAL PROPERTIES

Figure 2 shows the visible spectra of alcohol oxidase isolated from *P. pastoris.* The mayor absorption bands at 383 and 456 nm are typical of FAD-containing proteins. The unusually low absorbancy of the 456 band relative to the 383 band and its 398 nm shoulder are notable characteristics of yeast alcohol oxidases.Aged alcohol oxidase solutions have altered absorption spectra, with the wave length of the absorption maxima changing little but the relative absorbances of the two bands becoming more equal. The electron paramagentic resonance (EPR) of *P. pastoris* alcohol oxidase is typical for that found for anionic (red) protein-bound flavosemiquinone; i. e. , the peak distance corresponds to about 12 Gauss and is similar to that observed in the *Hansenula* and *Candida* enzyme (11). The radical can also be easily observed by light absorption. The small shoulder at about 400 nm, the intense band at about 370 nm and an increase in intensity in the 500 nm region are diagnostic for the anionic flavosemiquinone (5).The flavosemiquinone apparently is not needed during catalysis as the radical signal is unaffected by the presence of substrate or inhibiting levels of sodium azide. If the radical were a catalytic intermediate

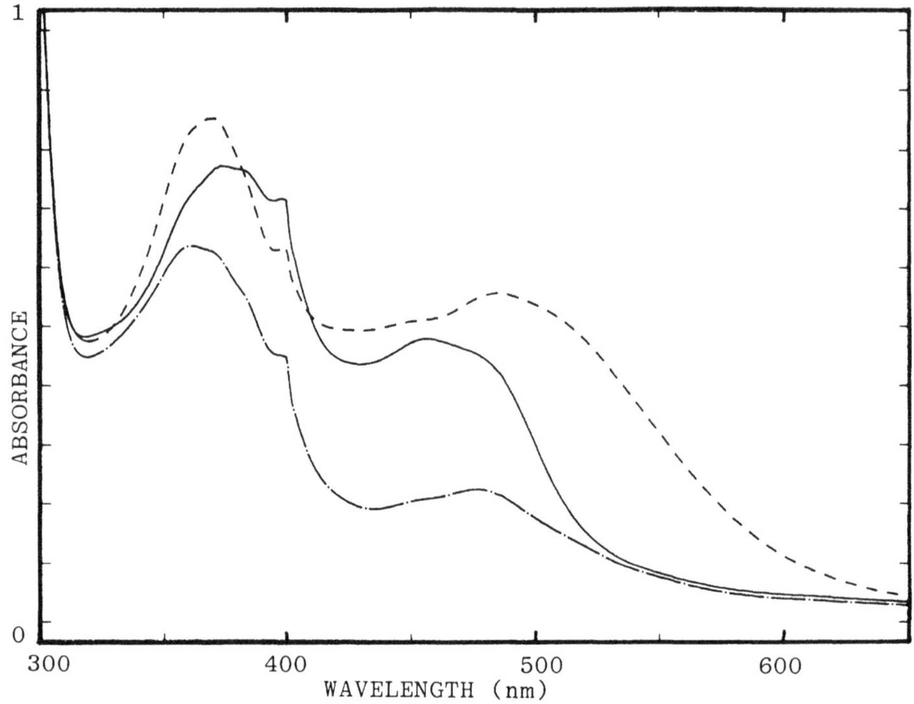

Figure 2. Spectra of *Pichia pastoris* (NRRL- Y- 4290) alcohol oxidase. Native enzyme (——), enzyme exposed to N_aN_3 (- - -) and enzyme in presence of methanol without air (—·—·—·).

stabilized by the protein, O_2^- might be released from the protein. We were unable to detect O_2^- in the medium during enzyme turnover. The data suggest that alcohol oxidase molecules which contain stable semiquinone are structurally different from the active molecules. It is not clear how the flavosemiquinone is formed.

5. AZIDE ANION BINDING

Concentrated solutions of the native, yellow form of alcohol oxidase turn wine red in color when exposed to low concentrations of sodium azide. As shown in Figure 2, the strongly bound ligand shifts in absorption maxima with the 383 nm peak being blue shifted to 367 nm and the 456 nm band being red shifted to 482 nm. The greatest change in absorption intensity is an increase at 520 nm.

The spectral shifts induced by sodium azide can be used to titrate the azide binding sites on the alcohol oxidase. Using high enzyme concentrations which flavor saturation of the azide binding sites, 7.7 binding sites per native alcohol oxidase were found. It is, therefore, likely that there is one spectra perturbing binding site per enzyme subunit. Analysis of the binding data by a Scatchard plot gave a straight line fit. From this, it was deduced that the association constant for sodium azide was 3×10^6 M^{-1} with 0.97 binding sites per enzyme subunit.

A binding site with this affinity should release its ligand upon extensive dialysis against a large volume of ligand-free buffer, and, in the process, change from a red colored alcohol oxidase back to the yellow enzyme. This has not been observed no matter how long the enzyme is dialyzed. Similarly, there was no loss of the red color when the enzyme-azide complex was sieved through a Sephadex 200 column preequilibrated with water or subjected to electrophoresis in 15% polyacrylamide gel. Clearly there is a red colored alcohol oxidase-azide complex which has an affinity constant several orders of magnitude greater than that deduced from the spectrophotometric titration experiments. Perhaps the initial binding of azide by alcohol oxidase induces a slow change in the binding site configuration which greatly increases the affinity of the site for azide anion.

Samples of the red colored alcohol oxidase freed of unassociated azide ions by dialysis or recrystallization have catalytic properties undistinguishable from the native, yellow form of the enzyme. The alcohol oxidase-azide complex could be disrupted by solvent conditions outside those needed for support of catalytic activity. Thus, red enzyme solutions turned yellow at pH values below 5.5 or temperatures above 50°C.

Typical of many flavo-enzymes, the yellow color was reversibly bleached when the enzyme was converted to the E-$FADH_2$ form by reaction with substrate in the absence of molecular oxygen (see Figure 2). Similar bleaching effects of the red form of alcohol oxidase were observed. In both cases, the visible absorption bands of the reduced enzyme were identical and were restored upon aeration. Solutions of the reduced flavin form of alcohol oxidase rapidly lose activity.

Other alcohol oxidases have been observed to turn form yellow to red when exposed to azide ions. Bringer, et al, (13) obtained the color change with purified enzyme from *Poria contigua* and we have observed like changes with the alcohol oxidase from *Hansenula polymorpha* and *Candida boidinii.* The red azide complexes of these two yeast enzymes appeared to be weaker than that from *P. pastoris* as their color return to yellow after dialysis overnight against azide-free buffer. A highly purified yellow colored alcohol oxidase preparation from an unknown yeast supplied by Toyo Jozo, Ltd., Japan, turned red very slowly (30 minutes) when exposed to sodium azide.

Free sodium azide is a strong inhibitor of *P. pastoris* Y-4290 alcohol oxidase. Double reciprocal plots of initial enzyme velocities against methanol concentrations in the presence of varying concentrations of sodium azide yielded a $K_i = 6 \times 10^{-6}$ M and $\alpha = 150$. The reciprocal plot approximated pure competitive inhibition because α is greater than one. These graphs indicated that sodium azide was a linear, mixed type inhibitor, system C1 (14). This type of inhibition can exist when the substrate and inhibitor bind to the enzyme at different sites or when the inhibitor binds at two different, mutually exclusive sites with binding at one site excluding the substrate. Interestingly, alcohol oxidase from *Pichia* sp. (NRRL- Y - 11328) apparently was not strongly inhibited by sodium azide (12). However, this alcohol oxidase from Y-11328 is different in other respects as it is reported to have four, rather than eight, subunits.

The nature of the red colored alcohol oxidase-azide complex is unknown. It could be a molecular complex with the oxidized flavin or be a transition complex in the active site similar to that proposed for azide with formate dehydrogenase from *C. boidinii* (12). The complex does not involve liganding to transition metals such as Fe, Zn and Mo (16). The levels of transition metals in alcohol oxidase (red or yellow forms) measured by plasma ion spectroscopy were an order of magnitude lower than that needed to form a unit molar ratio with the alcohol oxidase subunit.

6. FORMALDEHYDE ADDUCTS

Concentrated solutions or "dried" preparations of carefully purified alcohol oxidase exposed to air produced low levels of H_2O_2 in the absence of added substrate. This apparent autooxidation effect was first detected while developing a dip stick reagent pad for detection of alcohol in blood or saliva. A mixture of alcohol oxidase, horseradish peroxidase and dianisidine spontaneously formed color during the drying process. This unexpected and undesired color formation was prevented by titrating the regent mixture with a nonvolatile hydrogen peroxide scavenger such as dithiothreotol. It was later found that the oxidizable substrate of the reaction was formaldehyde reversibly bound to the alcohol oxidase. The reaction is as follows:

$$\text{E-N(H)-CH}_2\text{OH} \xrightarrow{\text{Spont.}} \text{E} + \text{HCHO} \xrightarrow[\text{O}_2]{\text{Oxidase}} \text{E} + \text{HCOOH} + \text{H}_2\text{O}_2$$

Table 3 shows the results of testing alcohol oxidase and two control proteins using chromotropic acid, a reagent specific for formaldehyde (10). Samples of alcohol oxidase have varying mounts of formaldehyde adduct, depending upon the history of the preparation. The adduct is sufficiently stable to survive the purification process, but unstable enough to dissociate in solutions of pure enzyme. The likely form of the adduct is a Shiff's base adduct with α -amino groups of the lysine residues. The formaldehyde might also react with free sulfhydryl groups of alcohol oxidase in an analogous way to that of amines. A failure in early attempts to react all lysine residues of freshly prepared alcohol oxidase with succinic anhydride is probably explained by the masking of reactive amines with the formaldehyde. Aged samples of alcohol oxidase, having less formaldehyde adducts, were more completely succinylated.

The Shiff's base is a labile, covalent bond and dissociation of the formaldehyde complex can easily occur in aqueous solution at temperatures and pH values where the enzyme is active. Assuming that all α -amino groups of alcohol oxidase are formaldehyde binding sites, it can be estimated that the effective concentration of bound formaldehyde would be in the 1- 10 mM range in concentrated protein solutions. And in crystals with 20% water of hydration, as might be found in peroxisomes containing crystalloid alcohol oxidase, the estimated concentration of formaldehyde binding sites is about 0.4 M. It is tempting to speculate that such a capacity to bind and release formaldehyde within the peroxisome would serve as a formaldehyde buffering system. This proposed function offers at least one explanation for the apparent large overproduction of alcohol oxidase in cells grown on methanol. In the cell, reduced

Table 3. Formaldehyde Content of Alcohol Oxidase and Two Control Proteins

Protein	Subunit MW	Moles Formaldehyde / Moles Protein	Percent Complexed Lysine Residues
Alcohol Oxidase	75,000	7.6	25%
Bovine Serum Albumin	93,000	0.0	0%
Glucose Oxidase	80,000	0.9	3%

glutathione is considered the normal reactant with formaldehyde produced by oxidation of methanol (1). The resultant formaldehyde-glutathione complex is the substrate of formaldehyde dehydrogenase. However, if steady-state glutathione concentrations were too low to handle a surge in formaldehyde level, the formaldehyde can be temporarily stored as the adduct in the peroxisome. The formation or disruption of formaldehyde adducts with amino groups on the alcohol oxidase may also modify the solubility or catalytic properties, as observed above. Finally, formaldehyde adducts with alcohol oxidase may mask sites from proteolytic attack by trypsin-like degradative proteases and, thus, constitute one possible regulatory mechanism for peroxisome degradation.

References

[1] Anthony, C. (1982) In The Biochemistry of Methylotrophs. pp. 269-295. Academic Press, London and New York.
[2] Blanchard, J. S. and W. W. Cleland. 1980. Biochem. 19, pp. 3543-3550.
[3] Bringer, S., B. Sprey and H. Sahm. 1979. Eur. J. Biochem. 101, pp. 563-570.
[4] Couderc, R. and J. Baratti. 1980. Agric. Biol. Chem. 44, pp. 2279-2289.
[5] Ehrenberg, A., F. Muller and P. Hemmerich. 1967. Eur. J. Biochem. 2. 286-293.
[6] Fujii, T. and K. Tonomura. 1972. Agr. Biol. Chem. 36, pp. 2297-2306.
[7] Fujii, T. and K. Tonomura. 1975. Agr. Biol. Chem. 39, pp. 2325-2330.
[8] Hopkins, T. R. 1985a. Amer. Biotech. Lab 3, pp. 32-35.
[9] Kato, N., Y. Omori, Y. Tani and K. Ogata. 1976. Eur. J. Biochem. 64, pp. 341-350.
[10] Keleti, G. and W. H. Lederer. 1974. In Handbook of Micromethods for the Biological Sciences. Van Nostrand Reinhold Co., New York, pp. 62-63.
[11] Mincey, T., G. Tayrien, A. S. Mildvan and R. H. Abeles. 1980. Proc. Natl. Acad. Sci. USA 7, pp.7099-7101.
[12] Patel, R. N., C. T. Hou, A. I. Laskin and P. Deleranko. 1981. Arch. Biochem. Biophys. 210, pp. 481-488.
[13] Sahm, H. and F. Wagner. 1973. Eur. J. Biochem. 36, pp. 250-256.
[14] Segal, I. H. 1975. In Enzyme Kinetics. Wiley-Interscience. New York, pp. 170-178.
[15] Tani, Y., T. Miya and K. Ogata. 1972. Agr. Biol. Chem. 36, pp. 76-83.
[16] Thauer, R. K., G. Fuchs and K. Jungermann. 1977. In Iron-Sulfur Proteins (Lovenberg, W., Ed.), Vol. III, p. 121. Academic Press, New York.
[17] Van Dijken, J. P., R. Otto and W. Harder. 1976. Arch. Microbiol. III. pp. 137-144.

GENETICS OF METHYLOTROPHIC YEASTS

J. M. CREGG
The Salk Institute Biotechnology/Industrial Associates Inc.
La Jolla, California 92037 USA

1. INTRODUCTION

Biochemical and physiological investigations of methylotrophic yeasts have produced a number of intriguing observations with regard to the ability of these yeasts to grow on methanol [1, 2]. At least two observations are of particular interest to geneticists and molecular biologists. The first is that many of the enzymes involved in methanol utilization are normally present only when the yeasts are growing on methanol. It should be most interesting to determine the mechanisms by which the expression of this family of genes appears to be regulated by methanol. The second observation is that some of the methanol-metabolizing enzymes, such as alcohol oxidase and dihydroxyacetone synthase [3,4], are sequestered in organelles called the peroxisomes or microbodies. We would like to understand the processes and signals involved in the compartmentalization of these enzymes in peroxisomes.

The application of genetics has great potential for contributing to our understanding of these and other aspects of methylotrophic yeasts. In recent years, significant advances have been reported in techniques for the genetic manipulation of these yeasts, both at the organismal (classical genetic) and molecular (molecular genetic) levels. The intention of this report is to review results of genetic experiments which have advanced our understanding of either methanol utilization by these yeasts or their genetics.

2. CLASSICAL GENETICS

2.1. Life Cycle

The four known genera of methylotrophic yeasts are divided into two groups [5, 6]. Species of the genera *Candida* and *Torulopsis* are asporogenous and therefore do not have a genetically definable life cycle. Thus, classical genetics of these species is limited primarily to the isolation and characterization of mutants. Species of the genera *Hansenula* and *Pichia* are ascomycetous. With some understanding of the life cycle of species in this second group, it should be possible to develop techniques for their genetic manipulation. Such genetic studies have been reported for two ascomycetous methylotrophic yeast species, *Pichia pinus* [7, 8, 9, 10, 11] and *Hansenula polymorpha* [12, 13]. Their life cycles appear to be quite similar (Figure 1). Both species can be maintained indefinitely as vegetative haploids. However, upon nutritional limitation, particularly for nitrogen, mating occurs and diploid cells are formed.

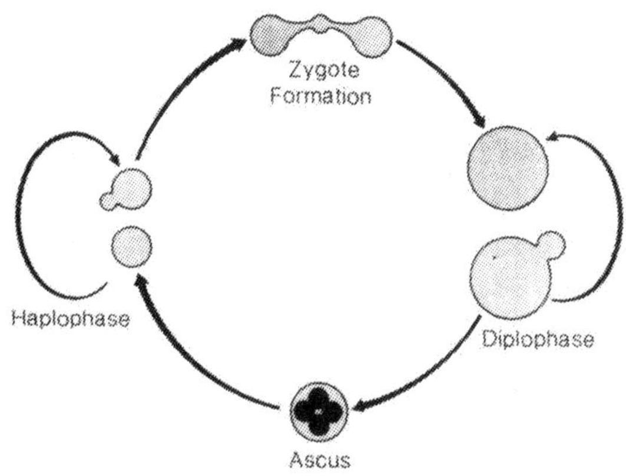

Figure 1. Life Cycle of *P. pinus* and *H. polymorpha.*

Since daughter cells originating from a single parent can mate in cultures of either *P. pinus* and *H. polymorpha,* both yeasts are, by definition, homothallic. In *P. pinus*, it has been shown that, as in *Saccharomyces cerevisiae* , two mating types actually exist and that mating occurs primarily between cells of opposite type [8]. It appears that, upon nutritional limitation, cells of *P. pinus* switch mating type at high frequency and proceed to mate with those of the opposite type [9]. This is an interesting contrast to homothallic strains of *S. cerevisiae,* which are known to switch mating type and mate under vegetative growth conditions [14], and to those of *Schizosaccharomyces pombe*, which are believed to switch type during vegetative growth, but to mate only in response to nutritional limitation [15]. Mutants of *P. pinus* have been isolated which appear to be heterothallic, *i. e.*, defective in mating type interconversion [9, 10]. After mating, strains of *P. pinus* and *H. polymorpha* can either be maintained as diploids by shifting them to a nonstarvation medium or allowed to proceed through meiosis and to the production of asci, each of which usually contains 4 haploid spores.

The key feature in the life cycle of these homothallic yeasts that permits their genetic manipulation is the physiological control of mating. Strains of *P. pinus* and *H. polymorpha* can be maintained as haploids, which is a great advantage in the isolation and phenotypic characterization of mutants. To initiate genetic characterizattions, selected pairs of mutant strains are first crossed by mixing and subjecting the strains to nutritional limitation. The cultures are shifted to a non-starvation medium supplemented with a combination of nutrients on which only diploids composed of two opposite strains can grow. Diploid strains are then shifted to a growth-limiting medium a second time and allowed to sporulate. Finally, asci are dissected, and spores are subjected to tetrad analysis using standard yeast genetic procedures. In both *P. pinus* and *H. polymorpha* , spore viability is high (>80%), and upon dissection of tetrads, genetic markers are most often observed at a ratio of two wild-type to two mutant spores (2:2), as expected for the meiotic segregation of alleles of nuclear genes.

2.2. Mutant Analysis

2.2.1. Auxotrophic Mutants.

Auxotrophic mutations have played a major role in studies on the genetic behaviour of *P. pinus* and *H. polymorpha*. Tolstorukov *et al.*, [7] reported the isolation of 84 *P. pinus* auxotrophs whose mutations represented 31 complementation groups in 11 different biosynthetic pathways. For *H. polymorpha*, Gleeson *et al.*, [12, 13] described a collection of 201 auxotrophs representing 40 different genes in 8 biosynthetic pathways. Interestingly, the auxotrophic collections of both of these yeasts showed a similar bias in the biosynthetic pathways affected [7, 12, 13, 16]. Approximately 50-75% of the mutants required either adenine or methionine, whereas mutants which were defective in many of the other biosynthetic pathways were not found at all. Gleeson*et al.*, [12, 13] suggested that this bias for mutants affecting specific biosynthetic pathways muight be due in part to the number of genes involved in a pathway, *i. e.*, the more genes, the greater the probability of isolating a mutant in that pathway, or the tendency to select mutants for further analysis which have the best characteristics, such as low reversion frequency and easily observed phenotype. Other possible explanations for this observed bias include aneuploidy, large numbers of duplicated genes, or mutational "hot-spots". This mutational bias is of concern, since a similar bias in the isolation of the other types of mutants, such as those defective in methanol utilization, might occur as well.

In *P. pinus*, 17 biosynthetic genes have been assigned to one of four linkage groups. Genetic linkage for some of these genes was demonstrated by tetrad analysis[7], but for most, linkage was determined from the analysis of aneuploid strains [11]. Such strains were derived from genetically marked diploid strains which were irradiated to induce mitotic chromosome loss ("haploidization"). A high frequency of co-loss ("segregation") of pairs of marked genes indicated genetic linkage.

2.2.2. Methanol-Utilization-Defective Mutants.

Until recently, results from the analysis of methanol-utilization-defective (Mut^-) strains were employed primarily to add to biochemical evidence for the existence of a particular enzymatic pathway for methanol metabolism. The first mutant strain of a methylotrophic yeast to be described was a Mut^- strain of *Candida boidinii*. In 1973, Sahm and Wagner [17] reported a *C. boidinii* mutant which could not grow on methanol and which was defective in alcohol oxidase, but not alcohol dehydrogenase activity, providing evidence for the former enzyme and not the latter was involved in methanol metabolism. Evidence in support of the operation of the dihydroxyacetone (DHA) pathway of formaldehyde fixation was provided by the analysis of Mut^- mutants of *H. polymorpha* and *C. boidinii* [18]. One of the *H. polymorpha* mutants was defective in triokinase and another had low levels of fructose 1, 6-bisphosphate (FBPase) activity. Neither of these *H. polymorpha* mutants was defective in the other enzymatic activity or alcohol oxidase. Thus, triokinase and FBPase, which are part of the DHA pathway, appeared to be required for methanol utilization.

A more systematic effort to isolate and characterize Mut⁻ strains has been reported for *P. pinus* and *H. polymorpha.* Popova and Pugovkina reported the isolation of 17 Mut⁻ strains of *P. pinus* [19]. Five of these strains were also defective in ability to utilize glycerol, ethanol, and xylose. Of nine mutants which were defective in methanol utilization alone, seven were subjected to genetic analysis and found to represent 5 complementation groups, termed *cut 1* through *cut 5.* Tetrad analysis revealed that two sets of complementation groups, *cut 2* and *cut 4,* and *cut 3* and *cut 5,* were linked. Gleeson *et al.* [12, 13] reported the isolation and charactetization of Mut⁻ mutants of *H. polymorpha.* About half of the mutants were unable to grow on ethanol, glycerol, maltose, and dihydroxyacetone as well. These mutants, called alternative carbon-utilization (*acu*) mutants, were not defective in respiration but potentially could be defective in carbon catabolite repression. Eighteen mutants which were defective only in methanol utilization were classified into 5 complementation groups, termed $C_1 1$ through $C_1 5$ Mutant strains in each complementation group were assayed for the presence of 11 of the 13 enzymes involved in the metabolism of methanol. One, $C_1 3$ was dihydroxyacetone synthase defective. Each of the remaining 4 had significant levels of activity for each enzyme. It was suggested that some of these mutants may not be defective in any of the methanol-pathway enzymes directly but are defective in other protein required for C_1 metabolism, such as those required for the proper functioning of peroxisomes [13].

3. MOLECULAR GENETICS

3.1. Methanol-Regulated Genes

The first evidence that the synthesis of methanol-pathway enzymes in yeasts may be controlled at the level of transcription was reported in late 1983 and early 1984. Roa and Blobel [20] and Roggenkamp *et al.* [21] used *in vitro* translation techniques to demonstrate that translatable mRNA for alcohol oxidase, catalase, and dihydroxyacetone synthase is present in methanol-grown, cells of *H. polymorpha.* In addition, by comparing the relative sizes of *in vivo* and *in vitro*-synthesized alcohol oxidase [20, 21] and catalase [21], it appeared thet neither of these enzymes, which are known to be located in peroxisomes, is synthesized as a larger precursor, as is typical of many transported proteins.

Definitive proof of transcriptional control and absence of cleaved amino-terminal signal sequences has been obtained as a result of the isolation and characterization of several methanol-regulated genes. Working with *Pichia pastoris,* Ellis *et al.* [22] described 3 such genes: one encoding alcohol oxidase *AOX1*, a second encoding a protein of 76,000 daltons which is now known to be dihydroxyacetone synthase (*DAS1*) (G. Thill, personal communication), and a third encoding a 40,000 dalton protein of unknown function. *AOX1* and *DAS1* mRNAs were detectable in *P. pastoris* cells grown on ethanol. Similar results were reported for transcription of *AOX (MOX)* and *DAS* mRNAs in *H. polymorpha* [23, 24]. Comparison of the alcohol oxidase DNA and protein sequences of *P. pastoris* and *H. polymorpha* clearly showed that neither *AOX* gene encodes a processed amino-terminal signal sequence.

Ledeboer *et al.* [24] and Janowicz *et al* [23] reported the entire nucleotide sequences of the *H. polymorpha AOX* and *DAS* genes, respectively including 5' and 3' non-encoding regions. Since expression of both of these genes appears to be coordinately regulated, any sequence homologies shared between their putative promotor regions would be reasonable candidates for sequences involved in methanol regulation. Although a number of interesting sequence features were noted in both the 5' and 3' regions flanking the two genes, including sequences implicated in transcription initiation and termination in *S. cerevisiae*, no sequence homology was observed between the 5' non-coding regions of the *AOX* and *DAS* genes. The identification of sequences involved in the regulation of C_1-pathway genes may require the molecular dissection of one or more of the methanol-regulated promoters, as performed on promoters from the galactose-utilization pathway genes of *S. cerevisiae* [25].

3.2. Transformation

The ability to transfer and maintain cloned DNA sequences of interest in an organism is essential to establish the function of those sequences. Transformation is useful for other types of experiments as well. For example, in methylotrophic yeasts, genes whose products are required for methanol utilization could be isolated by transforming Mut^- hosts with plasmid-borne DNA libraries and then selecting transformants with restored methanol-growth ability. As a second example, commercial exploitation of methylotrophic yeasts as hosts for heterologous gene expression obviously requires the ability to transform these yeasts.

Transformation has been reported for 2 methylotrophic yeast species, *P. pastoris* [26] and *H. polymorpha* ([27, 28, 29]; M. Gleeson, personal communication). In both yeasts, transformation is based on auxotrophic mutant hosts and vectors which contain a complementing biosynthetic gene. As selectable markers, previously isolated and readily available *S. cerevisiae* biosynthetic genes functioned sufficiently well in these methylotrophic yeasts to complement and support near wild-type growth rates. These methylotrophic yeast strains were transformed using modified versions of the spheroplast generation- ($CaCl_2$- polyethylene glycol) fusion procedure developed for *S. cerevisiae* [30, 31]. However, Roggenkamp *et al.*, [29] reported that, in their hands, higher transformation frequencies in *H. polymorpha* were obtained using the frozen cell method of Klebe *et al.*, [32]. In *P. pastoris,* it has been demonstrated that, as in *S. cerevisiae*, integration of plasmid and linear DNA fragments into the genome occurs most often by homologous recombination between sequences shared by vector and host [26, 33].

In both *P. pastoris* [26] and *H. polymorpha* [29], a plasmid can be maintained as a multi-copy autonomous element, if the plasmid contains a DNA fragment with autonomous replication sequence (ARS) activity. In both yeasts, ARS elements with the best plasmid-maintenance characteristics were obtained by selecting for the elements directly in the native hosts. It was apparent that DNA fragments with ARS activity in *S. cerevisiae* often did not support autonomous replication in these methylotrophic yeasts. For example, of four *P. pastoris* - and two *H. polymorpha* -selected ARS elements, none functioned in *S. cerevisiae* [26, 29, 33]. The DNA sequences of ARS fragments from *P. pastoris,* PARS1 and PARS2 (26), and from *H. polymorpha,* HARS1 [29], have been determined and compared to those of ARS elements from *S. cerevisiae* . Despite the fact that the methylotrophic yeast ARSs did not function in *S. cerevisiae* and vice-versa, at least two general similarities in sequence structure were noted. Each sequence was found to be rich in A+T base pairs and to contain one or more sequences with close (but not exact) homology to the 11 base pair *S. cerevisiae* consensus ARS [34].

3.3. Construction of Mutants by Gene Disruption

In *S. cerevisiae* , it is possible to examine the function of a cloned DNA sequence by constructing mutant strains which are defective in that sequence. This technique is referred to as site-specific gene disruption or gene replacement [35, 36]. Using *in vitro* recombinant DNA methods, one or more precise alterations are first introduced into a cloned DNA fragment.

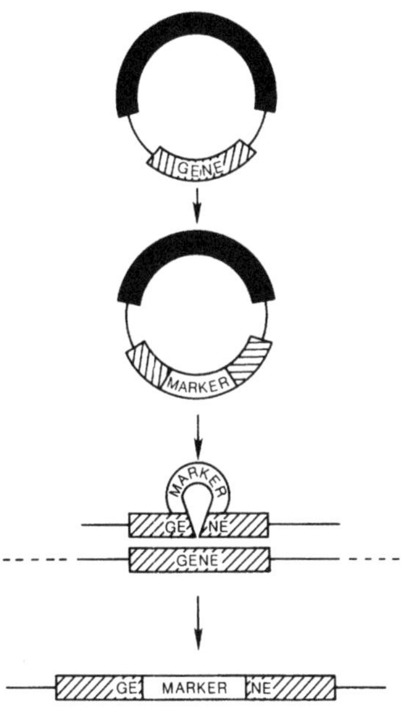

Figure 2. Site-specific gene disruption. The diagram depicts events resulting in the insertion of a modified host DNA fragment into a chromosome. The black box indicates sequences from pBR322. The box with hashed lines indicates a cloned host gene, and the thin lines represent sequences flanking the gene. The open box indicates a selectable marker gene.

In the cartoon shown in Figure 2, a gene is disrupted by the insertion of a selectable marker gene. A linear fragment which contains the altered sequence is then transformed into the host. In *S. cerevisiae,* DNA termini are recombinogenic and stimulate homologous recombination. Thus, in a high proportion of transformed cells, the *in vitro* modified DNA fragment is substituted for the native sequences, as shown at the bottom of Figure 2.

The gene disruption technique has been applied to *P. pastoris* ,and evidence has been obtained that the *P. pastoris* genome contains two functional alcohol oxidase genes [33]. Previously, Ellis *et al.* [22] identified twodifferent genomic fragments with homology to *AOX* cDNAs, one of which clearly contained a functional gene, *AOX1*. To determine whether the *P. pastoris* genome contained a second functional alcohol oxidase gene, *AOX2*, the two genomic fragments were modified into gene disruption vectors and inserted into the *P. pastoris* genome. As shown in Figure 3, KM71, a mutant strain in which *AOX1* had been disrupted, can still grow on methanol, although more slowly than PPF1, the fully methanol-utilization-competent parent strain (31 versus 5 hours per generation). In addition, cell-free extractsof the AOX1-defective strain contain approximately 15% of the alcohol oxidase activity of the parent. In a second experiment, KM7121, a mutant with disruptions in both *AOX1* and putative *AOX2* genes, was created.

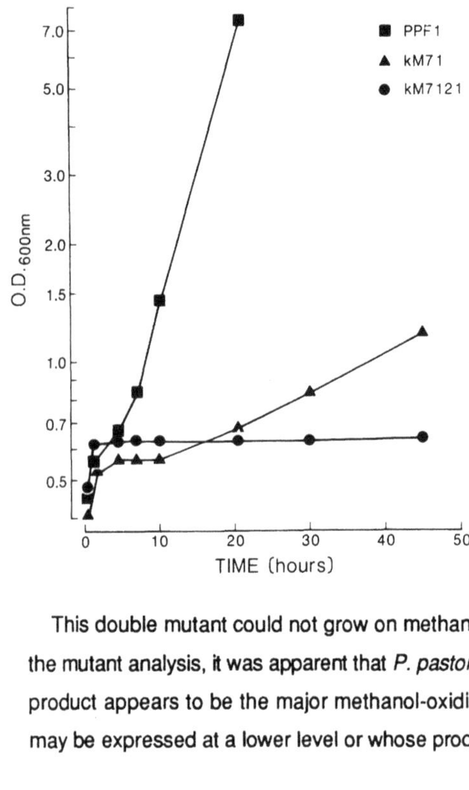

Figure 3. Growth of alcohol oxidase-defective mutants of *P. pastoris* in methanol. The times shown are those after shifting cultures from a medium containing glycerol to one containing methanol as a sole carbon and energy source. ■ , PPF1, fully methanol-utilization-competent strain; ▲ , KM71, *AOX1*- defective strain; ● , KM7121, *AOX1*- and *AOX2*- defective strain.

This double mutant could not grow on methanol at all and contained no alcohol oxidase activity. From the mutant analysis, it was apparent that *P. pastoris* contains two alcohol oxidase genes: *AOX1*, whose product appears to be the major methanol-oxidizing activity in methanol-grown cells, and *AOX2*, which may be expressed at a lower level or whose product may be less active on methanol.

3.4. Methanol-Regulated Expression of LacZ

The fusion of genes to the *E. coli lacZ* gene has become the method of choice for confirming the presence of putative promoters and studying their regulation [37]. Such *lacZ* fusion studies were recently described for the *AOX1* and *DAS1* genes of *P. pastoris* [38]. The fusion approach was particularly important for studies on the *P. pastoris AOX1* promoter, since the *P. pastoris* genome contains two *AOX* genes whose mRNAs and proteins are very similar in size and other physical properties [33]. Thus, the approach provided a means of examining the regulation of one gene independently of the other. For these studies, *AOX1-lacZ* and *DAS1-lacZ* fusion vectors were constructed which directed the synthesis of hybrid proteins composed of a small amino-terminal portion of a *P. pastoris* peptide (AOXI, 15 amino acids; DASI, 4 amino acids) fused to β- galactosidase ([38], G. Thill, personal communication). Regulation of these genes was then monitored by examining transformed cells for β- galactosidase.

Figure 4 shows the results from an experiment in which an *AOX1-lacZ* fusion strain was sequentially grown in a medium with different carbon sources and then assayed for both β- galactosidase and alcohol oxidase activities. In the *AOX1-lacZ* strain, no detectable activity for either enzyme was observed in cells grown on glucose. When shifted to medium without carbon source, the level of activity for both enzymes increased significantly. Finally, after addition of methanol, activity for both β-galactosidase and alcohol oxidase increased approximately 50-fold relative to the carbon starvation state. Thus, the putative *AOX1* promoter fragment appear to control the synthesis of β- galactosidase in a manner which was indistinguishable from that of its native product, alcohol oxidase. Similar results were obtained with the *DAS1-lacZ* fusion strain, with the important exception that little or no β-galactosidase activity was observed in carbon-starved cells.

These results on the regulation of methanol-controlled genes in *P. pastoris* are considerably different from those obtained from primarily biochemical studies of other methylotrophic yeasts, particularly *H. polymorpha.*

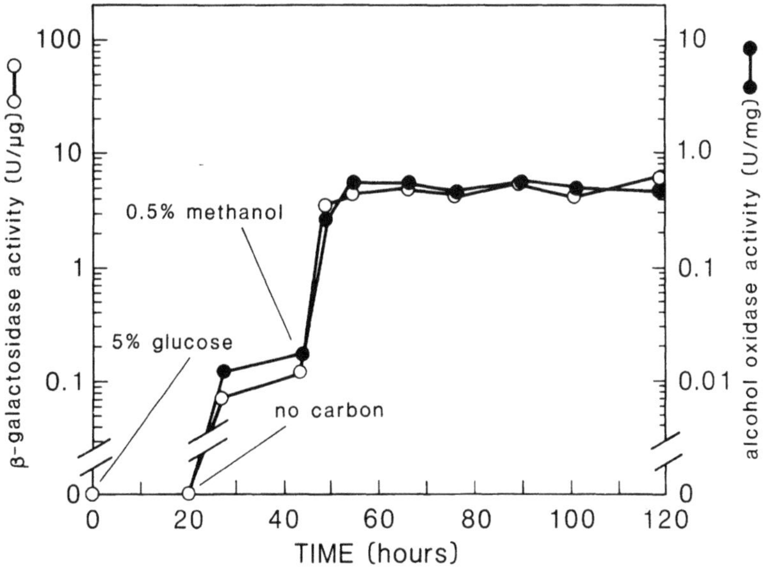

Figure 4. Expression of the AOX1-*lacZ* fusion in *P. pastoris.*The culture was shifted by centrifugation at the indicated times from a medium with glucose to the same medium with no carbon source, and finally to the medium with methanol. Samples were removed, and cell-free extracts were prepared for enzyme activity assays. A unit of β- galactosidase represents 1 nmole of O-nitrophenol produced per minute at 30°C. A unit of alcohol oxidase is 1 μmole of hydrogen peroxide produced per minute at 30°C.

In both *P. pastoris* and *H. polymorpha* , catabolic repression appears to be a major part of the regulatory mechanism, since most methanol-pathway enzymes are not present in cells grown on repressing carbon sources such as glucose, even when methanol is also present in the medium [38, 39, 40, 41]. When *H. polymorpha* is grown on derepressing carbon sources, such as glycerol, high levels of many C_1-pathway enzymes such as alcohol oxidase and catalase are observed [41]. In contrast, no *AOX1*- or *DAS1*- driven β- galactosidase or alcohol oxidase is observed in glycerol-grown cells of *P. pastoris* (data not shown), and only a small derepression response is observed upon carbon starvation. In *P. pastoris* , a high level of expression of these genes is observed only in response to methanol. Thus, in *H. polymorpha,* the pattern of C_1-pathway enzyme synthesis best fits the description of a repression/derepression mechanism, while in *P. pastoris* the regulation of C_1-pathway genes appears to involve both repression/derepression and induction mechanisms. Further investigations at the molecular level are required to determine the regulatory mechanisms involved in controlling the expression of these genes.

4. CONCLUSION

To date, the contribution of genetics to our understanding of methylotrophic yeasts has been quite small relative to that of biochemistry and physiology. Until recently, the lack of techniques for the genetic manipulation of these yeasts was a significant block towards progress. Of particular importance to classical genetics, methods for crossing and analyzing mating products have been described. For molecular genetic studies, the key technique of DNA-mediated transformation has been applied to these yeasts. As a result of these advantages in genetic techniques, rapid progress in studies on methylotrophic yeasts is anticipated.

ACKNOWLEDGEMENTS

I wish to thank my colleagues, Rich Buckholz, Mary Ellen Digan, Tom Gingeras, Greg Thill, and Juerg Tschopp, for their assistance and comments on this manuscript and for communicating unpublished results. I thank Martin Gleeson for citing the literature on the genetics of *Pichia pinus* . I thank Janice Doty for help in preparing this manuscript. Research on *Pichia pastoris* was supported by Phillips Petroleum Company.

REFERENCES

[1] Anthony, C. 1982. The Biochemistry of Methylotrophs. New York, Academic Press, Inc.
[2] Veenhuis M., van Dijken JP, Harder W. 1983. Adv. Microbial Physiol. 24, 1-82.
[3] Goodman JM. 1985. J. Biol. Chem. 260, 7108-7113.
[4] Douma AC, Veenhuis M., de Koning W., Harder W. 1985. Arch. Microbiol. 143, 237-243.
[5] Lee J-D, Komagata K. 1980. J. Gen. Appl. Microbiol. 26, 133-158.
[6] Barnett JA, Payne RW, Yarrow D. 1983. Yeasts: Characteristics and Identification. Cambridge, Cambridge University Press.
[7] Tolstorukov II, Dutova TA, Benevolenskii SV, Soom YaO. 1977. Genetika 13, 322-329.
[8] Tolstorukov II, Benevolenskii SV. 1978. Genetika 14, 519-526.
[9] Tolstorukov II, Benevolenskii SV. 1980. Genetika 16, 1335-1341.
[10] Benevolenskii SV, Tolstorukov II. 1980. Genetika 16, 1342-1349.

[11] Tolstorukov II, Efremov BD, Bliznik KM. 1983. Genetika 19, 897-902.
[12] Gleeson MA, Waites MJ, Sudbery PE. 1984. In Microbial Growth on C_1 Compounds. Crawford RL., Hanson RS, Eds. Washington, DC, American Society for Microbiology. Pp. 228-235.
[13] Gleeson Ma, The Genetics of the Methylotrophic Yeast, *Hansenula polymorpha* , Ph. D. thesis, University of Sheffield, 1986.
[14] Hicks JB, Herskowitz I. 1976. Genetics 83, 245-258.
[15] Beach DH. 1983. Nature 305, 682-688.
[16] Sanchez S, Demain AL. 1977. Eur. J. Appl. Microbiol. 4, 45-49.
[17] Sahm H, Wagner F. 1973. Eur. J. Biochem. 36, 250-256.
[18] O'Conner ML, Quayle JR. 1979. J. Gen. Microbiol. 113, 203-208.
[19] Popova IA, Pugovkina NA. 1982. Genetika 18, 916-923.
[20] Roa M, Blobel G. 1983. Proc. Natl. Acad. Sci. USA 80, 6872-6876.
[21] Roggenkamp R, Janowicz Z, Stanikowski B, Hollenberg CP. 1984. Mol. Gen. Genet. 194, 489-493.
[22] Ellis SB, Brust PF, Koutz PJ, Waters AF, Harpold MM, Gingeras TR. 1985. Mol. Cell. Biol. 5, 1111-1121.
[23] Janowicz ZA, Eckard MR, Drewke C, Roggenkamp RO, Hollenberg CP, Maat J, Ledeboer AM, Visser C, Verrips CT. 1985. Nucl. Acids. Res. 13, 3043-3062.
[24] Ledeboer AM, Edens L, Maat J, Visser C, Bos JW, Verrips CT, Janowicz Z, Eckart M, Roggenkamp R, Hollenberg CP. 1985. Nucl. Acids. Res. 13, 3063-3082.
[25] Giniger E, Varnum SM, Ptashne M. 1985. Cell 40, 767-774.
[26] Cregg JM, Barringer KJ, Hessler AY, Madden KR. 1985. Mol. Cell. Biol. 5, 3376-3385.
[27] Tikhomirova LP, Vel'kov VV. 1985. Dokl. Akad. Nauk. SSSR 282, 741-743.
[28] Makhina EN, Nesterova GF, Grishin AV. 1985. Mol. Genet. Mikrobiol. Virusol. 5, 28-31.
[29] Roggenkamp R, Hansen H, Eckart M, Janowicz Z, Hollenberg CP. 1986. Mol. Gen. Genet. 202, 302-308.
[30] Hinnen A, Hicks JB, Fink JR. 1978. Proc. Natl. Acad. Sci. USA 75, 1929-1933.
[31] Beggs JD. 1978. Nature (London) 275, 104-109.
[32] Klebe RJ, Harris JV, Sharp ZD, Douglas MG. 1983. Gene 25, 333-341.
[33] Cregg JM, Madden KR. 1978, in press. In Biological Research on Industrial Yeasts. Stewart GG, Russell I, Klein RD, Hiebsch RR, Eds. Boca Raton, Uniscience Series, CRC Press.
[34] Broach JR, Li Y-Y, Feldman J, Jayaram M, Abraham J, Nasmyth KA, Hicks JB. 1983. In Cold Spring Harbor Symp. Quant. Biol. Vol 47. Cold Spring Harbor, New York, Cold Spring Harbor Laboratory. pp. 1165-1173.
[35] Rothstein RJ. 1983. In Methods in Enzymology. Vol. 101, Part C. Wu R, Grossman L, Moldave K, Eds. New York, Academic Press, Inc. pp. 202-210.
[36] Winston F, Chumley F, Fink GR. 1983. In Methods in Enzymology. Vol. 101, Part C. Wu R, Grossman L, Moldave K, Eds. New York, Academic Press, Inc. pp. 211-228.
[37] Silhavy TJ, Beckwith JR. 1985. Microbiol. Rev. 49, 398-418.
[38] Tschopp JF, Brust PF, Cregg JM, Stillman CA, Gingeras TR. 1986, manuscript in preparation.
[39] van Dijken JP, Otto R, Harder W. 1976. Arch. Microbiol. 111, 137-144.
[40] Eggeling L, Sahm H. 1978. Eur. J. Appl. Microbiol. Biotechnol. 5, 197-202.
[41] Egli Th, van Dijken JP, Veenhuis M, Harder W, Fiechter A. 1980. Arch. Microbiol. 124, 115-121.

PROTEIN TOPOGENESIS IN METHYLOTROPHIC YEASTS

H.F. TABAK and G. AB*

Section for molecular Biology, Laboratory of Biochemistry, Biotechnology center, University of Amsterdam and *Laboratory of Biochemistry, University of Groningen, Groningen, The Netherlands.

1. INTRODUCTION

Eukaryotic cells contain many subcellular structures that serve to carry out various processes. It requires that each of these compartments is provided with specific sets of proteins necessary to perform such tasks. With minor exceptions concerning certain organelles, all proteins are synthesized in the cytoplasm from where they are distributed to their various locations (see Fig. 1). This process is referred to as protein topogenesis, protein traffic or sorting and its basic principles are coming into focus. All the information gathered up till now suggests that information needed for protein sorting is present in short blocks of adjacent amino acids (topogenic or addressing signals) in the precursor or mature protein [1-5]. The evidence disfavours conformational models in which the tertiary structure of the protein plays the major role in sorting. In that case topogenic information could consist of amino acids close to each other in space but far apart from each other in the linear protein chain and corresponding gene, raising formidable difficulties in analysing such signals. In many proteins important topogenic information is present in the N-terminal part of the precursor protein caled signal peptide. It is believed that the nature of the signal peptide determines to a large extent which of the two main routes in protein traffic is chosen (Fig. 1). The nascent N-terminal protein chain protruding from a ribosome engaged in translation is recognized in one group of proteins by the Signal Recognition Particle (SRP). It consists of 6 proteins and a 7S RNA and by binding to the signal peptide stops further translation [6]. Only after contact with a receptor (Docking protein) located in the endoplasmid reticulum (E.R.) translation is resumed, resulting in transfer of the protein into the lumen of the E.R. (Fig. 2). Insertion into the lumen of the E.R. is coupled to precursor processing and the signal peptide is cleaved from the mature protein by a specific protease (signal peptidase). From here it is further transported to the Golgi and delivered to various locations. Some proteins following this route are secreted, others remain in the plasma membrane and also lysosomal enzymes diverge from this traffic line. In the second major route the translational arrest characteristic for entry into the E.R. is lacking and complete precursor proteins are synthesized in the cytoplasm. Each protein carries its own topogenic information by which it is delivered to the various organelles. N-terminally located signal peptides are a general characteristic for proteins imported into mitochondria or chloroplasts and arrival is coupled in most cases to removal of the signal peptide [4,7]. Internal topogenic information is the property of proteins destined for import in nuclei [8-10] and microbodies [11].

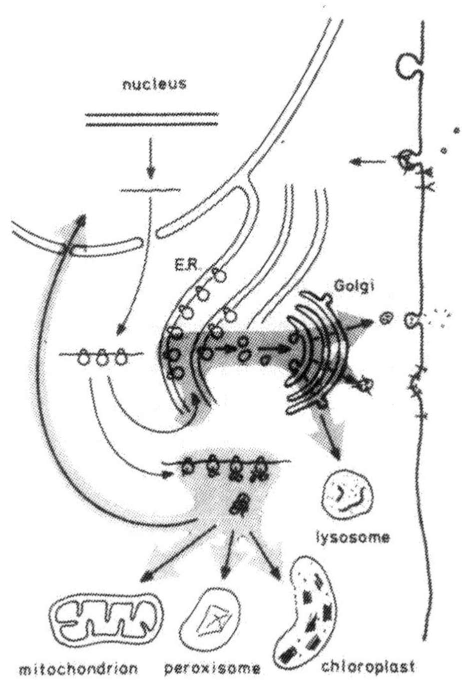

Figure 1.A schematic view of traffic lines in a eukaryotic cell going from site of synthesis to place of destination. The two main routes of protein traffic are distinguished from each other by the intensity of shading.

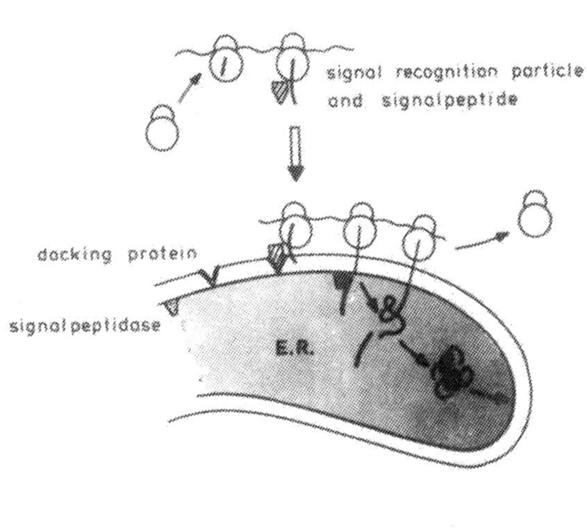

Figure 2.A schematic view of the signal hypothesis.

Cells can also acquire proteins that have been synthesized by other cells. Specific receptors present in the plasma membrane can sequester such proteins from the outside. Receptor-ligand complexes assemble in specific regions which are then internalized in a process called "receptor-mediated endocytis" [12]. In a less specific manner extracellular material can also be internalized without the intervention of specific receptors by "fluid-phase endocytosis".

After reaching a subcellular compartment the sorting process is not yet finished, since many proteins find a specific localization within them. Some are inserted in membranes with a specific orientation or have to pass through a membrane to be positioned in a second one. This can be illustrated by looking at import of mitochondrial proteins. Mitochondria are surrounded by two membranes which divide the organelle into the following compartments: outermembrane, intermembrane space, innermembrane and the inside of matrix space each characterized by specific sets of proteins. Again the evidence suggests that short blocks of adjacent amino acids carry the topogenic information necessary for sorting proteins within an organelle. For instance, a stretch of approximately 20 mostly hydrophobic amino acids serve as a "stop transfer" sequence, trapping a protein in a membrane [13].

Through a number of windows we can catch a glimpse of the complicated traffic networks along which proteins are efficiently sorted in a cell without formation of traffic jams. As windows to observe these activities many workers resort to specialized biological systems or to cells that are especially suitable for experimental manipulation. In this respect fungi have much to offer to those who are interested in this field.

2. PROTEIN TOPOGENESIS IN FUNGI

Saccharomyces cerevisiae displays all the characteristics of a eukaryotic cell as shown in Fig. 1 with the possible exception of receptor-mediated endocytosis. It has been developed into a workhorse comparable in versatility to that of *Escherichia coli* amongst prokaryotes. Recombinant DNA techniques can be fully exploited to study gene expression and function, including the powerfull possibility of exchanging an authentic wild-type gene for a manipulated version of the gene [14]. Combined with the highly developed classical techniques of nuclear and mitochondrial genetics, the tools are provided to thoroughly investigate the principles of protein topogenesis in this organism. Typical examples that illustrate these possibilities are: i) the isolation of many thermosensitive mutants blocked at the nonpermissive temperature in the various steps of the secretion pathway, leading to accumulation of proteins in the compartments they have to traverse on their way out [15]; ii) the *in vitro* construction of fusion proteins by manipulating the respective genes [16,17]. For instance the DNA encoding the signal peptide of a mitochondrial protein has been transplanted to the gene coding for a secretion protein replacing its signal sequence encoding DNA. After transforming yeast cells with the gene resulting from such DNA surgery, it is observed that the protein is diverted from the secretion pathway and is now, albeit with low efficiency, imported into mitochondria [18]. Such experiments offer strong clues to the nature and function of addressing signals and the possible influence of the remaining body of the protein in sorting. There is also a reverse side of the medal however. *S. cerevisiae* has a strong cell wall which makes it difficult to prepare clean homogenates from which organelles can be isolated in good yield. Such homogenates are full of proteases complicating protein purification.

Certain fungi from the genera *Hansenula, Pichia, Candida* and *Torulopsis* show dramatic morphological changes when shifted to a methanol-containing growth medium [19]. A single microbody grows and divides to such an extent that the whole cell becomes filled with them after some time (Fig. 3). The presence of a hydrogen peroxide-producing alcohol oxidase and katalase allows them to be regarded as peroxisomes according to the definitions spelled out by de Duve [20]. Similar but less

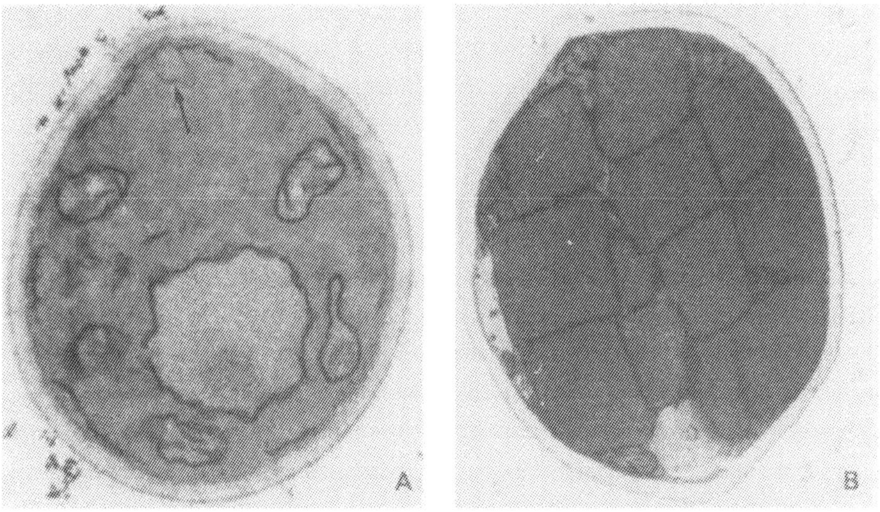

Figure 3.*Hansenula polymorpha* grown on glucose (Fig. 3A) or methanol (Fig. 3B). The photograph was kindly provided by M. Veenhuis, Laboratory of Microbiology, University of Groningen.

extensive proliferation of peroxisomes is seen when cells are dependent on methylated amines as source for their nitrogen requirement. Due to their spectacular biological specialization methylotrophic yeasts offer many attractive possibilities to study biosynthesis of peroxisomes. Thus far peroxisomes have been the Cinderella among the organelles with regard to studies in protein topogenesis. This is in part due to their manifestation in various disguise (peroxisome, glyoxysome, glycosome and perhaps hydrogenosome) and to difficulties in isolating the organelles in intact form. A few generalizations can be made however. For a detailed discussion the reader is referred to recent reviews [21,22].

Peroxisomes do not obtain their proteins from the E.R. as suggested in older literature. Proteins are synthesized in the cytoplasm and subsequently imported. No difference is found between the primary translation products made in *in vitro* systems and the mature proteins isolated from cells [11]. It rules out that topogenic information is present in the N-terminal part which is dispensed with after completion of import, as has been demonstrated to be the case for mitochondrial and chloroplast proteins. It is therefore not immediately obvious where to look for the signal peptide-like sequences responsible for import. Progress can be expected from several lines of research. Genes coding for peroxisomal proteins can be cloned and the amino acid sequences derived from them can be compared to reveal common conserved elements possibly involved in import. Genes can be modified by various techniques of site-directed mutagenesis and reintroduced into the cell to observe if import is affected. In this respect it is promising that DNA transformation systems have been worked out for *Pichia* and *Hansenula* enabling this approach [23,24]. Finally, it may be easier to isolate peroxisomes of good quality and in decent yield from methanol-induced methylotrophic yeasts. It would offer the possibility to set up in vitro systems for protein import like have been worked out for mitochondrial import studies.

It would also be worthwhile to reinvestigate the status of peroxisomes in *C. cerevisiae*. After the first visualization of peroxisomes in *S. cerevisiae* by electron microscopy [25], attempts to isolate and characterize the organelles with biochemical techniques, mostly failed. However, study of the peroxisomal import problem could benefirt enormously from the technical advantages afforded by the use of *S. cerevisiae*.

3. APPLICATIONS RESULTING FROM RESEARCH ON PROTEIN TOPOGENESIS

Research on protein topogenesis has spin off for application to problems in biotechnology. There is a growing interest in producing heterologous proteins in yeast. The occurrence of topogenic signals in short blocks of amino acids offers unique possibilities to transplant the DNA in which they are encoded to other genes. In this way we cannot only produce the heterologous protein but also impose upon it the information to seek out a specific location within the cell or to direct them into the secretory pathway for delivery in the medium. In this respect the huge peroxisomal organelles present in methanol-grown yeasts could offer save storehourse space to package heterologous proteins away from the proteolytic activities present elsewhere in the cell. To try out such scenario's it will first be necessary to establish the nature of the peroxisomal import signal.

REFERENCES

[1] Blobel G, Dobberstein B. (1975) J. Cell Biol. 67, 835-851.
[2] Blobel G, Dobberstein B. (1975) J. Cell Biol. 67, 852-862.
[3] Garif G, (1985) Ann. Rev. Cell Biol. 1, 403-407.
[4] Hay R, Bohni P, Gasser S. (1984) Biochim. Biophys. Acta 779, 65-87.
[5] Wickner WT, Lodish HF. (1985) Science 230, 400-407.
[6] Walter P, Blobel G. (1982) Nature 299, 691-698.
[7] Hurt EC, Van Loon APGM. (1986) Tr. Biochem. Sci. 11, 204-207.
[8] Hall MN, Hereford L, Herskowitz I. (1984) Cell 36, 1057-1065.
[9] Moreland RB, Nam HG, Hereford LM, Fried HM. (1985) Proc. Natl. Acad. Sci. USA 82, 6561-6565.
[10] Kalderon D, Roberts BL, Richardson WD, Smith AE. (1985) Cell 39, 499-509.
[11] Rao M, Blobel G. (1983) Proc. Natl. Acad. Sci. USA 80, 6872-6876.
[12] Goldstein JL, Anderson RGW, Brown MS. (1979) Nature 279, 679-685.
[13] Kaput J, Goltz S, Blobel G. (1982) J. Biol. Chem. 257, 15054-15058.
[14] Struhl K. (1983) Nature 305, 391-397.
[15] Schekman R. (1985) Ann. Rev. Cell Biol. 1, 115-143.
[16] Hurt EC, Pesold-Hurt B, Schatz G. (1984) EMBO J. 3, 3149-3156.
[17] Hurt EC, Pesold-Hurt B, Suda K, Oppliger W, Schatz G. (1985) EMBO J. 4, 2061-2068
[18] Van Steeg H, Oudshoorn P, Van Hall B, Polman JEM, Grivell LA. personal communication.

[19] Veenhuis M, Van Dijken JP, Harder W. (1983) Adv. Microbiol. Physiol. 24, 1-82.
[20] De Duve C. (1983) Sci. American 248, 52-62.
[21] Lazarow PB, Fujuki Y. (1985) Ann. Rev. Cell Biol. 1, 489-530.
[22] Borst P. (1986) Biochim. Biophys. Acta 866, 179-203.
[23] Roggenkamp R, Hansen H, Eckart M. Janowiz Z, Hollenberg CP. (1986) Mol. Gen. Genet. 202, 302-308.
[24] Cregg JM, Barringer KJ, Hessler AY, Madden KR. (1985) Mol. Cell. Biol. 5, 3376-3385.
[25] Hoffmann RB, Szabo A, Avers CJ. (1970) J. Bacteriol. 104, 581-584.

ENERGETICS OF C_1 METABOLISM

ENERGETICS OF C_1- COMPOUND METABOLISM

Henk W. VAN VERSEVELD and Rudolf K. THAUER
Biologisch Laboratorium, Vrije Universiteit, Amsterdam, The Netherlands and
Fachbereich Biologie, Philipps-Universität Marburg, Federal Republic of Germany

1. INTRODUCTION

The thermodynamics of C_1-metabolism, the mechanism of energy conservation, and growth yields will be discussed. Aerobic C_1-utilizers and anaerobic C_1-utilizers will be dealt with in separate chapters. The treatise is selective as is the literature review.

2. AEROBIC C_1- UTILIZERS

2.1. Thermodynamics

Aerobic organisms that metabolize C_1-compounds oxidize these compounds with O_2 to CO_2. The free energy change (ΔGo) associated with the redox process are sufficiently high to drive the phosphorylation of ADP for which in vivo generally - 80 to - 90 kJ/mol are required [1,2]. The oxidation of formate (- 241 kJ/mol), of CO (- 257 kJ/mol), of formaldehyde (- 501 kJ/mol), of methanol (- 693 kJ/mol), and of methane (- 817 kJ/mol) theoretically allow the formation of 3 mol ATP/mol formate or CO, of 6 mol ATP/mol formaldehyde, of 8 mol ATP/mol methanol, and of 10 mol ATP/mol methane. The experimental data indicate that these theoretical ATP yields are not always reached in vivo especially in the case of methane metabolism where the oxidation of CH_4 to the level of formaldehyde is, for mechanistic reasons, probably not linked with phosphorylation [3].

2.2. Mechanism of energy conservation

It is generally accepted that the driving force of ATP formation in aerobically grown cells is the proton motive force, $\Delta\mu_{H^+}$ or pmf. Although the proton gradient is widely accepted as intermediate between oxidation and phosphorylation the stoichiometry of proton or charge separation across the membrane in mitochondria or bacterial cytoplasmic membranes has been a long standing dispute. For a long time full agreement has existed about the stoichiometry of proton and charge separation during electron transfer from ubiquinone to cytochrome *c*. Per two electrons transfered, 4 protons are released at the cytoplasmic side of the mitochondrion or at the periplasmic side of the cytoplasmic membrane of a gram-negative bacterium and 2 positive charges are translocated across the membrane [see 4,5]. The $H^+/2e^-$ and $q^+/2e^-$ stoichiometries between NADH and ubiquinone and cytochrome *c* and oxygen have long remained controversial. Recently Wikström [6] showed, by using the pH-indicators neutral red and safranine to measure intramitochondrial alkalinisation, that 4 protons and 4 charges are translocated during transport of 2 electrons from NADH to ubiquinone. On the basis of experiments carried out by Wikström and Casey [7], Mitchell accepted that cytochrome oxidase translocates protons with a $\rightarrow H^+/2e^-$ stoichiometry of 2 and a $\rightarrow q^+/2e^-$ stoichiometry of 4 [8] at the same time proposing the protonmotive O loop and O cycle mechanisms by analogy with the action of ubiquinone as the conductor of hydrogen in proton motive Q loop and Q cycle mechanisms.

Another long standing dispute is the chemiosmotic delocalized, bulk to bulk phase membrane potential against localized proton circuits, which was proposed by Williams as early as 1961 [9]. An excelent compilation of data and arguments for and against localized proton flow pathways has recently been published by Ferguson [10]. A good example of data indicating that localized proton circuits could be involved in energy generation is given by Laszlo et al [11] with chloroplast thykaloids. They found an array of protein buffering groups residing either within the membrane matrix or in proteins at the membrane surface, not in equilibrium with the bulk aqueous phases and closely associated with the water-oxidizing apparatus.

Considerable information has been obtained on the magnitude of the proton motive force in nongrowing cells and cells growing in batch culture, recently reviewed by Kashket [12], indicating that the stoichiometry of the H^+-ATPase is 3 H^+/ATP. However, presentation of this value as definite would be premature, because a non-linear relationship between ΔG_p and $\Delta\mu_{H^+}$ was observed in vesicles from *Paracoccus denitrificans* and complete equilibration between these forces could be prevented by ATP utilization within cells [13].

Not many data concerning the mechanism of oxidative phosphorylation are collected using microorganisms growing on C_1-compounds. Only with *Paracoccus denitrificans* and *Methylophilus methylotrophus* the stoichiometry of proton translocation has been investigated. Methanol-limited chemostat-grown *Paracoccus denitrificans* has been used to investigate the proton translocating capacity of the cytochrome *c* to oxygen part of the electron transport chain [14]. The advantage of methanol-grown cells is that methanol via methanol dehydrogenase and PQQ directly donates its electrons to cytochrome *c*. Consequently methanol is a physiological electron donor for the oxidase part of the respiratory chain. It has been shown that cytochrome *c* oxidase (aa_3) of *Paracoccus denitrificans* is proton translocating with a stoichiometry of 2 $H^+/2e^-$. The $\rightarrow q^+/2e^-$ consequently is 4, like that found for the mitochondrial electron transport chain.

Methanol-grown *Methylophilus methylotrophus* show the presence of two major terminal oxidases, cytochrome aa_3 and *o*. In contrast to *Paracoccus denitrificans* the oxidase is not proton translocating and the $\rightarrow q^+/2e^-$ of the cytochrome *c* to oxygen segment of the respiratory chain is only 2. It has been speculated that cytochrome *o* is involved in the oxidation of NADH and flavin-linked substrates [15,16]. A more thorough treatment is given by Anthony and Jones in this session of the symposium.

A promising line of research is the isolation of different respiratory chain complexes and reconstitution into liposomes. Isolation and reconstitution enables one to establish the proton translocating properties and the influence of the different subunits on these properties in small parts of the electron transport chain as has been done extensively with the mitochondrial ubiquinone-cytochrome *c* and cytochrome *c* oxidase complexes. Especially the oxidases are of interest, because mostly C_1-compounds are physiological electron donors for the oxidase complex.

Cytochrome oxidases have been isolated from *Paracoccus denitrificans* [17], *Methylophilus methylotrophus* [18] and *Pseudomonas* AM1 [19]. The isolated complex of *Paracoccus denitrificans* is a two subunit aa_3-type capable of proton translocation when reconstituted in liposomes [20]. The *Methylophilus methylotrophus* complex is a four subunit *co*-type and was reconstituted with methanol dehydrogenase and cytochromes c_L and c_H as a methanol oxidase complex, showing the same sensitivity to inhibitors as during methanol oxidation in whole bacteria. The *Pseudomonas* AM1 complex is a two subunit aa_3-type. From *Paracoccus denitrificans* also the ubiquinol oxidase has been isolated [21] from cells cultured under different growth conditions including methanol-limitation (see abstract 5-16 of Bosma et al). Also the soluble *c*-type cytochromes have been isolated and characterized from methanol-grown cells (see abstract 5-16). Three additional types were found when compared with heterotrophically grown cells: a 30 kD, 22 kD and 9.2 kD cytochrome *c* with midpoint potentials of 160, 200 and 200 mV respectively. The 22 and 30 kD are also found in membranes of heterotrophically grown cells and are probably involved in electron transport from methanol to oxygen. The 9.2 kD is exclusively found in methanol grown cells and could therefore be part of the methanol dehydrogenase.

For future research the involvement of the pmf on regulatory processes is of interest. It is known that the pmf is inversely related to the specific growth rate [22,23]. However, in relaxed mutants of *Escherichia coli*, not showing the stringent response and the concomittant intracellular increase in regulatory nucleotides (pp(p)Gpp), the pmf is independent of the specific growth rate [23]. It is well known that the induction of enzymes for C_1-metabolism mostly occur under stringent situations [see e.g. 24]. Measurement of the pmf and the level of regulatory nucleotides most probably will give insight in regulatory mechanisms involved in C_1-metabolism.

2.3. Growth Yields

To extract definite conclusions about the energetics of growth from growth yield determinations, it is a prerequisite that close fitting mass and energy balances are presented [25-27]. In these balance equations an empirical formula of cell material is used, $CH_pO_nN_q$, and substrates and products are converted to formulae that contain one mol of carbon. This leads to equation 1 for growth.

$$CH_mO_l \text{ (substrate)} + a\,NH_3 + b\,O_2 \rightarrow y_c\,CH_pO_nN_q + z\,CH_rO_sN_t \text{ (product)} + c\,H_2O + d\,CO_2 \qquad (1)$$

in which y_c and z are the fractions of the substrate converted into biomass and product respectively. Carbon and redox balances are given by equations 2 and 3 respectively.

$$y_c + z + d = 1 \qquad (2)$$

$$b = 1/4(\gamma_s - y_c\gamma_b - z\gamma_p) \tag{3}$$

in which γ_s, γ_b and γ_p are the degree of reductance for organic substrate, biomass and product respectively. Under aerobic conditions the efficiency of growth is characterized by $Y_{substrate}$ and Y_{O2} (biomass yield on substrate and oxygen, g/mol). We like to emphasize again [see also 28] that these parameters are not independent as is already clear from equation 1. It is surprising that this point is not sufficiently realized by microbiologists. The relationship between these parameters is absent in many of the published experimental yield data, indicating that the author did not realize that the experimental determination of these parameters can be subject to large errors [see 29,30]. An extensive treatment of the use of carbon and energy balancing in experimental growth yield determinations is given by Babel and van Verseveld in this session of the symposium. Kelly et al will present the thermodynamic and biochemical interpretation of growth yield data of chemolithotrophic sulphur bacteria.

In this paper we only present an example showing the potentialities of growth yield determinations to deduce statements about the efficiency of energy transduction [31]. Equations 1-3 are macroscopic equations. From known biochemical pathways it is possible to formulate microscopic equations that include the synthesis and use of reduction equivalents and ATP [31]. This leads to assimilation equation 4 and dissimilation equation 5 for methanol grown *Paracoccus denitrificans.*

$$CO_2 + 0.25\ NH_3 + 2.04\ (NADH + H^+) \rightarrow CH_{1.8}N_{0.25}O_{0.48}\ \text{(biomass)} + 1.52\ H_2O + 2.04\ NAD^+ \tag{4}$$

$$CH_3OH + H_2O + PQQ + 2\ NAD^+ \rightarrow CO_2 + PQQH_2 + 2\ (NADH + H^+) \tag{5}$$

Using equations 4 and 5, equations for production of reduction equivalents per mol methanol (v_{H2}), ATP per mol methanol (v_{ATP}) and specific rate of ATP production, mol ATP per gram biomass per hour (q_{ATP}) can easily be constructed.

$$v_{H2} = 3 - (Y_{methanol}/MW_b)2.04 \tag{6}$$

in which MW_b is molecular weight of the biomass in equation 4 (25 g).

$$v_{ATP} = \{2 - (Y_{methanol}/MW_b)2.04\}P/2e^-(NADH) + P/2e^-(PQQH_2) \tag{7}$$

$$q_{ATP} = q_{2e^-}P/2e^-(overall) \tag{8}$$

in which q_{2e^-} is the specific rate of electron transport.

It has been unequivocally shown that methanol grown cells possess 3 sites of oxidative phosphorylation [14,32-34]. Considerable information about the magnitude of the proton motive force and the phosphate potential [12] makes a translocated charge/ATP stoichiometry of 3 very likely. Consequently the $P/2e^-$(NADH) is 3 for growth on methanol due to the translocation of 9 charges during electron transport of $2e^-$ [14]. During electron transport from $PQQH_2$ to oxygen 4 charges are translocated and consequently $P/2e^-(PQQH_2)$ will be 1,33. Table 1 shows growth parameters for aerobic growth on methanol at $\mu = 0.1\ h^{-1}$. The $P/2e^-$ ratios mentioned are overall values. With the aid of equations 6 and 7 the theoretical overall $P/2e^-$ ratio can be calculated as

$$P/2e^- = v_{ATP}/v_{H2} \qquad (9)$$

which yield a value of 2.21 for methanol-limited cells.

Table 1. Growth parameters of *Paracoccus denitrificans* for aerobic methanol- and formate-limited growth in the chemostat at $\mu = 0.1\ h^{-1}$. Y and q are expressed as g biomass/mol and mmol/g biomass.h. For further explanation see text.

Growth condition	Y_{sub}	q_{2e^-}	Y_{2e^-}	v_{H2}	v_{ATP}	theoretical $P/2e^-$	experimental $P/2e^-$
methanol	10.91	18.30	5.46	2.11	4.66	2.21	-
formate	2.84	22.85	4.38	0.77	1.79	2.33	1.77

Insertion of this value in equation 8 yield a q_{ATP} of 40.44 mmol/g.h.

Synthesis of biomass from formate also proceeds via the Calvin cycle and assimilation equation 4 is therefore also valid for growth on formate. The theoretical overall $P/2e^-$ is in this case equal to the $P/2e^-$ (NADH), because NADH is the only reduction equivalent formed. Since growth on formate did not induce the third site of energy coupling [33,34] the $P/2e^-$(NADH) is 2.33 due to the translocation of 7 charges during electron transport. On assumption that the q_{ATP} of 40.44 calculated from the methanol-limited culture is also valid for growth on formate, an experimental $P/2e^-$ of 1.77 is obtained with equation 8, that is substantially lower than the theoretical value of 2.33. However, one important difference between the two growth conditions has not yet been taken into account i.e., transport of formate across the cytoplasmic membrane requires energy. Assuming a symport of one net positive charge, a q_{ATP} of 11.74 mmol/g.h is needed for formate transport. Therefore the q_{ATP} for growth on formate will be 52.18 mmol/g.h. and insertion of this value in equation 8 yields an experimental $P/2e^-$ of 2.28, that is almost equal to the theoretical value of 2.33.

It will be clear from the above mentioned experiments that with close fitting carbon and redox balances it is possible to calculate differences in energy requiring systems. In this case the conclusion that formate enters the cell in symport with one positive charge (one proton) is only valid when these balances do fit.

3. ANAEROBIC C_1-UTILIZERS

3.1. Thermodynamics

Anaerobic bacteria use C_1-compounds as electron donor and/or electron acceptor in their energy metabolism. In contrast to the aerobic organisms, the free energy changes associated with these redox reactions are only very small. Examples are the conversion of CO and H_2O to CO_2 and H_2 (ΔGo = -20 kJ/mol), the fermentation of formate and S^o to CO_2 and H_2S (ΔGo = -37 kJ/mol), and the formation of methane from CO_2 and 4 H_2 (ΔGo = -131 kJ/mol). Although under standard conditions the free energy change associated with the latter reaction is large, it is only between -15 and -40 kJ/mol at the H_2 concentrations prevailing in the natural habitat of the methanogens ($pH_2 = 10^{-4}$ to 10^{-5} atm). The free energy available to anaerobic C_1-utilizers is thus only a fraction of that normally required for the phosphorylation of ADP. There is at present no indication that the phosphorylation potential is much lower and the thermodynamic efficiency of ATP formation is much higher than in other organisms, although this cannot be completely ruled out. It is more likely that in the reactions mentioned above only 1/3 mol ATP is formed per mol C_1-compound metabolized. Chemiosmotic coupling via vectorial protons with both a $\rightarrow$ 1 $H^+/2e^-$ stoichiometry and a $\rightarrow$ 3 H^+/1 ATP stoichiometry would make a fractional ATP yield of 1/3 possible [1,2,35].

3.2. Mechanism of energy conservation

The information on the mechanism of energy conservation in anaerobic bacteria metabolizing C_1-compounds is still scarce and controversial.

Blaut et al (this session of the symposium) have provided evidence that in *Methanosarcina barkeri* methane formation from methanol and H_2 (ΔGo = -112 kJ/mol) is coupled with the phosphorylation of ADP via a chemiosmotic mechanism. Protonophorous uncouplers and the proton translocating ATP-synthase-inhibitor N,N'-dicyclohexylcarbodiimide (DCCD) affected the measurable $\Delta\mu_{H^+}$, the intracellular ATP concentration, and the rate of methanogenesis in accordance with this mechanism [36,37]. The interpretation of the results is, however, not unambiguous, since under the experimental conditions less than 0.1 mol ATP/mol CH_4 were formed (deduced from the increase in the intracellular ATP concentration upon onset of methanogenesis). $P/2e^-$, $\rightarrow H^+/2e^-$ and $\rightarrow q^+/2e^-$ stoichiometries have not been reported.

Schönheit and Beimborn [38] have investigated ATP synthesis coupled to methanogenesis from CO_2 and H_2 in *Methanobacterium thermoautotrophicum*. They have found that this organism can synthesize ATP and form CH_4 in the apparent absence of $\Delta\mu_{H^+}$. Similar observations have been made by Lancaster [39] for *Methanococcus voltae* and by Sprott [40] for *Methanospirillum hungatei*. It should be noted, however, that under the experimental conditions less than 0.01 mol ATP per mol CH_4 were generated, which makes the interpretation of these results ambiguous. It is therefore premature to propose that ATP synthesis in these methanogens proceeds via the mechanism of substrate level phosphorylation rather than by the chemiosmotic mechanism [41]. Other explanations for these findings can be thought of [42,43].

Methane formation from H_2 and CO_2, from acetate, and from methanol by cell suspensions of methanogenic bacteria is dependent on the presence of Na^+ in the medium [44,45], whereas methane formation from formaldehyde and from H_2 and methanol is not [46]. A sodium proton antiporter is operative in these bacteria [47,48]. Its function in energy conservation is still unclear [49,50].

Autotrophic methanogens reduce CO_2 with H_2 to CO in an energy dependent reaction [51]. The bacteria can convert CO to CO_2 and H_2 and couple this reaction with the phosphorylation of ADP. A CO/CH_4 stoichiometry of 0.1 has been measured. Coupling appears to proceed via a chemiosmotic mechanism [52]. Similar conclusions have been drawn for the energy dependent interconversion of CO_2 and CO in *Acetobacterium woodii* [53].

Methanogenic bacteria contain a membrane associated ATP synthase/ATPase, which has not yet been well characterized [see 54,55]. The enzyme activity in *M. barkeri* and in *M. hungatei* is sensitive towards DCCD, whereas it is not in *M. thermoautotrophicum* and *M. voltae* [36,52,56,57]. Available information on the membrane associated ATPase in *Halobacterium* [58] and in *Sulfolobus* [59,60] suggests that the archaebacterial enzyme complex differs considerably from that found in energy transducing membranes of mitochondria, chloroplasts, and eubacteria.

The mechanism of energy conservation in anaerobic eubacteria that grow at the expense of CO_2 reduction to acetic acid has not yet been unravelled [see 53,61,62]. *Clostridium thermoaceticum* contains a proton translocating ATPase [63,64] which resembles that of *Escherichia coli.*

An interesting finding is that *Methanobacterium* species contain a membrane integrated corrinoid protein [65] rather than cytochromes [66]. The corrinoid protein may be involved as electron donor in methyl-CoM reduction to methane [67], which is the energy yielding step in methanogenesis. The methyl-CoM reductase has recently been shown to be membrane associated in *M. voltae* [68] and in *M. thermoautotrophicum* (H. Aldrich et al., unpublished results) by immunocytochemical techniques.

3.3. Growth yields

Y_{CH4} of methanogenic bacteria growing on H_2 and CO_2 have been shown to increase significantly when the growth rate is limited by the supply of H_2 [69,70]. Methanogenesis and phosphorylation of ADP appear to uncouple at high H_2-concentrations. Experiments with non-growing cells support this interpretation. It was recently found that ADP phosphorylation proceeds with an apparent ATP/CH_4 stoichiometry of 0.2 - 0.3 when the H_2 concentration is low (0.2% H_2, 20% CO_2, 80% N_2 in the gas phase) whereas the stoichiometry is only < 0.01 when the H_2 concentration is high (80% H_2, 20% CO_2 in the gas phase) (P. Schönheit and B. Kaesler, unpublished results).

References

[1] Thauer RK, Jungerman K, Decker K. (1977) Bacteriol. Rev. 41, 100-180
[2] Thauer RK, Morris G. (1984) In: The Microbe 1984: Part II Prokaryotes and Eukaryotes (Eds. Kelly DP, Carr NG), pp. 123-168, Cambridge University Press, Cambridge.

[3] Anthony C. (1982) The Biochemistry of Methylotrophs, Academic Press, London.
[4] Wikström M, Krab K (1980) Curr.Top.Bioenerg. 10, 51-101
[5] Wikström M, Krab K, Saraste M (1981) Annu.Rev.Biochem. 50, 623-655
[6] Wikström,M (1984) FEBS Lett. 169, 300-304
[7] Wikström,M, Casey,R (1985) FEBS Lett. 183, 293-298
[8] Mitchell,P, Mitchell,R, Moody,AJ, West,IC, Baum,H, Wrigglesworth,JM (1985) FEBS Lett. 188, 1-7
[9] Williams,RJP (1961) J.Theor.Biology 1, 1-13
[10] Ferguson,SJ (1985) Biochim.Biophys.Acta 811, 47-95
[11] Laszlo,JA, Baker,GM, Dilley,RA (1984) J.Bioenerg.Biomembr. 16, 37-51
[12] Kashket,ER (1985) Annu.Rev.Microbiol. 39, 219-242
[13] McCarthy,JEG, Ferguson,SJ (1983) Eur.J.Biochem. 132, 417-424
[14] van Verseveld,HW, Krab,K, Stouthamer,AH (1981) Biochim.Biophys.Acta 635, 525-534
[15] Dawson,MJ, Jones,CW (1981) Eur.J.Biochem. 118, 113-118
[16] Carver,MA, Jones,CW (1984) Arch.Microbiol. 139, 76-82
[17] Ludwig,B, Schatz,G (1980) Proc.Natl.Acad.Sci.USA 77, 196-200
[18] Froud,SJ, Anthony,C (1984) J.Gen.Microbiol. 130, 2201-2212
[19] Fukumori,Y, Nakayama,K, Yamanaka,T (1985) J.Biochem. 98, 493-499
[20] Solioz,M, Carafoli,E, Ludwig,B (1981) J.Biol.Chem. 257, 1579-1582
[21] Berry,EA, Trumpower,BL (1985) J.Biol.Chem. 260, 2458-2467
[22] Lolkema,JS, Hellingwerf,KJ, Konings,WN (1982) Biochim.Biophys.Acta 681, 85-94
[23] van Verseveld,HW, Braster,M, Kashket,ER, Stouthamer,AH (1986) FEBS Lett. 202, 319-342
[24] van Verseveld,HW, Arbige,M, Chesbro,WR (1984) Tr.Biotechnol. 2, 8-12
[25] Erickson,LE, Minkevich,IG, Eroshin,VK (1978) Biotechnol.Bioengin. 20, 1595-1621
[26] Roels,JA (1980) Biotechnol.Bioengin. 22, 2457-2514
[27] van Verseveld,HW, de Hollander,JA, Frankena,J, Braster,M, Leeuwerik,FJ, Stouhamer,AH (1986) Ant.van Leeuwenhoek 52, 325-342
[28] Stouthamer,AH (1984) In: Proceedings of the Third European Congress on Biotechnology Vol.4, 223-239
[29] Stouthamer,AH (1979) In: Microbial Biochemistry (Ed.Quayle,JR), pp 1-47, University Park Press, Baltimore, USA
[30] Heijnen,JJ, Roels,JA, Stouthamer,AH (1979) Biotechnol.Bioengin. 21, 2175-2201
[31] Boogerd,FC (1984) PhD Thesis, Vrije Universiteit, Amsterdam, The Netherlands
[32] van Verseveld,HW, Stouthamer,AH (1978) Arch.Microbiol. 118, 13-20
[33] van Verseveld,HW, Stouthamer,AH (1978) Arch.Microbiol. 118, 21-26
[34] van Verseveld,HW, Boon,JP, Stouthamer,AH (1979) Arch.Microbiol. 121, 213-223
[35] Daniels L, Sparling R, Sprott D. (1984) Biochim. Biophys. Acta 768, 113-163
[36] Blaut M, Gottschalk G. (1984) Eur. J. Biochem. 141, 217-222
[37] Blaut M, Gottschalk G. (1985) Tr. Biochem. Sc. 10, 486-489
[38] Schönheit P, Beimborn D. (1985) Eur. J. Biochem. 148, 545-550
[39] Crider BP, Carper SW, Lancaster Jr JR. (1985) Proc. Natl. Acad. Sci. 82, 6793-6796
[40] Sprott GD, Shaw KM, Jarrell KF. (1985) J. Biol. Chem. 260, 9244-9250
[41] Lancaster Jr JR. (1986) FEBS Lett. 199, 12-18
[42] Mountfort DO, Mörschel E, Beimborn D, Schönheit P. (1986) J. Bacteriol. in press
[43] Krämer R, Schönheit P. (1986) Arch. Microbiol. submitted
[44] Perski HJ, Moll J, Thauer RK. (1981) Arch. Microbiol. 130, 319-321
[45] Perski HJ, Schönheit P, Thauer RK. (1982) FEBS Lett. 143, 323-326
[46] Blaut M, Müller V, Fiebig K, Gottschalk G. (1985) J. Bacteriol. 164, 95-101.
[47] Schönheit P, Beimborn D. (1985) Arch. Microbiol. 142, 354-361
[48] Schönheit P, Beimborn D. (1986) Arch. Microbiol. in press
[49] Schönheit P, Perski HJ. (1983) FEMS Microbiol. Lett. 20, 263-267
[50] Carper SW, Lancaster Jr JR. (1986) FEBS Lett. 200, 177-180
[51] Eikmanns B, Fuchs G, Thauer RK (1985) Eur. J. Biochem. 146, 149-154
[52] Bott M, Eikmanns B, Thauer RK (1986) Eur. J. Biochem. in press
[53] Diekert G, Schrader E, Harder W. (1986) Arch. Microbiol. 144, 386-392
[54] Doddema HJ, Hutten TJ, van der Drift C, Vogels GD. (1978) J. Bacteriol. 136, 19-23
[55] Roth R, Duft R, Binder A, Bachofen R. (1986) System. Appl. Microbiol. 7, 346-348
[56] Sprott GD, Jarrell KF. (1982) Can. J. Microbiol. 28, 982-986
[57] Mountfort DG. (1978) Biochem. Biophys. Res. Commun. 85, 1346-1351
[58] Kristjansson H, Hochstein LI. (1985) Arch. Biochem. Biophys 241, 590-595
[59] Wakagi T, Oshima T. (1985) Biochim. Biophys. Acta 817, 33-41
[60] Lübben M, Anemüller S, Schäfer G. (1986) in: Archaebacteria '85 (Eds. Kandler O, Zillig W) Gustav Fischer Verlag, Stuttgart, New York.

[61] Wood HG, Ragsdale SW, Pezacka E. (1986) Tr. Biochem. Sc. 11, 14-18
[62] Fuchs G. (1986) FEMS Microbiol. Rev. , in press.
[63] Ivey DM, Ljungdahl LG. (1986) J. Bacteriol. 165, 252-257
[64] Mayer F, Ivey DM, Ljungdahl LG. (1986) J. Bacteriol. 166, 1128-1130
[65] Schulz H, Fuchs G. (1986) FEBS Lett. 198, 279-282
[66] Kühn W, Fiebig K, Hipper H, Mah RA, Huser BA, Gottschalk G. (1983) FEMS Microbiol. Lett. 20, 407-410
[67] Ankel-Fuchs D, Thauer RK. (1986) Eur. J. Biochem. 156, 171-177
[68] Ossmer R, Mund T, Hartzell PL, Konheiser U, Kohring GW, Klein A, Wolfe RS, Gottschalk G, Mayer F. (1986) Proc. Natl. Acad. Sci. in press.
[69] Schönheit P, Moll J, Thauer RK. (1980) Arch. Microbiol. 127, 59-65
[70] Fardeau M-L, Belaich J-P. (1986) Arch. Microbiol. 144, 381-385

ENERGY METABOLISM IN CHEMOLITHOTROPHS

D.P. KELLY, Julie MASON and Ann P. WOOD

Department of Biological Sciences, University of Warwick, Coventry CV4 7AL, England

1. INTRODUCTION

The chemolithotrophic bacteria obtain energy from the oxidation of inorganic substrates. The best known are those which oxidize hydrogen, ammonia, nitrite, ferrous iron, and sulphur and its reduced compounds. Other oxidations probably yielding metabolic energy to some bacteria include U(IV) to U(VI), Cu(I) to Cu(II) and those of some selenium and antimony compounds.

Many chemolithotrophs can also be autotrophs, using CO_2 as the source of all the carbon required for growth. In many cases, however, they are facultative heterotrophs, as is seen for the immensely diverse hydrogen-oxidizing bacteria, and some of the sulphur- and iron-oxidizing bacteria. Others can exhibit chemolithotrophic heterotrophy or mixotrophy, in which energy, carbon or both may be obtained from inorganic and organic sources and carbon dioxide.

The boundary between previously distinct groups of chemolithotrophs has become blurred as more work has been done on their metabolic capacity and more new types have been isolated. Examples of this are seen in the fact that the ammonia-oxidizers are able to oxidize C_1-compounds such as methane, methanol and methylamine, formate is oxidized by *Nitrobacter,* and that the methanotrophs are themselves capable of ammonia oxidation [1,2]. The extreme thermophile *Sulfolobus ambivalens,* which uses aerobic sulphur oxidation as a source of energy is also able to grow by generating energy from hydrogen-dependent sulphur reduction [3], thereby blurring the classical division of dissimilatory sulphur metabolism into reductive and oxidative processes unique to different types of bacteria. Some of the aerobic hydrogen-oxidizing chemolithotrophs have been shown to be able to obtain energy to support growth using the oxidation of inorganic sulphur compounds [4], an attribute previously ascribed to the specialized groups such as the thiobacilli, *Sulfolobus* and some of the filamentous sulphur bacteria. The growth yields produced by some of these are quite comparable to those seen with thiobacilli, as is the aerobic chemolithotrophic growth in the dark of some normally anaerobic phototrophic bacteria, when grown with inorganic sulphur compounds as the sole source of energy [5,6].

The blurring of the distinctions between the chemolithotrophs is itself paralleled by the demonstration that some sulphur-oxidizing thiobacilli are also able to grow as methylotrophic autotrophs, using the oxidation of methanol, methylamine, formaldehyde or formate as sources of energy [7]. This methylotrophy/chemolithotrophy boundary has most recently been crossed by a strain of *Thiobacillus thioparus* that grows autotrophically on dimethyl sulphide, apparently obtaining energy from the oxidation of the one-carbon units as well as the sulphide in the molecule [8]. Comparable levels of ribulose bisphosphate carboxylase have been found in this organism during growth on either thiosulphate or dimethyl sulphide (Neil Smith and D.P. Kelly, unpublished data).

2. ELECTRON TRANSPORT AND ENERGY-CONSERVING SYSTEMS IN CHEMOLITHOTROPHS

Apart from the oxidation of hydrogen, all the oxidations used by known chemolithotrophs have electrode potentials less electronegative than the $NAD^+/NADH + H^+$ couple. In all cases this means that energy-dependent NAD^+ reduction, at the expense of reduced cytochrome *c* oxidation has to occur, and in *Nitrobacter* even the initial reduction of cytochrome *c* is not a thermodynamically favourable process [1,2,9].

Considerable progress has been made in recent years in identifying the electron transport chain components of chemolithotrophs, and energy conservation by means of essentially similar chemiosmotic mechanisms, dependent on establishing transmembrane proton gradients and maintaining a protonmotive force, have been demonstrated for organisms obtaining energy from ammonia, nitrite, thiosulphate or ferrous iron [1,2,9-14].

In some of the sulphur- and nitrogen-oxidizing chemolithotrophs, numerous cytochromes are found [2,13,15]. Some of these are unusual, such as the high molecular weight cytochrome c_{551} and $c_{552.5}$ found in *Thiobacillus versutus*: these also each have two redox centres of widely different midpoint potential, the physiological significance of which is uncertain [15]. In this organism, these cytochromes form part of a multienzyme complex that catalyzes the oxidation of thiosulphate to sulphate without the detectable production of free intermediates. This system is the most thoroughly known mechanism so far elucidated in the thiobacilli [16-18], and has now been shown to be located periplasmically in this organism thus showing that this multicomponent complex is part of a 'standard' chemiosmotic energy-conserving system for oxidative phosphorylation.

3. ENERGY METABOLISM OF SULPHUR-OXIDIZING CHEMOLITHOTROPHS

These bacteria come from diverse bacterial groups and represent a biochemically complex problem in energy relations. In recent years it is the sulphur-chemolithotrophs that have received the most intense study of their growth energetics and with which we have worked at Warwick. The remainder of this article will be devoted to a summary of these recent studies.

Most work has been done on members of the *Thiobacillus* genus, but other genera are gradually being studied. These include *Thiomicrospira, Thiosphaera, Sulfolobus, Thermothrix* and the filamentous bacteria such as *Beggiatoa*. I reviewed some aspects of sulphur-dependent energy generation in 1982 [18] and shall now deal mainly with data obtained since that time.

It is now well established that most, if not all, sulphur chemolithotrophs generate energy by oxidative phosphorylation and that some also obtain ATP from substrate-level processes with adenylyl sulphate (APS) as an intermediate. Many of them fix CO_2 by the Calvin Cycle, but *Sulfolobus* almost certainly uses a 'reverse carboxylic acid cycle' similar to that seen in *Chlorobium*. It is probable that the latter process is less energy-demanding than the Calvin Cycle, although the energy needed for the generation of reductant for the carboxylation of succinyl CoA and acetyl CoA in *Sulfolobus* is not yet known. Certainly in the case of the Calvin Cycle sulphur-chemolithotrophs up to 80% of the energy from the sulphur-oxidations is expended for only two processes: the energy-dependent reduction of NAD^+ and the ATP-demand of the Calvin Cycle.

The uncertainties of the relative contribution of oxidative and substrate level phosphorylation for ATP synthesis, the uncertainty of the point of entry of electrons to the respiratory chain from all steps in the sulphur-oxidation pathway, the uncertainty of the nature of the intermediates and electron-generating reactions of the pathway itself and of the amount of energy required to drive NAD^+ reduction, all combine to make direct biochemical analysis of the energy metabolism of the sulphur bacteria very difficult. Our approach has been to study the overall energy-conserving process in terms of its quantifiable expression as fixation of CO_2 and the production of biomass during growth on precise amounts of known substrates. This, coupled with comparison of the free energy efficiencies of the growth processes, enables comparison of the apparent energy conservation exhibited by different bacteria, and allows some assessment of the total ATP production as well as the relative importance of oxidative and substrate-level phosphorylation processes.

4. SOME OBSERVATIONS ON CHEMOSTAT GROWTH YIELDS OF THIOBACILLI AND OF *THERMOTHRIX THIOPARA*.

All the sulphur chemolithotrophs grown under energy limitation in chemostat culture, with the oxidizable sulphur compound as the limiting nutrient, show an increase in yield (biomass produced per mol of energy-substrate oxidized) with increasing growth rate (=dilution rate, D), as was described for some heterotrophs by Pirt (20). Graphical plots of the reciprocals of the yields and dilution rates (1/Y x 1/D) or of the specific rates of substrate oxidation (mmol sulphur substrate oxidized/ g of organisms/hour) against D for steady state culture conditions (q x D) each enable the calculation of theoretical maximum (or 'true') growth yield (Y_{max}) values [17,20,21). The q x D plot provides a more accurate assessment of Y_{max} [22] and is illustrated in Fig. 1 and 2. By the q x D plot the Y_{max} is given by the slope of the line and it is immediately evident that there is considerable variation among different organisms using the same substrate. It is also evident that some species exhibit very similar Y_{max} values, even though the pH values for their optimum growth may be five units apart, as in the case of *Thiobacillus versutus* and *T. ferrooxidans* (Fig.1). Similarly for any one organism the Y_{max} value does not vary greatly over a range of temperature or pH around the optimum (e.g. pH 5.5-7.6 for *T.*

neapolitanus or pH 6.8-8.4 for *T. versutus)* even though observed yield values in the chemostat may be lower at "extreme" temperatures (e.g. 17°C for T. *versutus* or 40°C for *T. neapolitanus*).

Earlier studies on Y_{max} (see Ref 18 for review) suggested that values for the thiobacilli and *Thiomicrospira* were all fairly similar, with the exception of *Thiobacillus denitrificans,* which gave higher yields than the others. Y_{max} for growth on thiosulphate in the data illustrated (Fig. 1 and 2) range from around 8 g dry wt/mol for *T. versutus, T. neapolitanus* and *T. ferrooxidans* (and probably also *T. thiooxidans, T. acidophilus),* through 11 g/mol for *Thiobacillus tepidarius* and the newly isolated

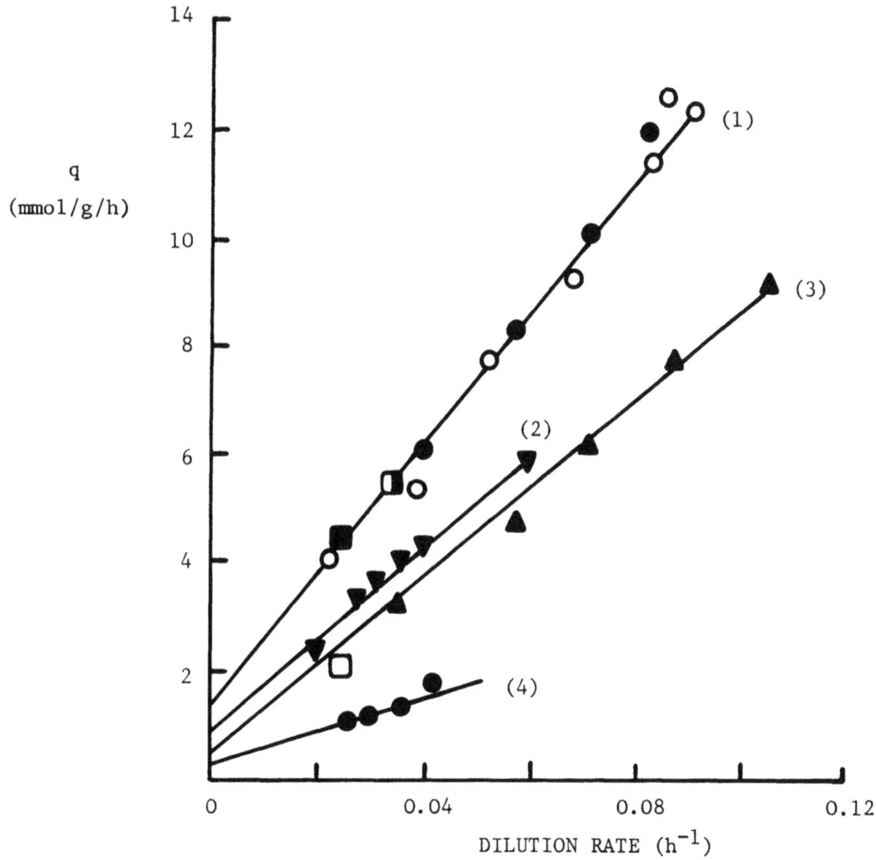

Figure 1. Relationship of dilution rate and specific rate of consumption of energy-limiting substrate in steady state chemostat cultures of some sulphur-chemolithotrophs. Specific oxidation rate (q) is given as nmol sulphur substrate oxidized/h/g dry wt of bacteria in the steady state. Results are shown for (1) *Thiobacillus versutus* (pH 7.6;○), *Thiobacillus ferrooxidans* (pH 2.5;●), *Thiobacillus thiooxidans* (pH 2.5;□) and *Thiobacillus acidophilus* (pH 3.0;■), each limited by thiosulphate; (2) *Thiobacillus ferrooxidans* (pH 2.5) with tetrathionate-limitation; (3) *Thiobacillus thiooxidans* (pH 2.5) with tetrathionate-limitation (▲) and *Thiobacillus acidophil;us* (pH 3.0) with tetrathionate-limitation (□); (4) *Thermothrix thiopara* (pH 7.0, 72°C) with tetrathionate-limitation (●).

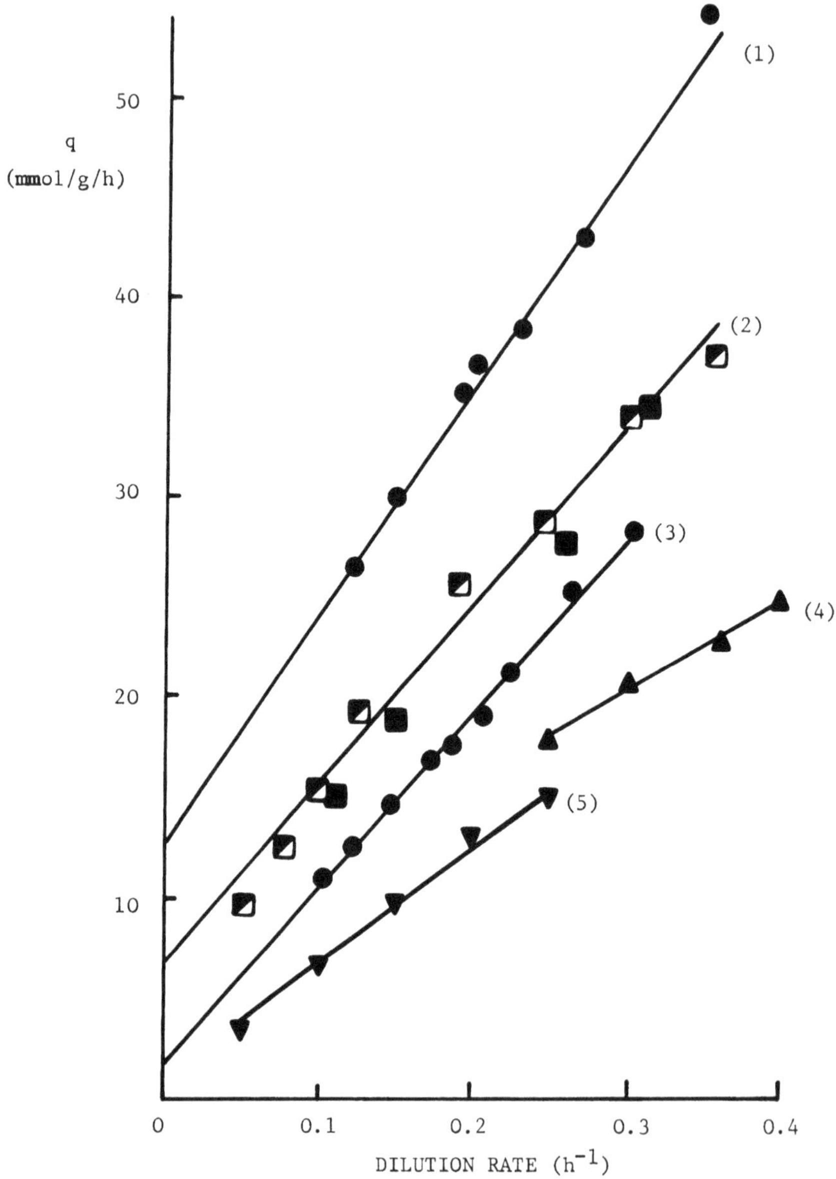

Figure 2. Specific rates of thiosulphate or trithionate oxidation over a range of dilution rates in steady state chemostat cultures of (1) *Thiobacillus neapolitanus* (thiosulphate-limited); (2) *Thiobacillus tepidarius* limitedby thiosulphate (◪) or trithionate (■); (3) *Thiobacillus aquaesulis* (pH 7.6, 43°C) limited by thiosulphate; (4) *Thermothrix thiopara* (pH 7.0, 72°C) limited by thiosulphate; (5) *Thermothrix thiopara* limited by trithionate.

Table 1. 'True growth yields' (Y_{max}) for some sulphur chemolithotrophs grown in aerobic chemostat culture under energy-limitation by the inorganic reduced sulphur respiratory substrate.

Organism	Culture Temp	pH	Y_{max} (g dry wt/g mol S-compound) $Na_2S_2O_3$	$Na_2S_3O_6$	$K_2S_4O_6$
Thiobacillus versutus	30	7.8	7.8	NG[a]	NG
T. neapolitanus	30	7.0	8.6	8.8	15.9
T. thiooxidans	30	2.5	-	-	12.1
T. ferrooxidans	30	2.5	8.2	-	11.9
T. acidophilus	30	3.0	-	-	16.0
T. tepidarius	43	7.0	10.7	11.4	20.5
T. aquaesulis	43	7.6	11.3	-	-
T. denitrificans	30	7.0	14.7	-	(27.7)[b]
Thiomicrospira pelophila	25	6.9	7.0[c]	-	-
Thermothrix thiopara	72	7.0	22.8	17.1	31.5

[a] NG = no growth on this substrate. [b] Value calculated from the $Na_2S_2O_3$ Y_{max} relative to the Y_{max} values of anaerobic nitrate-limited cultures of 11.4 ($S_2O_3^{2-}$) and 21.5 ($S_4O_6^{2-}$). [c] From Ref 18

Table 2. Comparative observed growth yields (g dry wt/g mol sulphur compound oxidized) for several sulphur chemolithotrophs at single growth rates (=dilution rates) in the chemostat with CO_2 as sole carbon source.

Organism	Dilution rate (h^{-1})	Observed yield Na_2S	$Na_2S_2O_3$	$Na_2S_3O_6$	$K_2S_4O_6$
Thiobacillus acidophilus	0.025	-	5.5	7.0	12.5
T. thiooxidans	0.034	-	6.2	-	10.3
T. tepidarius	0.150	6.9	7.4	8.8	14.8
T. aquaesulis	0.110	-	9.6	9.6	18.7
T. neapolitanus	0.050	4.9	5.0	-	-
Thiomicrospira denitrificans[a]	0.030	-	7.7	-	-
Tms. denitrificans (anaerobic + NO^-_3)[a]	0.030	5.9	5.2	-	-
Thermothrix thiopara	0.050	-	15.4	13.8	31.3

[a] From Ref 24. Unless indicated, all cultures were aerobic.

T. acquaesulis (23; Ann Wood and Don Kelly, unpublished) to the extremely high value of 22.8 for *Thermothrix thiopara* growing at 72°C (Julie Mason and Don Kelly, in preparation). Similar relative Y_{max} values are seen for chemostat cultures energy-limited by trithionate or tetrathionate (Fig. 1 and 2; Tables 1 and 2).

Recent work has shown that chemostat cultures of *Thiobacillus versutus, T. tepidarius, T. denitrificans* and *Thiomicrospira denitrificans* (6,23,24; Table 2) each give similar relative growth yields on sulphide or thiosulphate, as would be expected from the fact that both oxidations involve the transfer of eight electrons. Using the newly isolated chemolithotrophic marine strain of *Beggiatoa,* Nelson *et al* have shown its yield in sulphide-gradient cultures to be about 8.4 g dry wt/mol sulphide (25). This indicated the *Beggiatoa* to be similar in chemolithotrophic energy efficiency to the other species discussed here, whose aerobic Y_{max} values on sulphide would be in the range 8 - 15 g/mol.

Values of Y_{max} for different species can be used to estimate gross thermodynamic efficiency and to gain some idea of the production and use of ATP for growth (18). Based on free energy data (e.g. Ref 21), taking the energy required to fix CO_2 to the level of (CH_2O) as 495 kJ/mol and the free energy made available by thiosulphate or tetrathionate oxidation to sulphate as 936 and 1654 kJ/mol respectively, the thiobacilli show a thermodynamic growth efficiency of 14-30%. Differences between yields reflect differences in the contribution of substrate-level phosphorylation to ATP synthesis and may indicate whether electrons from the oxidations enter the respiratory chain at cytochrome *b* or *c*. Comparison of the Y_{max} and observed yields gives some idea of whether substrtate-level phosphorylation occurs at all. In the case of *Thiobacillus neapolitanus,* where substrate-level phosphorylation was suggested to exist (18), the Y_{max} and Y data suggest otherwise (Table 1 and 2). As sulphide, thiosulphate and trithionate are all eight-electron oxidation substrates they should only allow different growth yields if they also support substrate-level phosphorylation, which should then be in the ratio 1:2:3 for the three substrates. As the yields on all three are not significantly different, energy conservation may be entirely chemiosmotic. In contrast, the observed yields for *T. acidophilus* (Table 2) would be consistent with the operation of a substrate-level phosphorylation process.

5. GROWTH YIELDS FOR *THERMOTHRIX THIOPARA* AND *SULFOLOBUS*

T. thiopara was previously reported (26) to give Y_{max} values on thiosulphate (at 65° and 70°) of 41.4 and 24.3 g dry wt/mol respectively. We have not been able to confirm the former value, growing the organism either at 65° or at the optium temperature of 72° at which we found Y_{max} of 18.0 and 22.8 respectively. A value of 41.4 g/mol (at an assumed carbon content of 50% of the dry wt; 26) would represent a thermodynamic efficiency of 91%. We find the carbon content of *T. thiopara* to be only about 29% (from numerous assays of chemostat cultures on all three sulphur substrates the range was 23-33% carbon), thereby reducing the apparent efficiency of a 41.4 g/mol yield to a still unlikely 53%. Our Y_{max} value of 22.8 thus represents a thermodynamic efficiency of 29%, which makes *T. thiopara* comparable in growth efficiency terms to *Thiobacillus denitrificans.*

No Y_{max} values have yet been reported for the extreme thermoacidophile *Sulfolobus*, which we have cultured under wholly autotrophic conditions on tetrathionate as sole energy source at 70°C. Our best estimates of mean autotrophic yield for four strains in batch culture at pH 3.0 (specific growth rate range 0.017-0.028 h^{-1}) are around 7 g dry wt/mol $K_2S_4O_6$ with a carbon content of 42% (A.P. Wood, D.P. Kelly and P.R. Norris, submitted for publication). This is within the range seen in batch culture for some thiobacilli or for *T. ferrooxidans* in the chemostat at D, 0.02 h^{-1} where Y = 8.3, and indicates that *Sulfolobus* is not comparable in yield with the high-yielding thermophile *Thermothrix* (Y = 22.8 at D, 0.026 h^{-1}). Recently a sulphur oxygenase (converting sulphur to sulphite) was purified from *Sulfolobus* (T. Emmel, personal communication 1986), which was proposed to be central to the sulphur oxidation pathway in the organism. The energetic disadvantages of such a non-energy-conserving oxygenase in the thiobacilli were discussed previously (18). If such an oxygenase operates in the *Sulfolobus* for the conversion of the sulphane groups of tetrathionate as well as of elemental sulphur to sulphite, a considerable proportion of the free energy theoretically available from the conversion of $S_4O_6^{2-}$ to 4 SO_4^{2-} could not be metabolically conserved. Regarding tetrathionate as approximately equal thermodynamically to two sulphur and two sulphite moieties, an oxygenase reaction would effectively convert $S_4O_6^{2-}$ to 4 SO_3^{2-} without conserving any energy as ATP. The oxidation of the four sulphite ions to sulphate makes available 1008 kJ overall, or 61% of the total free energy released in the conversion of tetrathionate to sulphate (1654 kJ/mol). Thus about 39% of the energy of the reaction could not be used for growth by *Sulfolobus*. It remains to be seen from more detailed growth and biochemical studies whether any common oxidation mechanisms exist in the thiobacilli, *Thermothrix* and the archaebacterium *Sulfolobus*.

6. CONCLUSION

Chemostat culture is proving a powerful tool for obtaining a direct estimate of relative amounts of energy actually retained for growth on various inorganic substrates by diverse sulphur-chemolithotrophs. The data it gives are of value in themselves and as indicators of the kind of biochemical study that needs to be pursued subsequently in order to test the predictions that follow from growth studies.

7. ACKNOWLEDGEMENTS

The original work included in this article was supported by the Natural Environment Research Council, Science and Engineering Research Council and Biogen S.A. (Geneva).

REFERENCES

[1]. Wood PM (1986) In: Nitrification (Ed Prosser J.), in press. Special publication of the Society for General Microbiology.

[2]. Suzuki I (1984) In: Microbial growth on C_1 compounds (Ed Crawford RL, Hanson RS), 42-48. Washington, American Society for Microbiology.
[3]. Segerer A, Stetter KO, Klink F (1985) Nature 313, 787-789.
[4]. Friedrich C, Mitrenga G (1981) FEMS Microbiol Lett 10, 209-212.
[5]. Kämpf C, Pfennig N (1980) Arch Microbiol 127, 125-137.
[6]. Kelly DP, Kuenen JG (1984) In: Aspects of microbial metabolism and ecology (Ed Codd GA), 211-240. London, Academic Press.
[7]. Kelly DP, Wood AP (1984) In: Microbial growth on C_1 compounds (Ed Crawford RL, Hanson RS), 324-329. Washington, American Society for Microbiology.
[8]. Kanagawa T, Kelly DP (1986) FEMS Microbiol Lett 34, 13-19.
[9]. Aleem MIH, Sewell DL (1984) In: Microbial chemoautotrophy (Ed Strohl WR, Tuovinen OH), 185-210. Columbus, Ohio State University Press.
[10]. Hooper AB (1984) *ibid* (Ref 9) pp 133-167.
[11]. Cox JC, Brand MD (1984) *ibid* (Ref 9) pp 31-46.
[12]. Hooper AB, DiSpirito AA, Olson TC, Andersson KK, Cunningham W, Taaffe LR. (1984) In: Microbial growth on C_1 compounds (Ed Crawford RL, Hanson RS), 53-58. Washington, American Society for Microbiology.
[13]. Ingledew WJ (1982) Biochim Biophys Acta 683, 89-117.
[14]. Wheelis M (1984) Arch Microbiol 138, 166-169.
[15]. Lu WP, Poole RK, Kelly DP (1984) Biochim Biophys Acta 767, 326-334.
[16]. Lu WP, Swoboda BEP, Kelly DP (1985) Biochim Biophys Acta 828, 116-122.
[17]. Kelly DP (1985) Microbiol Sci 2, 105-109.
[18]. Kelly DP (1982) Phil Trans R. Soc Lond B 298, 499-528
[19]. Lu WP (1986) FEMS Microbiol Lett 34, 313-317.
[20]. Pirt SJ (1965) Proc Roy Soc Lond B 163, 224-231.
[21]. Kelly DP (1978) In: Companion to microbiology (Ed Bull AT, Meadow PM),363-386.
[22]. Fieschko J, Humphrey AE(1984) Biotechnol Bioeng 26, 394-396.
[23]. Wood AP, Kelly DP (1986) Arch Microbiol 144, 71-77.
[24]. Timmer-ten Hoor A (1981) Antonie van Leeuwenhoek 47, 231-243.
[25]. Nelson DC, Jörgensen BB, Revsbech NP (1986) Appl Env Microbiol, in press.
[26]. Brannan DK, Caldwell DE (1983) Appl Env Microbiol 45, 169-173.

ENERGY METABOLISM OF AEROBIC, METHYLOTROPHIC BACTERIA

C. ANTHONY and C.W. JONES

Department of Biochemistry, University of Southampton, Southampton, U.K. and Department of Biochemistry, University of Leicester, Leicester, U.K.

1. INTRODUCTION

This review will be limited to consideration of aerobic electron transport in methylotrophic bacteria growing on methylamine and methanol by way of quinoprotein dehydrogenases, and the synthesis of ATP that is coupled to this process. We will concentrate on those aspects of the subject that have seen the most development since last reviewed [1-5].

The oxidation of methanol and methylamine by methylotrophs is unusual in a number of respects. Both these substrates are C_1 compounds and both are more reduced than cell material. Thus, for every molecule of glucose synthesised (for example), the oxidation of six substrate molecules is required. Furthermore, methanol and methylamine are together unique in being the only organic substrates whose oxidation does not involve the membrane quinones, iron-sulphur proteins and mid-chain *b*-type cytochromes. The dehydrogenases for methanol and methylamine oxidation are both unusual, neither being coupled to NADH production and both being quinoproteins that pass electrons to the electron transport chain at a relatively high redox potential. One consequence of this is that the percentage of electron flow to oxygen during growth on methanol and methylamine from NADH or flavoproteins is very low (0 - 30%) compared with that during oxidation of conventional substrates (100%) [1]. Between 50% and 70% of electron flow to oxygen is always by way of the electron transport chains from methanol or methylamine, and what is special about energy metabolism in methylotrophic bacteria is the nature of the dehydrogenases for initial oxidation of these growth substrates, their electron acceptors and their relationships to the oxidases and ATP synthesis.

2. RESPIRATION

2.1. Respiratory chain components

There is no evidence to suggest that the electron transport chain for oxidation of NADH in methylotrophs is different in composition and function to those in other bacteria; they contain ubiquinone, iron-sulphur proteins and mid-chain *b*-type cytochromes. All methylotrophs having the quinoprotein dehydrogenases for oxidation of methanol and methylamine contain high concentrations of two different *c*-type cytochromes and an oxidase which is either cytochrome aa_3 or an *o*-type oxidase, or they may be able to synthesise both, the growth conditions determining which is predominant (see below).

2.1.1. Soluble *c*-type cytochromes.

The two types of soluble cytochrome *c* found in most methylotrophs have a single polypeptide chain which bears a single haem, having histidine and methionine as the two axial ligands. They both have a high midpoint redox potential, absorption maxima at about 550nm and are low spin in both oxidised and reduced states. They are labelled according to their isoelectric points [see ref. 2 for a recent review of these cytochromes].

Cytochrome c_H is smaller, has a higher pI, higher midpoint redox potential, only two cysteine residues (binding the haem) and is relatively less sensitive to proteolytic digestion. Preliminary work on the amino acid sequence of this cytochrome indicates that it is similar to other soluble *c*-type cytochromes that mediate electron transfer between the cytochrome *bc* complex and cytochrome oxidase [R.P. Ambler, personal communication]. It is always periplasmic and easily dissociates from membranes.

Cytochrome c_L is an unusual cytochrome *c* that is probably unique to methylotrophs. It differs from cytochrome c_H in being much larger, with a lower pI and redox potential. It has an extra cysteine residue and is relatively more sensitive to proteolytic digestion. Although it is also located in the periplasmic fraction a considerable amount is not so easily removed from membranes; after removal, this cytochrome c_L appears to be identical to the readily-soluble cytochrome. As there is no identifiable cytochrome c_1 in methylotrophs it is possible that the firmly bound cytochrome c_L fulfils this function. Cytochrome c_L is the electron acceptor for methanol dehydrogenase and the special role of this cytochrome in methanol oxidation has been confirmed recently by Nunn and Lidstrom [6,7] who have shown that mutants lacking this cytochrome (but containing cytochrome c_H) are unaltered except in their ability to oxidise methanol. The cytochrome c_L is probably regulated together with methanol dehydrogenase in the same operon in *Pseudomonas* AM1 [D.N. Nunn and C. Anthony, unpublished results].

Besides its large size and its function as electron acceptor for methanol dehydrogenase, cytochrome c_L is unusual in being rapidly autoreduced at high pH values. When reacting with methanol dehydrogenase the pK for this autoreduction is diminished so that it occurs at the pH value found in the periplasm (pH 7). Whether or not the autoreduction process is involved in electron transfer between the dehydrogenase and the cytochrome is not yet certain but some recent evidence suggests that it is. *Acetobacter methanolicus* grows on methanol at pH 4.0 which will also be the pH value of its periplasm. If the autoreduction phenomenon is important in catalysis then it might be expected that it would occur at a much lower pH in this acidophile than in a typical methylotroph. This is indeed the case. Autoreduction occurs in the absence of methanol dehydrogenase at about pH 7 but it occurs at pH 4 in the presence of methanol dehydrogenase [E.J. Elliott and C. Anthony, unpublished results].

2.1.2. Blue copper proteins (cupredoxins).

These are small, water soluble proteins that contain one copper atom per polypeptide chain, the ligands to the copper being two histidines, one cysteine and one methionine residue. It has been suggested that these proteins should be called cupredoxins by analogy with the iron-sulphur proteins, the ferredoxins [8]. The cupredoxins act, like cytochrome *c*, as soluble carriers in electron transport having redox potentials that are similar to those of the high potential cytochromes *c* with which they are able to react. Amino acid sequence data suggest four classes but these do not necessarily correlate with any known function of the proteins [9]. The proposed classes are plastocyanin, pseudoazarin, amicyanin and azurin, which differs from the other three in having a disulphide bridge.

At least two cupredoxins occur in methylotrophic bacteria (Table 1). The only one with a known function was first isolated from *Pseudomonas* AM1 and was called amicyanin [5,10], and this acts as the electron acceptor for methylamine dehydrogenase. Unfortunately (in terms of ease of discussion) the cupredoxin which functions as electron acceptor for methylamine dehydrogenase in *Methylomonas* J (an obligate methylotroph) falls into the azurin sequence class, and it has been suggested that the name amicyanin should be reserved for cupredoxins in the same sequence class as amicyanin from the facultative methylotroph *Pseudomonas* AM1 [9]. Although logical, this poses problems because the amino acid sequence of a protein is necessarily one of the last aspects of a new protein to be characterised. To avoid confusion we shall avoid the issue and refer (when in doubt) to 'amicyanin' for cupredoxins that react with methylamine dehydrogenases and 'azurin' for other cupredoxins.

Table 1. THE BLUE COPPER PROTEINS (CUPREDOXINS) OF METHYLOTROPHS

Organism	Mol. Wt.	pl	Em_7	λ_{max} (mM^{-1} cm^{-1})
	'AMICYANIN'			
Pseudomanas AM1 [5,9,10]	11,700	9.3	280mV	596nm (4.5)
Methylomonas J [5,10]	-	7.7	-	613nm
Organism 4025 [11,12]	11,500	5.3	294mV	620nm (3.8)
Paracoccus denitrificans [13,14]	15,000	4.8	-	595nm (4.6)
Thiobacillus versutus [37]	13,800	4.7	261 mV	596nm (4.0)
	'AZURIN'			
Pseudomonas AM1 [5,9,10]	13,000	9.4	310mV	593nm
Methylomonas J [5,10]	-	9.6	-	620nm
Organism 4025 [11,12]	12,500	9.4	323mV	625nm (16.0)
Paracoccus denitrificans [15]	13,800	4.6	230mV	595nm (1.5)

NB: The proteins are grouped here according to their function; 'amicyanins' are electron acceptors for methylamine dehydrogenase and 'azurins' are other cupredoxins also found in these methylotrophs. On the basis of sequence data [9], it has been suggested that the 'azurin' from *Pseudomonas* AM1 should be called pseudoazurin; and that the 'amicyanin' from *methylomonas* J is a true azurin.

Table 1 summarises the properties of known cupredoxins from methylotrophs. Like the soluble *c*-type cytochromes, the cupredoxins of methylotrophs are located in the periplasm [12,13].

2.1.3. Cytochrome oxidases.

Methylotrophic bacteria produce either cytochrome aa_3 or an *o*-type oxidase called cytochrome *co*.In *Pseudomonas* AM1 the predominant oxidase is cytochrome aa_3; this has been completely purified and shown to be similar to other oxidases of this type [16]. *Methylophilus methylotrophus* is able to synthesise two oxidases. Cytochrome aa_3 is only observed in bacteria grown under carbon-limited conditions, whereas the cytochrome *co* is always present, but is induced 10-fold during growth in carbon-excess conditions [17,18]. This oxidase has been solubilised and completely purified, and shown to consist of two identical cytochrome *b* (or cytochrome *o*) subunits plus two identical cytochrome *c* subunits. The cytochrome *c* is completely distinct (although similar in size) from the soluble *c*-type cytochromes of this organism [19,20, H.M. Ashworth and C. Anthony, unpublished results]. Although first purified and characterised in a methylotroph it is probably similar to the *o*-type oxidases found in other bacteria able to form *c*-type cytochromes.

2.2. Interaction of electron transport components

The dehydrogenases for methanol and methylamine both occur in the periplasm or loosely attached to the outer surface of the periplasmic membrane. This is also the location of the soluble *c*-type cytochromes and the cupredoxins which are all able to react rapidly with each other. Specificity of interaction is conferred by the dehydrogenases and, to some extent, the oxidases. Methanol dehydrogenase interacts specifically with cytochrome c_L, but the rate of electron transfer with the pure proteins is (inexplicably) very low [3,21].

By contrast, methylamine dehydrogenase is able to react very rapidly with amicyanin, its specific electron acceptor, whose synthesis is induced during growth on methylamine or other methylated amines [5,11-14]. The amount of 'amicyanin' present varies markedly from one methylotroph to another, and in some bacteria (e.g. *Methylophilus methylotrophus)* it is very difficult to detect. In such organisms the 'amicyanin' may be replaced by cytochromes. The second site of specificity in the electron transport chains for methanol and methylamine is in the oxidases. Both the solubilised and purified cytochrome aa_3 from *Pseudomonas* AM1 [16] and cytochrome *co* from *M. methylotrophus* [19] react specifically with the small cytochrome c_H. Although cytochrome *co* does bind this cytochrome and cytochrome c_L with similar affinities, it oxidises cytochrome c_H 50 times faster than cytochrome c_L, and this is also observed with cytochrome *co* reconstituted into lipid vesicles [H.M. Ashworth and C. Anthony, unpublished results]. The observation that 'azurin' is also oxidised by the cytochrome *co* in lipid vesicles raises the possibility that electron transport during the oxidation of methylamine may be independent of soluble cytochrome *c*. Possible routes for electron flow from methanol and methylamine in a 'generalised' methylotroph is shown below (the dashed lines indicate reactions that can occur, although they may not be physiologically important):

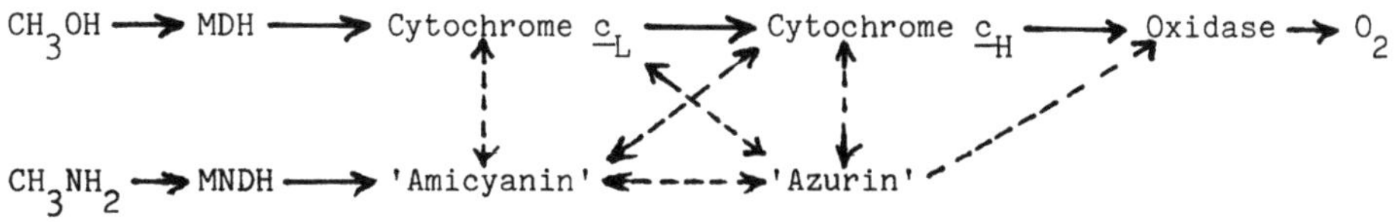

2.3. Environmental regulation of synthesis of 'methanol oxidase' of Methylophilus methylotrophus

In this obligate methylotroph, the concentration of methanol dehydrogenase is maximal during growth in continuous culture at very low concentrations of methanol [18], thus indicating that its synthesis is repressed by either methanol or a product of its catabolism. This may be formaldehyde because it is also repressed during batch growth on methylamine or trimethylamine. The cytochrome oxidase aa_3 is also derepressed under methanol-limited growth conditions. By contrast, cytochrome oxidase *co* is derepressed when oxygen is limiting, thus largely replacing cytochrome aa_3 during oxygen-limited growth. Neither of the two cytochromes *c* is markedly affected by the nature of the nutrient (carbon or oxygen) limitation, but both cytochromes decrease in concentration as a function of growth rate [20,22]. The methanol oxidase system is thus precisely tuned to the concentrations of methanol and oxygen in the culture such that the latter is able to maintain *in situ* rates of respiration and methanol consumption that satisfy the energy demands of the cell and increase linearly as a function of the imposed growth rate.

The addition of saturating concentrations of methanol to whole cells prepared from methanol-limited cultures leads to a substantial overflow of formaldehyde, since the rate of production of the latter under these conditions exceeds the capacity of the ribulose monophosphate cycle to remove it [J.A. Greenwood, S.M. Burton and C.W. Jones, unpublished]. ^{13}NMR spectroscopy indicates that the spillage of formadehyde is accompanied by the overflow of dihydroxyacetone [H. Santos, D.L. Turner and C.W. Jones, unpublished], which arises from the pyruvate produced by the assimilatory cycle. As formaldehyde and dihydroxyacetone are both non-phosphorylated and have identical oxidation states, they may serve as alternative overflow products. Spillage is not observed with cells prepared from methanol-excess cultures, since the methanol dehydrogenase is repressed and the rate of methanol oxidation is therefore relatively slow.

2.4. Spatial organisation of respiratory chains

Methanol and methylamine dehydrogenases, the cupredoxins azurin and amicyanin, cytochrome c_H and some of the cytochrome c_L are either in the periplasm or are loosely attached to the external surface of the periplasmic membrane [12,14,23-29]. The extent to which these hydrophilic redox carriers are bound to the membrane is probably determined by their concentrations, since a higher proportion of the total methanol dehydrogenase is unbound following growth under methanol-limitation compared with oxygen-limitation, presumably because the amount of enzyme grossly exceeds the number of available binding sites.

By contrast, the hydrophobic cytochrome oxidases a_3 and *co* are firmly membrane-bound. Little is known about the arrangement of cytochrome oxidase aa_3 in the membrane although, by analogy with higher organisms, it might be expected to occupy a transmembrane orientaion. Radiolabelling studies have indicated that cytochromes *c* and *o* are both partially exposed on the periplasmic side of the membrane (J.A. Greenwood and C.W. Jones, unpublished), but as this technique does not unambiguously detect the exposure of proteins on the cytoplasmic side, the possibility that cytochrome oxidase *co* is also a transmembrane protein cannot be discounted. Only when the exact sites of oxygen reduction by the two cytochrome oxidases have been determined will it be clear whether overall electron transfer from methanol or methylamine to oxygen is a scalar or an inwardly-directed vectorial process.

3. ENERGY TRANSDUCTION

3.1. Proton translocation

Respiration-linked proton translocation has been most extensively investigated in whole cells of *M. methylotrophus* using oxygen-pulse and initial-rate techniques to measure both $\rightarrow H^+/0$ and $\rightarrow$charge/0($\rightarrow K^+0$) quotients. Three proton- and/or charge-translocation sites have been identified in cells grown under methanol-limited conditions: site 1 (NADH-Q reductase; $\rightarrow H^+/2e^- = 2$ and $\rightarrow$charge/$2e^- = -2$); site 2 (QH_2-cytochrome *c* reductase, $\rightarrow H^+/2e^- = 4$ and $\rightarrow$charge/$2e^- = -2$); and site 3 (cytochrome *c* oxidase; $\rightarrow H^+/2e^- = 0$ and $\rightarrow$charge/$2e^- = -2$). NADH oxidation thus involves all three sites ($\rightarrow H^=/0=6$ and $\rightarrow$charge/0=-6), whereas the oxidation of methanol to formaldehyde involves only site 3 plus methanol dehydrogenase ($\rightarrow H^+/O=2$ and $\rightarrow$charge/O=-2) [30,31]. Similar results have been obtained with cells grown under oxygen-limited conditions, where cytochrome oxidase aa_3 is largely replaced by cytochrome oxidase *co* (M.J. Dawson and C.W. Jones, unpublished).

The protonmotive properties of the methanol oxidase system is most simply explained either by a redox arm mechanism (with the two protolytic reactions on oppisite sides of the membrane and linked by vectorial electron transfer) or by a proton pump mechanism (with the two protolyctic reactions on the same side of the membrane and linked by scalar electron transfer through a proton-pumping oxidase); it has not yet proved possible to distinguish experimentally between these two mechanisms. Interestingly, whole cells of *Paracoccus denitrificans* exhibit an $\rightarrow H^+/0$ quotient of 4 for the oxidation of methanol, a result which indicates a combined redox arm/proton pump mechanism (with the two protolytic reactions linked by vectorial electron transfer through a proton-pumping cytochrome oxidase aa_3) [32]. Methylotrophs may thus contain two different types of cytochrome oxidase aa_3 which differ from each other either in their capacity to pump H^+ or in the sidedness of their protolytic reaction

3.2. ATP SYNTHESIS

Respiratory chain phosphorylation has been examined in detail in *M. methylotrophus* by measuring the changes in the intracellular concentrations of adenine nucleotides and/or inorganic phosphate over the first few seconds following the addition of oxidisable substrate to whole cells grown under methanol-limited conditions. After correcting for the utilisation of nascent ATP to synthesise other nucleoside triphosphates, which is reflected in the failure of the cells to exhibit respiratory control, Pi/0 quotients of approximately 0.5 and 1.4 nmol/ng-atom 0 respectively are obtained for the oxidation of methanol (or related substrates) and NADH generated from formate [33]. Comparably low Pi/0 quotients have also been reported for the oxidation of methanol or NADH by membrane vesicles prepared from *Ps.*AM1 [34]. In both cases the values obtained are proportional to the whole cell $\rightarrow H^+/0$ and/or $\rightarrow$charge/0 quotients and yield an average $\rightarrow H^+$/ATP (= $\rightarrow H^+$/Pi) quotient of approximately 4 ng-ion H^+/nmol.

Neither pyrophosphate synthesis nor reversed electron transfer, both of which could divert energy from ATP synthesis, and thus yield an erroneously low Pi/0 quotient, occurs at a significant rate, although rapid H^+ leakage through the membrane at high values of the protonmotive force cannot be entirely ruled out. Furthermore, a $\rightarrow H^+$/ATP quotient of more than 3 has been determined for the ATP synthases of *P. denitrificans* and *M. methylotrophus* by comparing values of Δp and ΔGp obtained with membrane vesicles and whole cells respectively [35,36]. It must be concluded, therefore, that the oxidation of methanol and endogenous NADH are accompanied by the synthesis of 0.5-0.7 and 1.4-2.0 ATP respectively in *M. methylotrophus* and *Ps.* AM1, and perhaps by significantly higher amounts in *P. denitrificans.*

Energy conservation *in vivo* has also been examined by measuring the steady-state concentrations of adenine nucleotides and inorganic phosphate in whole cells of *M. methylotrophus* [36]. The extent of intracellular energisation with various substrates is proportional to the rate of proton and/or charge translocation and reflects the relative energy conservation properties of sites 1,2 and 3,, outlined above. Interestingly, however, the artificial substrate ascorbate-TMPD elicits substantial ATP synthesis in cells grown under methanol-limited conditions but not in cells from oxygen-limited cultures [J.A. Greenwood and C.W. Jones, unpublished]. This suggests that cytochrome oxidases aa_3 and *co* may exhibit different energy conservation properties under certain experimental conditions in spite of their apparently similar charge translocation properties and their failure to elicit significant changes in growth yield.

The topography, function and regulation of these oxidases clearly remains a major problem of interest in the study of energy transduction in methylotrophs.

REFERENCES

[1] Anthony C. (1982) The Biochemistry of Methylotrophs. Academic Press, London.
[2] Anthony C. (1986) Adv. Microbial Phys. 27, 113-210.

[3] Beardmore-Gray M and Anthony C. (1984) In Microbial growth on C_1 compounds (R.L. Crawford and R.S. Hanson, eds), pp. 97-105. American Society for Microbiology, Washington.
[4] Jones CW, Carver MA, Dawson MJ and Patchett RA. (1984) In Microbial Growth on C_1 compounds (R.L. Crawford and R.S. Hanson eds), pp. 134-146. American Society for Microbiology, Washington.
[5] Tobari J. (1984) In Microbial Growth on C_1 Compounds (R.L. Crawford and R.S. Hanson eds), pp. 106-112. American Society for Microbiology, Washington.
[6] Nunn DN and Lidstrom ME. (1986) J. Bacteriol. 166, 581-590.
[7] Nunn DN and Lidstrom ME. (1986) J. Bacteriol. 166, 591-597.
[8] Adman ET. (1985) In Metalloproteins, part I (P.M. Harrison, ed), Macmillan, London.
[9] Ambler RP and Tobari J. (1985) Biochem. J. 232, 451-457.
[10] Tobari J.and Harada Y. (1981) Biochem. Biophys. Res. Commun. 101, 502-508.
[11] Lawton SA and Anthony C. (1985) Biochem. J. 288, 719-726.
[12] Lawton SA and Anthony C. (1985) J. Gen. Microbiol. 131, 2165-2171.
[13] Husain M and Davidson VL. (1985) J. Biol. Chem. 260, 14626-14629.
[14] Husain M, Davidson VL and Smith AJ. (1986) Biochemistry 25, 2431-2436.
[15] Martinkus K, Kennelly PJ, Rea T and Timkovitch R. (1980) Arch. Biochem. Biophys. 199, 465-472.
[16] Fukumori Y, Nakayama K and Yamanaka T. (1985) J. Biochem. 98, 493-499.
[17] Cross AR and Anthony C. (1980) Biochem. J. 192, 429-439.
[18] Greenwood JA and Jones CW. (1986) J. Gen. Microbiol. 132, 1247-1256.
[19] Froud SJ and Anthony C. (1984) J. Gen. Microbiol. 130, 2201-2212.
[20] Froud SJ and Anthony C. (1984) J. Gen. Microbiol. 130, 3319-3325.
[21] Beardmore-Gray M, O'Keeffe DT and Anthony C. (1983) J. Gen. Microbiol. 129, 923-933.
[22] Patchett RA and Jones CW. (1986) A. van Leeuwenhoek. In Press.
[23] Alefounder PR and Ferguson SJ. (1981) Biochem. Biophys. Res. Commun. 98, 778-784.
[24] Jones CW. Kingsbury SA and Dawson MJ. (1982) FEMS Microbiol. Letts. 13, 195-200.
[25] Burton SM, Byrom D, Carver MA, Jones GDD and Jones CW. (1983) FEMS Microbiol. Letts. 17, 185-190.
[26] Kasprzak AA and Steenkamp DJ. (1983) J. Bacteriol. 156, 348-353.
[27] Quilter JA and Jones CW. (1984) FEBS Letts. 174, 167-172.
[28] Carver MA, Humphrey KM, Patchett RA and Jones CW. (1984) Europ. J. Biochem. 138, 611-615.
[29] Carver MA and Jones CW. (1984) Arch. Microbiol. 139, 76-82.
[30] Dawson MJ and Jones CW. (1981) Biochem. J. 194, 915-924.
[31] Dawson MJ and Jones CW. (1981) Europ. J. Biochem. 118,113-118.
[32] Van Verseveld HW, Krab K and Stouthamer AH. (1981) Biochim. Biophys. Acta. 635, 525-534.
[33] Patchett RA, Quilter JA and Jones CW. (1985) Arch. Microbiol. 141, 95-102.
[34] Netrusov AI and Anthony C. (1979) Biochem. J. 178, 353-360.
[35] McCarthy JEG and Ferguson SJ. (1983) Europ. J. Biochem. 132, 417-424.
[36] Dawson MJ and Jones CW. (1982) Arch. Microbiol. 133, 55-61.
[37] Van Houwelingen T, Canters GW, Stobbelaar G, Duine JA, Frank, Jan, J, and Tsugita A. (1985) Eur. J. Biochem. 153, 75-80.

ENERGY METABOLISM IN METHANOGENS

G. GOTTSCHALK, M. BLAUT, A. JUSSOFIE, F. MAYER, V. MÜLLER AND R. OßMER
Institut für Mikrobiologie der Georg-August-Universität, D-3400 Göttingen FRG.

1.INTRODUCTION

Although there is a wide evolutionary divergence among the various genera of methanogenic bacteria, all organisms within this group share a common reaction: the reduction of methyl-coenzyme M to methane. This reaction is associated with a free energy change, large enough to be coupled to ATP synthesis:

$$CH_3\text{-coenzyme M} + H_2 \rightarrow CH_4 + \text{coenzyme M} \qquad \Delta G^{0'} = -85\ \text{kJ/reaction}$$

That the reduction of methyl-coenzyme M is indeed coupled to ATP formation could be concluded from a study of the simplest type of methane fermentation.

2.RESULTS AND DISCUSSION

2.1.Methanogenesis from Methanol + H_2

This fermentation consists of only two steps: the methyl group of methanol is transferred to coenzyme M -a reaction in which a protein-bound methyl-cobamide is an intermediate [1]; subsequently methyl-coenzyme M is reduced with H_2 by the methylreductase system [2]. Two organisms have been shown to gain energy for growth at the expense of this fermentation, *Methanosphaera stadsmanae* [3] and *Methanosarcina barkeri* [4].

When the bioenergetics of methane formation from methanol and H_2 was studied using cell suspensions of *M. barkeri* it could be unequivocally shown that ATP is synthesized in the course of this fermentation by a chemiosmotic mechanism [5].

The evidence is the following:

1. Addition of the protonophore tetrachlorosalicylanilide (TCS) to a cell suspension producing methane form methanol + H_2 results in the dissipation of the electrochemical transmembrane H^+ gradient and in a drastic decrease of the intracellular ATP content; methane formation, however, continues at a slightly increased rate (Fig. 1A).
2. The ATPase inhibitor N,N -dicyclohexylcarbodiimide (DCCD) prevents ATP formation, inhibits methanogenesis, but does not interfere with the electrochemical proton potential (Fig. 1B).

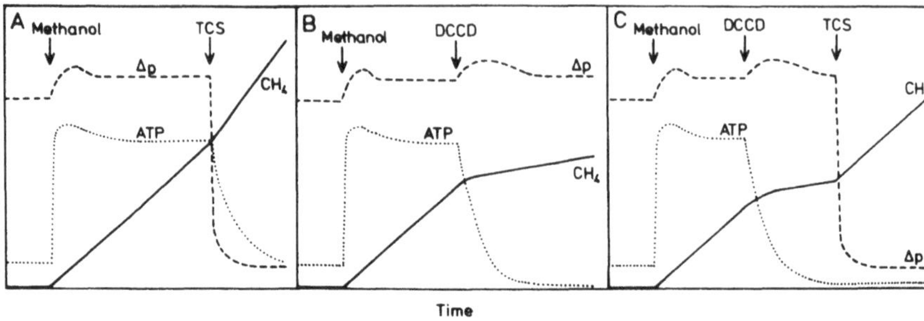

Figure 1. Formation of methane from methanol + H_2 by cell suspensions of *M. barkeri.* A) Effect of the protonophore TCS. B) Effect of DCCD, an inhibitor of the ATP synthase. C) Effect of DCCD followed by TCS.

3. The inhibition of methane formation by DCCD is relieved by TCS (Fig. 1C).

In analogy to results obtained with mitochondrial respiration it can, therefore, be concluded that methanogenesis from methanol + H_2 gives rise first to an electrochemical proton potential and then to ATP synthesis. Thus, substrate level phosphorylation as the mechanism of ATP synthesis is excluded.

2.2.Localization of component C of the methylreductase system in cells of methanogenic organisms.

On the basis of the above results, two models for the flow of electrons from an electron donor system (e.g. H_2) to methyl-coenzyme M can be envisaged; they are depicted in Fig. 2.

a) Electron transfer from the donor system to membrane-associated component C -the enzyme at which the reduction of methyl-CoM takes place [2]- results in proton translocation.

b) Electrons are transferred via membrane-bound components to a soluble redox carrier which funtions as electron donor in methyl-coenzyme M reduction.

Since the redox reactions in the membrane give rise to the generation of an electrochemical proton gradient, the redox potential of the hypothetical soluble electron donor must be considerably more positive than the one of H_2 or factor 420 (E'_0 = -420 mV and -340 mV, respectively). It must be in the neighbourhood of the one of methyl-coenzyme M/methane (E'_0 = + 26 mV). There is no experimental evidence for the presence of such a soluble carrier system in *M. barkeri* or other methanogenic organisms. On the other hand, component C has always been recovered as a soluble enzyme after disruption of cells of methanogenic bacteria [6] which argues against model a in Fig. 2. However, component C could be loosely associated with the membrane. Therefore, an immuno-electron microscopic approach was taken to localize component C in cells of methanogenic bacteria.

First, *Methanococcus volta* was studied with the help of antibodies raised against the three types of subunits of component C. Using the protein A gold technique, an unequivocal localization of

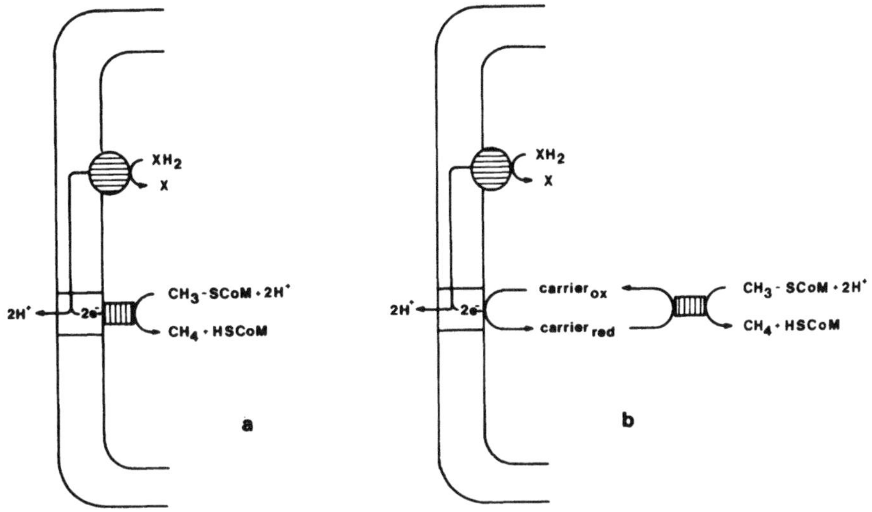

Figure 2. Tentative scheme for the involvement of the methylreductase in the generation of an electrochemical H^+ gradient. a) Methylreductase membrane-associated. b) A redox carrier mediates between membrane and methylreductase.

component C in the vicinity of the cytoplasmic membrane could be observed [7]. The typical picture which was seen is evident from Fig. 3. A comparable labeling pattern has also been obtained with *Methanobacterium thermoautotrophicum,* but rarely. In most of the ultrathin sections inspected a considerable percentage of the label was also present in the cytoplasm. This could be due to enzyme overproduction or to an insufficient fixation of the enzyme molecules by formaldehyde/glutaraldehyde in

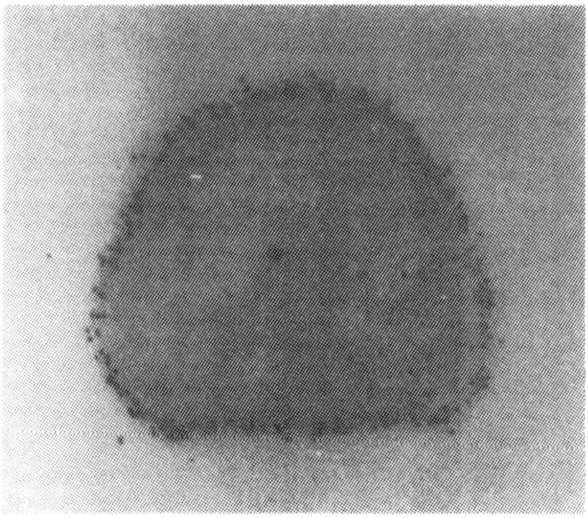

Figure 3. Immunocytochemical localisation of component C in *M. voltae.* Ultrathin sections were treated with antibodies against the subunits of component C and subsequently with colloidal gold-labelled protein A.

the thermophilic organism [7]. Although more organisms have to be studied, notably *M. barkeri*, the immuno-electron micriscopic investigations have indicated that an electron transfer might occur from membrane-bound components directly to component C.

2.3.ATP-independent methylreductase activity of vesicle preparations.

When cells of methanogenic organisms are disrupted in a conventional manner (French press or ultrasonication) component C appears in the soluble fraction and ATP in non-stoichiometric amounts is required for methylreductase activity [6]. Would subcellular fractions prepared from protoplasts also show an ATP requirement of the methylreductase reaction? Protoplasts were prepared by pronase treatment of the two as yet unclassified strains AJ2 and Gö1 (the latter strain was provided by F. Widel). These protoplasts were stable in buffer containing 0.5 M sucrose. They produced methane from methanol and H_2 and were energetically "intact". This was concluded from the finding that methanogenesis could be inhibited by DCCD and that this inhibition could be relieved by TCS [8]. When these protoplasts were broken with a gentle passage through a French press a vesicle preparation resulted that exhibited an ATP-independent methylreductase activity, in contrast to cell extracts prepared from whole cells (Fig. 4). So, a different form of the enzyme might be present, and it can be expected that further study of vesicle preparations will help to elucidate the role of the methylreductase system in energy-conserving reactions of methanogenic bacteria.

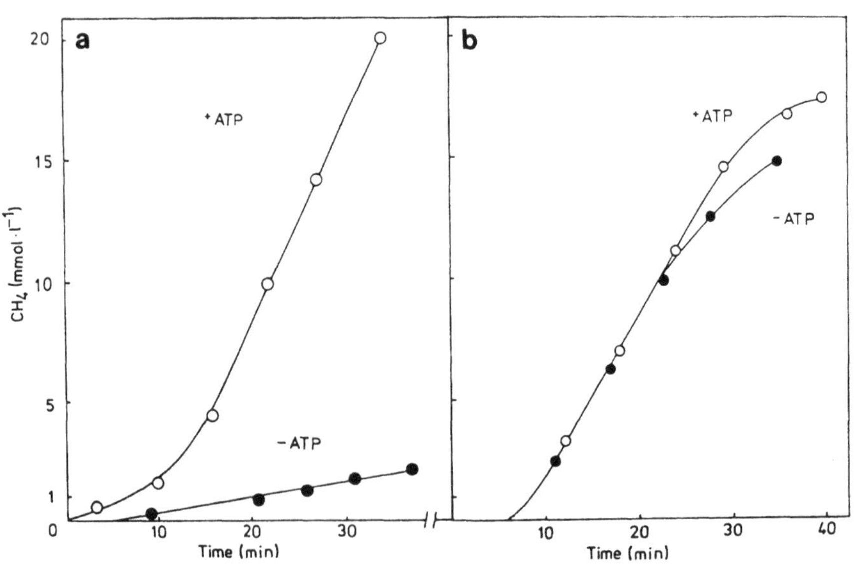

Figure 4. Effect of ATP on methylreductase activity in subcellular fractions of strain Go1. a|)Extract prepared from whole cells. b) Vesicles prepared from protoplasts; addition of an ATP trap (glucose + hexokinase) had no effect.

It should be mentioned in this connection that purified methylreductase of *M. thermoautotrophicum* will produce methane from methyl-coenzyme M in the absence of ATP if cobalamine is present and dithiothreitol is used as electron donor [9]. It was also shown that an ATP-activated reaction mixture could be passed through a Sephadex G-25 column without loss of methylreductase activity [10]. In addition, methane formation from $CO_2 + H_2$ by membrane fractions of *M. thermo-autotrophicum* and *M. ruminantium* has been reported [11,12].

2.4.Role of sodium ions

Growth of methanogenic bacteria depends on the presence of sodium ions in the growth medium [13]. Although it was tempting to speculate that Na^+ was involved in energy conservation, this apparently is not the case. Methanogenesis from methanol + H_2 as well as ATP synthesis in this fermentation have been shown not to require Na^+ [14]. However, one reaction specifically dependent on Na^+ has recently been identified [14]. Cell suspentions of *M. barkeri* produce methane from methanol or formaldehyde in the presence or absence of H_2 according to the following equations:

$$CH_3OH + H_2 \rightarrow CH_4 + H_2O$$

$$HCHO + 2\,H_2 \rightarrow CH_4 + H_2O$$

$$4\,CH_3OH \rightarrow 3\,CH_4 + CO_2 + 2\,H_2O$$

$$2\,HCHO \rightarrow CH_4 + CO_2$$

Only the conversion of methanol to methane and carbon dioxide requires Na^+. It follows that Na^+ must be involved in the oxidation of methanol to the level of formaldehyde. This reaction step is also uncoupler-sensitive. Growing cells of *M. Barkeri* maintain a transmembrane Na^+ gradient (Table 1) with the help of a very active Na^+/H^+ antiporter [15]. It is speculated that it is the Na^+ gradient that finally drives methanol oxidation. If the electrons of methanol oxidation to formaldehyde are to be channelled into the methylreductase system at the level of the redox potential of H_2 (or factor 420), then reverse electron transfer would be required; $HCHO/CH_3OH$: E'_0 = -182 mV; H^+/H_2: E'_0= -420 mV. This reverse electron transfer could be driven by Na^+ influx as tentatively depicted in Fig. 5.

Table 1. Intracellular concentration of Na^+ as dependent on the extracellular one in *M. barkeri* cells producing methane from methanol + H_2.

Na^+_e (mM)	Na^+_i (mM)	Na^+_i/Na^+_e	$\Delta\tilde{\mu}_{Na^+}$ (mV)
5	1.9	0.398	-24
10	3.6	0.360	-27.08
15	5.5	0.366	-26.6
20	5.8	0.291	-32.7
25	7.1	0.284	-33.3

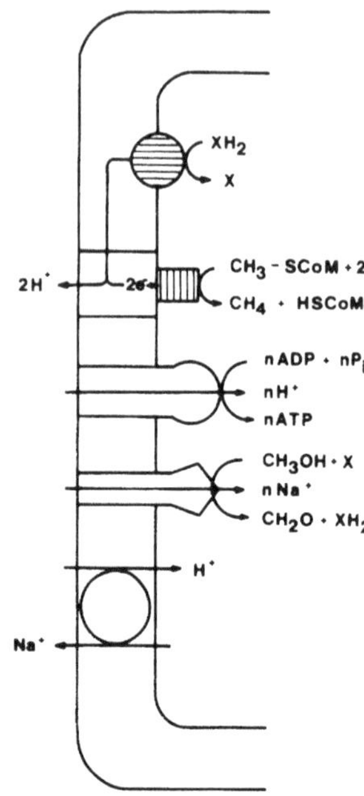

Figure 5. Tentative scheme of the role of Na+ in *Methanosarcina barkeri*.

3.CONCLUDING REMARKS

Conclusive evidence has been obtained that the last reduction step in methanogenesis as carried out by *M. barkeri* is coupled to ATP synthesis by a chemiosmotic mechanism. Since this step is inherent in methanogenesis *per se* it is tempting to assume that ATP is generally synthesized in methanogenic bacteria by ATP synthase and that the protonmotive force required is generated in the course of electron transfer from appropriate donors to methylcoenzyme M. However, controversial views have been published recently [16]; results on the effect of uncouplers on *M. thermoautotrophicum* were published that are currently not understood [17]. Consequently, a greater variety of organisms has to be studied with respect to the location of the methylreductase and the properties of the enzyme system.

Very interesting are the results on the effect of sodium ions. The Na^+/H^+ antiporter which in addition to *M. barkeri* has previously demonstrated to be present in *M. thermoautotrophicum* [18] might generally be responsible for the maintenance of a Na^+ gradient across the cytoplasmic membrane, and this gradient might drive endergonic reactions. This would be the opposite to some decarboxylase reactions in which exergonic decarboxylations are used to build up Na^+ gradients [19].

ACKNOWLEDGEMENTS

The work from the laboratory of the authors was supported by the "Deutsche Forschungsgemeinschaft". The antibodies against component C were obtained from A. Klein, Marburg, FRG.

REFERENCES

[1] Van der Meijden P, Heythuysen HJ, Pouwels A, Houwen FP, van der Drift C, Vogels GD. (1983). Arch Microbiol 134, 238-242.
[2] Wolfe RS. (1985). Tr Biochem Sci 10, 396-399.
[3] Miller TL, Wolin, MJ. (1985). Arch Microbiol 141, 217-222.
[4] Müller V, Blaut M, Gottschalk G. (1986). Appl Environ Microbiol, in press.
[5] Blaut M, Gottschalk G. (1984). Eur J Biochem 141, 217-222.
[6] Gunsalus RP, Wolfe RS. (1978). J Bacteriol 135, 851-857.
[7] Ossmer R, Mund T, Hartzell PL, Konheiser U, Kohring GW, Klein A, Wolfe RS, Gottschalk G, Mayer F. (1986). Proc Natl Acad Sci USA, in press.
[8] Jussofie A, Mayer F, Gottschalk G. (1986). Arch Microbiol, in press.
[9] Whitman WB, Wolfe RS. (1983). J Bacteriol 154, 640-649.
[10] Ankel-Fuchs D, Thauer RK. (1986). Eur J Biochem 156, 171-172.
[11] Sauer FD, Erfle JD, Mahadevan S. (1980). Biochem J 190, 177-182.
[12] Sauer FS, Erfle JD, Mahadevan S. (1979). Biochem J 178, 165-172.
[13] Perski HJ, Schönheit P, Thauer RK. (1982). FEBS Lett 143, 323-326.
[14] Blaut M, Müller V, Fiebig K, Gottschalk G. (1985). J Bacteriol 164, 95-101.
[15] Müller V, Blaut M, Gottschalk G. (1986). Eur J Biochem, in press.
[16] Lancaster JR. (1986). FEBS Lett 199, 12-18.
[17] Schönheit P, Beimborn D. (1985). Eur J Biochem 148, 545-550.
[18] Schönheit P, Beimborn D. (1985). Arch Microbiol 142,354-361.
[19] Dimroth P. (1985). Ann NY Acad Sci 447, 72-85.

THEORETICAL LIMITS OF GROWTH YIELDS AND AN ANALYSIS OF EXPERIMENTAL DATA

Wolfgang BABEL and Henk W. VAN VERSEVELD

Academy of Sciences of the GDR, Institute of Biotechnology, Leipzig, GDR and
Biologisch Laboratorium, Vrije Universiteit, Amsterdam, The Netherlands

Substrates for heterotrophic growth are carbon and energy sources at the same time. Hence it follows that, depending on the carbon/energy ratio, growth may be carbon- or energy-limited. From a physical-energy point of view, hydrocarbons and alcohols are energy excess substrates [1-4], i.e, firstly growth obtained on these substrates should be carbon-limited and secondly, substrate carbon might be completely incorporated into the "cell molecule" without losing carbon in the form of CO_2 for the purpose of energy generation (Table 1). However, during synthesis of amino acids and other cell constituents carbon gets lost as CO_2 due to unavoidable decarboxylations and a 100% carbon conversion efficiency is never possible. Theoretical calculations [5] show that, almost independent of cell composition, the upper limit of carbon incorporation into the "cell molecule" amount to about 85% with glycolytic substrates and to approximately 75% with non-glycolytic, gluconeogenetic or C_2 substrates (Table 2). Substrates are defined as glycolytic when a C_3 compound is formed as central precursor for the synthesis of all cell components without losing carbon, and as non-glycolytic when CO_2 is released along this way. According to this definition, methane and methanol belong to the first

Table 1. Upper limit (100%), carbon metabolism determined (85 and 75%), experimental (exp) and calculated (calc) growth yields of bacteria and yeasts on various substrates. Bacterial and Yeast "cellformula" are respectively $C_4H_8O_2N$ and $C_6H_{10}O_3N$.

	Bacteria					Yeasts				
substrate	$Y_{100\%}$	$Y_{85\%}$	$Y_{75\%}$	Y_{exp}	Y_{calc}	$Y_{100\%}$	$Y_{85\%}$	$Y_{75\%}$	Y_{exp}	Y_{calc}
methane	1.59	1.35	-	0.9	0.58[1](36.5%) 0.72[2](45%)					
methanol	0.8	0.68	-	0.5	0.45[1](56.7%) 0.6[2] (75%)	0.75	0.64	-	0.4	0.41(55.2%)
ethanol	1.11	-	0.83	0.8	0.83 (75%)	1.04	-	0.78	0.77	0.78 (75%)
acetate	0.85	-	0.64	0.4	0.41 (47.7%)	0.8	-	0.6	0.35	0.39(49.2%)
glucose	0.85	0.72	-	0.5	0.52 (61.4%)	0.8	0.68	-	0.5	0.51(63.4%)
hexadec-ane	1.81	-	1.35	1.1	1.35 (75%)	1.7	-	1.27	0.9	1.27 (75%)

Y : g dry weight per g substrate. % in parentheses : carbon conversion efficiency.

[1] Serine pathway-ICL+ variant, methanol → formaldehyde + 1 ATP.

[2] Hexulose phosphate pathway-FBP variant, TK, TA, TK

Table 2. Carbon-metabolism determined maximum carbon conversion efficiencies (%) of the synthesis of individual cell constituents and of yeast biomass with different elementary compositions

	methanol via DHA pathway	glucose via EMP pathway	acetate
Polysaccharides	100.	100.	75.
protein	84.7	84.7	77.2
lipid	67.9	67.9	98.1
RNA	81.3	81.4	81.3
DNA	89.1	83.1	83.1
|1|	84.3	84.3	76.5
|2|	83.7	83.6	77.4
|3|	86.3	86.2	n.d.
|4|	82.5	82.5	n.d.
|5|	86.2	86.3	n.d.

Macromolecular cell composition (given as % of dry weight)

	Polysaccharides	protein	lipid	RNA	DNA
|1|	20	55	7	12	3
|2|	35	40	15	6	1
|3|	35	40	7	12	3
|4|	27	40	15	12	3
|5|	33	50	7	6	1

category and long-chain hydrocarbons and ethanol to the second. Hence, the experimental upper limit of growth yield for bacteria on methane and methanol is 1.35 and 0.68 g/g respectively and for yeasts on methanol 0.64 g/g (Table 1) Most of the experimentally obtained growth yields reported for both bacteria and yeasts are markedly smaller than these upper limits. Since a carbon conversion efficiency of 85% is practically never reached, even in methanol-limited cultures, and differences in "cell formula" cannot explain these discrepancies either, methane and methanol should be considered as energy-deficient. Some reported high yield values on methanol and methane [6-9] appear to be hardly realistic.

The differences in carbon conversion efficiency between bacteria (and yeasts as well) using the serine- and hexulose phosphate-pathway can be explained by carbon catabolism [10,11], i.e. the sequence from formaldehyde to the central C_3 precursor is responsible for the differences, provided that the cell composition of both bacterial groups is similar, the efficiency of energy generation does not markedly differ, and that in both groups the oxidation of methanol to formaldehyde yields 1 ATP. When the Y_{ATP} concept [12] is used for calculating theoretical Y_{calc} (see Table 1) this results in values very much in agreement with experimental values. From the comparison between calculated and experimental yield values it can be deduced that the energetic inefficiencies of the methanol-oxidizing enzymes are responsible for the relatively low carbon conversion efficiency (see Table1). When the PQQ-linked methanol dehydrogenase of bacteria and the methanol oxidase of yeasts are replaced by a NAD^+-linked dehydrogenase, methanol could be converted by nearly 85% into biomass (Table 3).

The energy-deficient character and the metabolism-determined upper limit of carbon conversion of a substrate can be determined experimentally by adding an energy-donating substrate to the substrate to be evaluated. Using the transient-state cultivation technique, the upper limit of carbon conversion efficiency for glucose with *Hansenula polymorpha* [15] and for acetate with *Acinetobacter calcoaceticus* [16] has been determined. In the case of *Hansenula polymorpha* formate was used as an energy-donating substrate to glucose; in *Acinetobacter calcoaceticus* the energy necessary for the complete assimilation of acetate carbon was generated by the oxidation of glucose to gluconic acid by means of the PQQ-linked glucose dehydrogenase.

Table 3. Theoretical growth yields of bacteria and yeasts on methanol as the sole carbon and energy source, calculated for different energetic efficiencies of methanol oxidation.

prosthetic group of methanol oxidizing enzyme	Y_{calc} (g/g)	carbon conversion efficiency (%)
Bacteria [13] *HuP pathway KDPG-TK, TA, TK*		
PQQ (1 ATP)	0.53	67.1
NAD (2 ATP)	0.64	80.5
Yeasts [14] *DHA pathway*		
FAD (0 ATP)	0.41	55.2
FAD (0 ATP)[1]	0.45	60.5
NAD (2 ATP)	0.62	82.9

[1] Formaldehyde is thought to be oxidized via the TCA cycle, P/O ratio: 3 for mitochondrial and 2 for cytoplasmic NADH.

In Table 4 the results are summarized together with other literature data [17-21].

Table 4. Experimental growth yields (Y, g/g) and carbon conversion efficiencies (CCE, %) of single substrates and substrate mixtures.

Organism	Y	CCE	Y	CCE	Y	CCE
	glucose		formate		glucose/formate	
H. polymorpha [15]	0.52	61.4	0	0	0.7	82.3
C.utilis CBS 621 [17]	0.5	59.6	0	0	0.69	82.2
	acetate		glucose		acetate/glucose	
A.calcoaceticus [16]	0.4	46.9	0	0	0.62	72.9
	mannitol		formate		mannitol/formate	
P.denitrificans [18]	0.52	62.2	0.01	12.7	0.7	83.8
	glucose		formate		glucose/formate	
B.natriegens [19]	0.45	52.9	0	0	0.52	61.1
	methanol		formate		methanol/formate	
Pichia pastoris [20]	0.38	50.7	0	0	0.45	61.3
Pichia pinus MH4 [21]	0.37	49.3	0	0	0.55	73.3

For glucose as a glycolytic substrate the upper limit of carbon conversion efficiency actually amounts to about 85% as theoretically predicted and for acetate to approximately 75%. About 85% of carbon conversion efficiency was also reached for mannitol in *Paracoccus denitrificans* [18]. However, the situation concerning its use of formate is not as well-defined as for yeasts, because *P.denitrificans* is capable of assimilating formate autotrophically via the Calvin cycle. That 83.8% is obtained supports the view that in the presence of mannitol autotrophic CO_2 fixation is restricted or does not play any role at all as has been found for growth on mannitol/methanol [22]. If formate was assimilated as CO_2 the theoretically possible value of 65.8% for mannitol (i.e. 0.55 g/g) as the sole carbon and energy source can only be exceeded when the simultaneous assimilation of mannitol plus formate was accompanied by an increase in the P/O quotient. The results obtained with *Saccharomyces cerevisiae* (see [15]), *Beneckea natriegens* and *Pichia pinus* are remarkable. Although formate is oxidized NAD-dependently, the growth yield cannot be raised to the metabolic upper limit of 85%. With *Beneckea natriegens* this is easy to explain, if assimilation of glucose via the hexose monophosphate pathway is assumed [23]. With *Saccharomyces cerevisiae* and *Pichia pinus* the catabolic sequence does not give an explanation. *Saccharomyces cerevisiae* assimilates glucose via the EMP pathway and *Pichia pinus* methanol via the DHA pathway as other methylotrophic yeasts do. Indications for another assimilatory sequence for methanol in yeasts is not reported up to now. Probably yeasts possess at least two dehydrogenases, an energy-generating enzyme and a formate-detoxifying system. In *Saccharomyces cerevisiae I*, e.g., the capacity of the energy-yielding formate dehydrogenase is not sufficient to assimilate glucose carbon maximally. Formate added beyond this capacity is only oxidized to CO_2 without providing biologically useful energy. To what extend the different localizations in the membrane of the various dehydrogenases are of importance cannot be estimated. Probably this aspect does not play any role in bacteria (but see the contribution of Thauer and van Verseveld in this session). However, it cannot be excluded that in *Beneckea natriegens* the 85% carbon conversion efficiency is not obtained because the capacity of the energy-donating formate dehydrogenase is exhausted too early.

If the growth yield can be improved by adding an energy source it is difficult to imagine that such an effect can also be reached by adding an extra carbon source. Since methanol is energy-deficient, methane should be even more energy-deficient, provided it is assimilated like methanol and the oxidation of methane to methanol is a reducing equivalents consuming process, yielding no biologically useful energy. Consequently growth yield values of 0.9 g/g or even higher on methane as only carbon and energy source should be wrong and not succeptible to improvement by adding CO_2, although this seems to be suggested by the heterotrophic CO_2 fixation involved in the assimilation loop of the serine pathway and by the assumption that always more methane (or methanol) is oxidized to CO_2 via formate than is required for giving energy for carbon assimilation. Deviating experimental data reported occasionally [24-26] would hardly stand up to critical investigations due to the relative energy deficit of methane (and methanol, too). However, this statement does not exclude an increase in growth yield when two energy-deficient substrates are mixed which both simultaneously provide carbon and energy for growth. Examples are methane plus acetate [27,28] and methanol plus glucose [29]. In the first case the effect could be caused by a more economic use of the substrates, e.g., by an immediate use of acetate as a carbon skeleton for certain amino acids and for the formation of long-chain fatty acids by bypassing oxidative phosphorylation. Acetate cannot provide biologically useful energy, because the TCA cycle does not operate in obligately methylotrophic bacteria.

In the mixture of methanol plus glucose the auxiliary substrate effect can be explained as follows. In glucose grown *H. polymorpha* the biological useful energy is made available by means of the TCA cycle. For this purpose and for the synthesis of macromolecular cell constituents, especially proteins, lipids and nucleic acids, glucose must be metabolized to a C_3 compound (e.g. phosphoglycerate). However the "carbohydratic" cell components can be synthesized directly (growth yield is 0.52 g/g). During growth of yeasts on methanol the energy needed for growth is generated by oxidizing formaldehyde via formate to CO_2. The TCA cycle is dispensable as a generator of reducing equivalents and ATP (growth yield is 0.38 g/g). With the simultaneous assimilation of both substrates (Figure 1) glucose can serve as a precursor for "carbohydratic" components and as a C_2 donor for the fixation of formaldehyde after the oxidative decarboxylation to xylulose-5-phosphate. The growth yield on the mixture is higher than theoretically expected. This so-called auxiliary substrate effect mainly seems to result from putting or

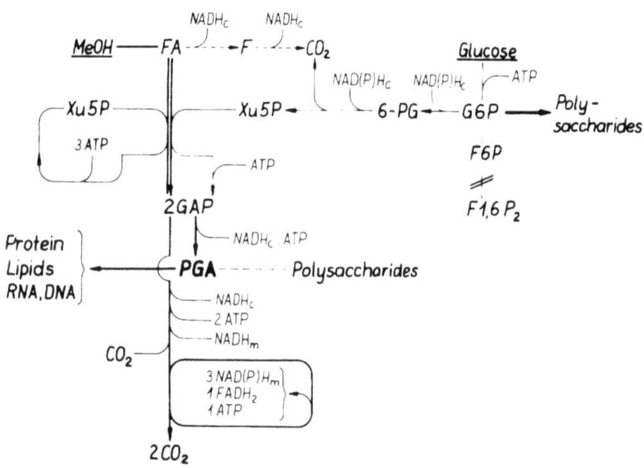

Figure 1. Metabolic scheme for simultaneous assimilation of methanol plus glucose in *Hansenula polymorpha*.

keeping the TCA-cycle into operation with the consequence that mitochondrial NADH instead of the cytoplasmic NADH of the (formaldehyde- and) formate dehydrogenase generates the biologically useful energy with a higher efficiency (P/O_{NADH}(mitochondria) = 3). At the present a complete picture of the regulatory network is not known, but the properties of the phosphofructokinase and the formate dehydrogenase, and the increased activities of some TCA cycle enzymes (Table 5) during growth on the mixture are in agreement with this model [32, cf. 33].

With ethanol as the sole carbon and energy source the experimental growth yields of yeasts and bacteria correspond with the carbon metabolism- determined upper limit, which suggests that ethanol is not energy-deficient. The growth yield increase obtained by Feiler et al [29] for *Candida utilis* growing on a mixture of ethanol and sucrose rather indicates that ethanol is even energy-excess.

Table 5. Influence of growth substrate on the specific activities of some enzymes in crude extracts of *Hansenula polymorpha*.

Substrate	α-ketoglutarate decarboxylase	formate dehydrogenase	phospho-fructokinase
methanol	0.004	0.235	0.035
glucose	0.015	0.023	0.125
glycerol	0.016		
MeOH/glucose (6:1)	0.052		
MeOH/glucose (1:5)	0.057		
MeOH/glycerol (5:1)	0.030		
MeOH/glycerol (1:5)	0.045		

Specific activity in U/mg. Numbers in parentheses correspond to the mixing proportion in g (w/w).

To get evidence *Hansenula polymorpha* MH20 was grown continuously on a mixture of ethanol plus formate. The experiments aimed at improving growth yield on ethanol in the presence of formate were performed with the transient-state cultivation technique. After having reached steady state conditions during cultivation on ethanol, formate was added to this culture in a linearly increasing concentration gradient generated in the medium reservoir flasks as described elsewhere [31]. In the presence of formate which was fed in differently steep concentration gradients no improvement of growth yield beyond 0.77 g/g could be obtained (Müller and Babel, unpublished). Hence, ethanol is either a carbon/energy balanced or even a real energy excess substrate.

Hydrocarbons (hexadecane) are reducing equivalents-excess substrates. Whether or not they are energy excess as well cannot be answered definitely. Reliable experimental results do not exist up to now. That a 75% carbon conversion efficiency is not reached and a yield of 1 g/g hardly exceeded may be ascribed to the fact that reducing equivalents are transferred into ATP via ETP with a low efficiency, that reducing equivalents are wasted by the cytochrome P-450 system or that the reducing equivalents of the ß-oxidation system cannot be used because of lacking transhydrogenases or because of its localization (peroxisomes). However, which of these aspects prevail remains to be investigated.

As has been shown, the Y_{ATP} concept allows to predict growth yields for any substrate. For this purpose the following must be known :

1. Close-fitting carbon- and energy balances.
2. The elementary composition of the biomass as a fictive "cell molecule"
3. The carbon catabolism pathway to the central precursor.
4. The ATP-yield during oxidation of substrate
5. The P/O quotient for the different reducing equivalents.

This approach also enables us to take into account the type of nitrogen source, e.g. ammonia, nitrate, and the terminal electron acceptor of the respiratory chain. On this basis alone the carbon metabolism-determined upper limit of carbon conversion efficiency cannot be estimated experimentally. This becomes possible when the auxiliary substrate concept is introduced and the transient-state cultivation technique is used as was demonstrated.

The real growth yield depends on the growth rate. On the basis of the Y_{ATP}^{MAX} concept [34] and the Pirt-equation the influence of the growth rate and the maintenance coefficient can be taken into account [5, 34], provided the Pirt-equation is valid.The validity of the Pirt-equation has during very slow growth been questioned [36-38] i.e in that area of growth when maintenance demands are significant. The Pirt-equation [39]

$$r_s = m_s x_t + (1/Y_{xsm}) r_x \qquad (1)$$

in which m_s, Y_{xsm}, r_s, r_x and x_t are, respectively maintenance coefficient, maximal growth yield corrected for maintenance, rate of substrate consumption, rate of biomass production and biomass concentration can be written as

$$r_x = (r_s - m_s x_t) Y_{xsm} \qquad (2)$$

In fed batch or recycling experiments with 100% biomass feedback, r_s has a constant value and integration of equation 2 yields

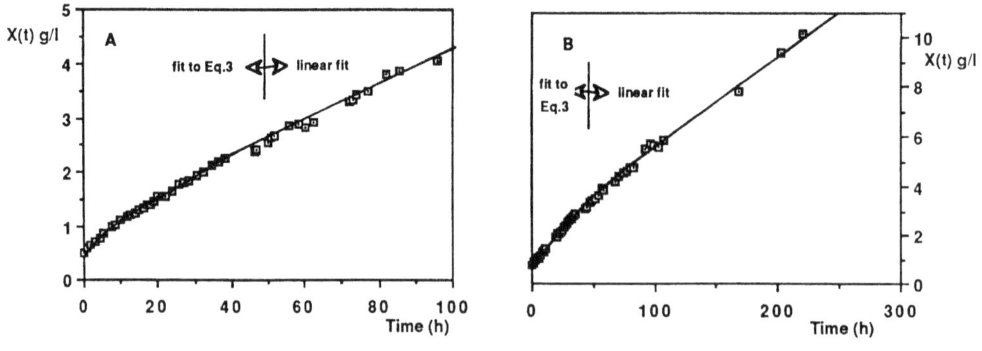

Figure 2. Biomass concentration (x_t) in methanol-limited recycling experiments with *Paracoccus denitrificans* (A) and *Hyphomicrobium* X (B). r_s = 5 and 4.5 mmol/liter.hour for respectively A and B.

$$x_t = r_s/m_s + (x_0 - [r_s/m_s]) e^{-m_s Y_{xsm} t} \qquad (3)$$

It will be clear from equation 3 that when $t \rightarrow \infty$, $x_t \rightarrow r_s/m_s$ via a logarithmic curve. It has been shown that at specific growth rates lower than 0.01 h^{-1} biomass increase deviates from behaviour expected from equation 3 [38] and the growing cells show all characteristics of the stringent response [37,40]. This deviation from equation 3 has also been observed in methanol-limited *Paracoccus denitrificans* and *Hyphomicrobium* X. The growth pattern in the recycling fermentor is shown in Figure 2. In both experiments of figure 2 growth can be described with equation 3 up to time is 50 h in the recycling experiment. The remaining data could be described with linear growth as has been described before with heterotrophic growth of *Paracoccus denitrificans* [38]. The fitting procedure, using equation 3, has been described before [38] and yielded fitted growth parameters summarized in Table 6.

Table 6. Growth parameters of *Paracoccus denitrificans* and *Hyphomicrobium X* grown in a methanol-limited recycling fermentor . r_s = 5 and 4.5 mmol/liter.hour respectively. m_s and Y_{xsm} (and Y) are, respectively, expressed as mmol methanol/g dry weight.hour and g dry weight/ mol methanol.

Parameter	organism	type of fitting-procedure	fitted value
m_s	*P.denitrificans*	all data (Eq.3)	0.55
Y_{xsm}			9.8
m_s	id.	data of 0 - 50 hour (Eq.3)	1.30
Y_{xsm}			13.7
Y	id.	data > 50 hours, linear	6.9
m_s	*Hyphomicrobium* X	all data	0.23
Y_{xsm}			10.7
m_s	id.	data of 0 - 50 hour (Eq.3)	0.66
Y_{xsm}			15.4
Y	id.	data > 50 hours, linear	8.3

As has been described before [38], it was impossible to describe all data of both experiments with equation 3. The obtained Y_{xsm} values were lower than the obtained Y-values in the preceding chemostat experiment. In accordance with the assimilation pathways, calvin cycle for *Paracoccus denitrificans* and serine pathway for *Hyphomicrobium* X, all yield data for *Hyphomicrobium* X were substantially higher than those estimated for *Paracoccus denitrificans.*

A striking observation is the change in cell shape during a recycling experiment of *Hyphomicrobium* X (Figure 3.)

It has been observed that the deviation from equation 3 is coupled to a change from normal mother and daughter cells (figure 3A) to an increasing amount of Y-shaped cells as has been observed before during growth in the presence of high metal-ion concentrations and shift from growth on methanol to growth on methylamine [41]. Most probably stringent growth induces this change in cell-shape.

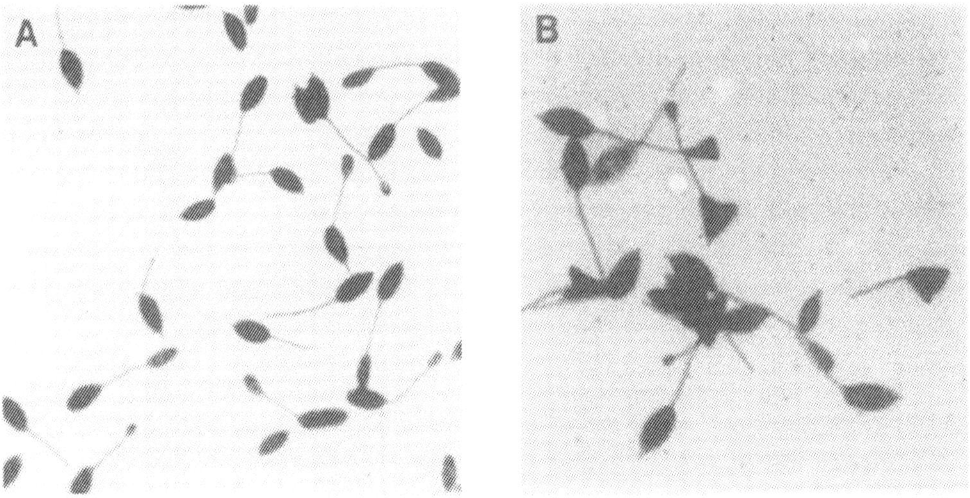

Figure 3. Cell shape of *Hyphomicrobium* X. **A**. At the start of the recycling experiment, enlargement : 1300 x. **B**. 71 hours after the start of the recycling experiment, enlargement : 1900 x.

In our opinion the recycling fermentor is a very good tool for investigating regulation of growth at very slow specific growth rates.

References

[1] Linton JD, StephensonRJ (1978) FEMS Microbiol. Lett. 3, 95-98
[2] Babel W (1979) Z.Allg.Mikrobiol. 19, 671-677
[3] Roels, JA (1980) Biotechnol.Bioengin. 22, 2457-2514
[4] Westerhoff HV, Hellingwerf KJ, van Dam,K (1983) Proc.Natl.Acad.Sci. 80, 305-309
[5] Babel W, Müller RH (1985) Appl.Microbiol.Biotechnol. 22, 201-207
[6] Papoutsakis E, Lim HC (1981) Ind.Eng.Chem.Fundam. 20, 307-314
[7] Krug ELR, Lim HC, Tsao GT (1979) Ann.Rep.Ferment.Proc. 3, 141-195
[8] Bewersdorff M, Dostalek M (1971) Biotechnol.Bioengin. 13, 49-62
[9] Klass DL (1968) Gas Wärme International 17, 249-254
[10] Anthony C (1978) J.Gen.Microbiol. 104, 91-104
[11] BabelW, Müller RH (1985) Acta Biotechnol. 5, 333-338
[12] Stouthamer AH (1973) Ant. van Leeuwenhoek 39, 545-565
[13] Anthony C (1983) Acta Biotechnol. 3, 261-268
[14] Babel W (1986) Acta Biotechnol. 6, 263-267
[15] Babel W, Müller RH, Markuske KD (1983) Arch.Microbiol. 136, 203-208
[16] Müller RH, Babel W (1986) Arch.Microbiol. 144, 62-66
[17] Bruinenberg PM, Jonker R, van Dijken, JP, Scheffers, WA (1985) Arch.Microbiol. 142, 302-306
[18] van Verseveld HW Stouthamer, AH (1980) FEMS Microbiol. Lett. 7, 207-211
[19] Linton JD, Griffiths K, Gregory M (1981) Arch.Microbiol. 129, 119-122
[20] Hazeu W, Donker RA (1983) Biotechnol.Lett. 5, 399-404
[21] Müller RH, Sysoev OV, Babel W (1986) Appl.Microbiol.Biotechnol. 23, in press
[22] van Verseveld HW, Boon JP, Stouthamer AH (1979) Arch. Microbiol. 121, 213-223
[23] Babel W, Müller RH (1985) J.Gen.Microbiol. 131, 39-45
[24] Babij T, Ralph BJ, Rickard PAD (1975) In: Proceedings of the 1st Int.Symp. on Microbial Growth on C1-compounds, Tokyo (Japan), pp 213-220
[25] Malashenko YR, Gagarina VA (1978) In: Abstracts of the 12th Int.Congr.Microbiol., Münich (FRG), No. F13
[26] Malashenko YR, Romanovskaya VA, Sokolov IN, Krystab TP, Lyndvychenko OS (1980) Ukrainian Biochem.J.(Kiev) 52, 159-163

[27] Patel RN, Hoare SL, Hoare DS, Taylor BF (1977) Appl.Environm.Microbiol. 34, 607-610
[28] Wadzinski AM, Ribbons DW (1975) J.Bacteriol. 123, 380-381
[29] Müller RH, Markuske KD, Babel W (1983) Z.Allg.Mikrobiol. 23, 375-384
[30] Feiler E, Becker U, Schneider J (1980) In: Abstracts 2nd Symp. of Socialist Countries on Biotechnol., Leipzig (GDR), Dec. 2-5
[31] Müller RH, Markuske KD, Babel W (1985) Biotechnol.Bioengin. 27, 1599-1602
[32]Babel W, Müller RH, Markuske KD (1986) In: 11th Intern.Spec.Symp.Yeasts "Regulation of Transport and Metabolism in Yeasts, Basic and Biotechnological Aspects" Lisboa (Portugal) p 75
[33] Egli T, Lindley ND, Quayle JR (1983) J.Gen.Microbiol. 129, 1269-1281
[34] Stouthamer AH (1977) Symp.Soc.Gen.Microbiol. 27, 285-315
[35] Harder W, van Dijken JP (1976) In: Microbial Production and Utilization of Gases (eds. Schlegel HG, Gottschalk G, Pfennig N) E.Goltze KG, Göttingen, pp 403-418
[36] van Verseveld HW, Chesbro WR, Braster M, Stouthamer AH (1984) Arch Microbiol. 137, 176-184
[37] van Verseveld HW, Arbige M, Chesbro WR (1984) Tr.Biotechnol. 2, 8-12
[38] van Verseveld HW, de Hollander JA, Frankena J, Braster M, Leeuwerik FJ, Stouthamer AH (1986) Ant.van Leeuwenhoek 52, 325-342
[39] Pirt SJ (1965) Proc.Roy.Soc.B 163, 224-231
[40] Arbige M, Chesbro WR (1982) J.Gen.Microbiol. 128, 693-703
[41] Harder W, Attwood yMM (1978) Adv. Microbiol.Physiol. 17, 303-359

GENETICS AND MOLECULAR BIOLOGY OF C_1 UTILIZERS

THE MOLECULAR GENETICS OF C_1 UTILIZING MICROORGANISMS AN OVERVIEW

B.W. HOLLOWAY, P.P. KEARNEY# and B.R. LYON

Department of Genetics, Monash University, Clayton, Victoria 3168, Australia and #The Galton Laboratory, Department of Genetics and Biometry, University College of London, London NW1 2HE, UK

The availability of recombinant DNA techniques has enabled the succesful genetic analysis and manupilation of a range of C_1 utilizing microorganisms. It has resulted in the identification of genes of interest on both plasmids and the chromosome; enabled the linkage of chromosomal genes to be determined; established the function and regulatory patterns of genes essential for utilization of C_1 compounds and provided information on the evolution of methanogenic bacteria.

1. INTRODUCTION

It did not take long for microbiologists in general to appreciate the sophistication of biological analysis which genetics brought to the study of enterobacteria and to seek the advantages of the genetic approach for a wider range of organisms. The genetic analysis of bacteria is now undergoing an interesting transitional phase. The identification and mapping of bacterial genes was previously dependent upon a trilogy of exchange systems - conjugation, transduction and transformation. As a result, the many benefits of genetic analysis were limited to about thirty species of bacteria [see 1]. In the case of C_1 utilizing organisms, the search for traditional means of genetic analysis was frustrating and to a large extent unsuccesfull as mutants of the type needed for selection of recombinants were difficult to obtain, chromosome mobilizing plasmids were rare and of low efficiency and transducing phages were absent [2].

The development of recombinant DNA technology has to a large extent overcome these difficulties. These techniques are applicable to all microorganisms although there is always the capacity for refinement to allow for particular properties of some genera and species. For many, such as a range of C_1 utilizing organisms, they are now the most succesful experimental approach to obtain genetic information and their achievements have been dramatic [3].

New procedures have been devised by which mutations can be induced, genomic fragments isolated, the distinction made between chromosomal and plasmid genes, gene products isolated in surrogate hosts, and gene function identified by complementation of selected genomic fragments. All these procedures are being applied to an increasing variety of C_1 utilizing organisms and the genetic data so obtained continues to play a vital role in the understanding of the unusual biochemistry and biology of these organisms, as well as the providing a basis for their directed manupilation for industrial purposes.

2. MUTATION

Auxotrophic mutants have a special place in the tools of trade of the microbial geneticist. In most organisms they are easy to isolate and characterize and are most effective as markers to select rare recombinants. For reasons which are still not completely understood, auxotrophs are difficult to isolate in most C_1 utilizing organisms. In addition, chemical mutagens such as ethyl methane sulfonate have proved to be more toxic to methylotrophs than is the case with many other bacteria [4] and indeed the effective use of chemical mutagens for mutagenesis of C_1 utilizing organisms is not common although nitrous acid is effective in *Methylobacterium* AM1 [5]. Both the difficulty of isolating auxotrophs and the ineffectiveness of chemical mutagens could be related to the permeability properties of C_1 utilizing organisms. Fortunately for the isolation of mutants UV has been shown to be an effective mutagen in both *Methylobacterium* AM1 [5] and *Methylophilus methylotrophus* AS1 [4].

The development of transposon mutagenesis has made the isolation of mutants very much easier in C_1 utilizing organisms. The most widely used transposon is Tn5 [6], which has a low specificity of insertion and hence mutations can be obtained in a wide range of genes. Tn5 encodes to kanamycin and streptomycin, and fortunately these resistances are well expressed in the majority of C_1 utilizing bacteria so far tested. As well, the restriction map of Tn5 is known and distinctive and this means that in any subsequent restriction enzyme mapping of the mutated gene the site of the mutation can be precisely determined. There are a variety of 'suicide' plasmids which can act as vectors of Tn5 so that insertion mutants can be obtained in either the intact organism [7-10] or on cloned chromosomal fragments of C_1 utilizing bacteria carried by cosmids [5,11]. As pointed out by de Vries [3], mutants affecting C_1 metabolism are being isolated and characterized in an increasing number of different organisms.

3. CLONING

The essence of recombinant DNA technology is the ability to isolate and characterize a fragment of the entire genome of an organism. It is worthwhile to recall that the use of plasmid and cosmid vectors is only an extension, albeit more efficient and embracing, of the classical microbial genetic experiments using F prime plasmids and transduction. For those experiments the choice of vector and the range of recipient hosts for the bacterial chromosome fragment was limited. Now there are a multitude of vectors derived from various plasmids, or in the case of cosmids, a hybrid between the bacteriophage λ and a plasmid.

The choice of vector is all important, depending upon the type of genetic analysis to be undertaken. If expression of the genes to be studied occurs at high level in *Escherichia coli*, then all the cloning and expression work can be carried out with vectors whose host range is limited to that organism. An example of this is the analysis of ribulose bisphosphate carboxylase oxygenase (RUBPC/O) of *Rhodopseudomonas sphaeroides* [12]

Tabita et al. [14] have shown that the form II RUBPC/O of *Rhodopseudomonas sphaeroides* produced from an *E.coli* strain in which this gene had been cloned, was functionally identical to the enzym produced in *R.sphearoides*. Thus with a need to produce variants of this enzyme with more favorite carbon dioxide fixation kinetic properties or to investigate in more detail the genetic organization of this function, it has been possible to carry out the site specific mutagenesis of the chromosomal region involved in *E.coli* where such techniques have been perfected, rather than having to do them in the genetically less facile *R.sphearoides*.

Complementation of functions carried by the genome fragments can sometimes only be detected in the original source of the DNA, so in these cases a vector with a wide host range must be used. The IncP-1 plasmids have been of particular value in this respect. These plasmids, first isolated in *Pseudomonas aeruginosa* have a bacterial host range which includes most gram negative genera. Plasmids of this group have been used to construct cloning plasmids and cosmids [11] with the same wide host range and these have been of special significance for C_1 utilizing organisms [13]. Similarly, the IncQ group plasmids, most derived from RSF010, have provided the basis for wide host range cloning vehicles. One difficulty with these vectors is that, in contrast to the IncP-1 vectors which have low copy numbers, the IncQ plasmids have high copy numbers. At least for some situations in *Pseudomonas* [AF Morgan, unpublished data], bacterial chromosome cloned in these vectors was unstable when transferred back to the original host. This problem has not been observed with IncP-1 vectors, which have been succesfully used with *Methylobacterium* AM1 [5,15], *Rhodopseudomonas sphearoides* [16], *Methylophilus methylotrophus* [4,17], *Methylobacterium organophilum* [11] and as described in this paper, *Methylophilus viscogenes.* One of the difficulties in the use of wide host range vectors is that native restriction enzymes may minimize the transfer of function DNA fragments between unrelated bacteria. If restriction deficient mutants can be found, as for example was done in *Paracoccus denitrificans* [3], then subsequent genetic analysis is facilitated.

4. COMPLEMENTATION

While recombinant DNA technology is highly effective in fragmenting the genome of an organism, the analysis of the function of the genes under investigation depends on a test of gene product activity, usually the repair of a defective function in either a surrogate organism or in the organism from which the DNA fragment has been derived. This procedure is commonly referred to as complementation, although the term is used with a slightly different meaning to the purpose for which this genetic term was originally devised.

The purposes of complementation are various. They include identification of a function in a genome library, localization of that function on a particular fragment of that library, determining the linkage relationships of the gene encoding the function and organizational and regulatory aspects of the relevant genes. Such tests can be applied equally well to genes located on either the bacterial chromosome or a plasmid. Once a gene function has been identified on a particular clone, it can then be manipulated and studied. *Tn5* or site specific mutagenesis combined with restriction endonuclease mapping can be used to identify specific genes which can then be subcloned and sequenced.

In the work of Reeve and his associates on the genetic structure of the two related methanogenic archaebacteria, *Methanococcus voltae* and *Methanococcus vannielii*, the genome of this archaebacteria was cloned in *E. coli* [18]. By complementation, a gene in *M. vannnielii* was identified which is functionally equivalent to the *his*I region of the *his*I/E gene of *E. coli* and to the *his*4A region of the *his*4 gene of *Saccharomyces cerevisiae* [19]. The sequences of these genes have been determined and from the degree of similarity it can be concluded that there are regions in all three organisms which are derived from a common ancestral gene, thus providing molecular evidence for the evolutionary origin of this archaebacterium.

In *Methylobacterium* AM1, Nunn and Lidstrom [5, 20] have used a genomic clone bank using the broad host range cosmid pVK100 to complement a range of mutants all of which have the common feature of being unable to grow on methanol. Ten complementation groups have been identified and related to individual regions of the gene bank. Each of these complementation groups has been characterized with respect to function, providing the first comprehensive genetic study of methanol dehydrogenase activity.

It is known that C_1 utilizing organisms commonly carry plasmids although there is little evidence to support the view that genes concerned with C_1 metabolism are located on plasmids. However, in the case of the *Hox* genes of *Alcaligenes eutrophus*, it was possible to determine whether or not they were found on megaplasmid DNA or located on the chromosome were involved. A range of aerobic facultatively autotrophic hydrogen bacteria were characterized with respect to their hydrogen oxidizing enzyme system (Hox). In the case of *Alcaligenes eutrophus* by means of plasmid curing, genetic transfer, and the analysis of mutations isolated by either chemical or transposon mutagenesis, it has been possible to localize structural and regulatory *Hox* genes on megaplasmids ranging from size 400-500 kilobase pairs [8, 9, 21]. While most of the genes were shown to be located on megaplasmid DNA the lack of transfer of some markers suggests that there is a chromosomally located regulatory gene. By means of cloning techniques, it was possible to confirm that *Hox* genes are clustered together on the megaplasmid, with at least six structural genes being identified.

5. CHROMOSOME MAPPING IN METHYLOTROPHS

The comprehensive genetic map available in *Escherichia coli* has been one of the reasons why this organism is favoured as a host for recombinant DNA technology procedures . The map constructed by the traditional procedures of conjugation and transduction is now being refined by restriction enzyme mapping and sequencing. It follows that the success of genetic investigation of C_1 utilizing organisms is going to depend on the contruction of genetic maps for them. Aspects of mapping have already been discussed above. The usual procedure for gene mapping is to first obtain mutants then to determine the spatial relationship of those mutants. The difficulties of isolating auxotrophs in methylotrophs has meant that these markers are essentially unavailable for map construction.

However by means of R68.45, Tatra and Goodwin [22] were able to determine the linkage relationships in *Methylobacterium* AM1 of antibiotic resistance markers and genes required for C_1 metabolism and were able to use auxotrophic markers isolated following N-methyl-N'-nitro-N-nitrosoguanidine treatment for the selection of recombinants in these crosses. As an alternative means of gene location, Moore et al. [17] used complementation mapping with wide host range prime plasmids to identify and map auxotrophic markers in the obligate methylotroph *Methylophilus methylotrophus* AS1. Prime plasmids are hybrids of an entire plasmid genome with an inserted segment of bacterial chromosome. Various primes carrying *M. methylotrophus* AS1 chromosome were constructed and transferred individually to mutants of *Pseudomonas aeruginosa.* Complementation of a mutant function in this surrogate host indicates that the isofunctional gene is carried on that individual prime. Complementation of two functions by the same prime plasmid indicates that the two *M. methylophilus* genes involved are on the same chromosomal fragment and hence linked. By this procedure, 28 primes of *M. methylophilus* were shown to carry 20 identifiable auxotrophic markers, individual primes carried from 1 to 8 different loci and four groups of linked markers were identified - the first chromosome map of any methylotroph.

Further work has shown that not all regions of methylotroph chromosome are accessible to mapping with this technique. To overcome this difficulty complementation mapping can also be carried out using genome banks constructed with wide host range cosmids. This procedure has been used to compare chromosomal gene arrangement in two obligate methylotrophs - *Methylophilus methylotrophus* AS1 and *Methylophilus viscogenes. M viscogenes* is an obligate methylotroph isolated from soil. The linkage data obtained by this method has been compared to that already established with the prime plasmids of *M. methylotrophus.* Complementation of markers has been demonstrated using both *P. aeruginosa* and *E. coli* as recipients.

The results of a series of complementations are summarized in Table 1. There is no implied gene order in the linkage groups identified, but in some cases gene order has been established by the availability of cosmids which can complement various combinations of the markers involved. The markers shown in the table are those complemented by fragments of methylotroph chromosome isolated as either prime plasmids or on the cosmid vector. The prime plasmids were derived from pMO172, an enhanced chromosome mobilizing plasmid similar to R68.45 [17]. The cosmid banks were derived using pLA2917 [11] and the host strain *E. coli* S17-1 [23], a strain which posseses plasmid mobilization genes integrated within the chromosome. By cloning in *E. coli* S17-1, it was possible to test complementation by biparental crosses rather than triparental matings in which one parent carries a plasmid capable of mobilizing the cosmid.

Table 1. Linkage groups in *Methylophilus methylotrophus* AS1 and *Methylophilus viscogenes* as determined by complementation mapping using prime plasmids (bold) and cosmid banks. While the presence of particular genes has been established in the various groups, the listing of the various genes does not imply that they are so ordered. The genotype symbols are the same as those used for *E.coli* [25].

M.methylotrophus	*M.viscogenes*
met-28 trpC trpD trpE pyrE	
metA met-28 trpC trpD trpE pyrE	met-28 trpC trpD pyrE
leu-8 proB pur-136	
proB pur-136	proB pur-136
pyrB proC pur-66	pyrB proC
pur-66 trpA trpB trpF argF argG leu-10 met-9011	
pheA trpA trpB trpF argG met-9011 purF	trpA trpF argF argG leu-10 met-9011
metE metC metB thr-48	
	hisIV hisIIA hisIIB hisIII
	hisV leu-8 pyrD
	pyrF phe-2

Complementation of auxotrophic markers by the methylotroph prime plasmids and cosmids was extensive. In *M. methylotrophus* for complementation in both *E. coli* and *P. aeruginosa*, 45 out of 66 (68%) of auxotrophs were complemented and in *M. viscogenes* of 34 *P. aeruginosa* markers tested, 65% were complemented. This supports the view that the genomic banks constructed in these organisms were comprehensive and should encompass all the genome.

Conclusions which can be drawn from these results include:

1. Gene arrangements in methylotrophs can be identified by complementation mapping with cosmids and plasmid primes.
2. Linkage arrangements of auxotrophs in *M. Methylophilus* and M. *viscogenes* show considerable similarity.
3. The similarity of the linkage results with both cosmids and plasmids suggests that internal rearrangements occurring during the isolation of these DNA fragments are uncommon.
4. Some arrangements of genes are conserved in the two species and show similarities to gene arrangements found in *Pseudomonas* [24]. For example, the (*trp*C, *trp*D *trp*E) cluster found in both

species is also found as a cluster in *P. aeruginosa* and *P. putida.* Likewise the *trp*F *leu*-10 *met*-9011 genes are clustered in both methylotrophs and in both *P. aeruginosa* and *P. putida..*

5. The fact that up to 7 auxotrophic markers can be found on the one cosmid spanning 25 kb of the methylotroph chromosome makes it likely that there is clustering of auxotrophs in the overall genome arrangement of these two bacteria.

It should be noted that linkage relationships derived by complementation mapping are determined with wild type alleles not, as is the case with most other ways of mapping, by using mutant alleles. Other markers, not shown above, including methanol dehydrogenase genes and genes on the pathway of hexose metabolism have also been shown to be complemented. Thus, complementation of genes in surrogate organisms by cosmid banks of C_1 utilizing organisms can be used to identify gene functions in those organisms and hence is an approach to the better understanding of the range of metabolic functions occurring in these organisms.

6. ACKNOWLEDGEMENTS

Work in the author's laboratory has been supported by the Australian Research Grants Scheme, Imperial Chemical Industries (U.K.), the Celanese Research Company and the Celgene Corporation.

REFERENCES

[1] Ball C (ed.) (1984). Genetics and Breeding of Industrial Microorganisms, 1-203, CRC Press, Boca Raton.
[2] Holloway BW (1984) In: CT Hou (ed), Methylotrophs: Microbiology, Biochemistry and Genetics. CRC Press, Boca Raton.
[3] de Vries GE (1986) FEMS Microbiol. Rev. 39, 235-258
[4] Kearney PP (1986) The biochemical and genetic analysis of Methylophilus methylotrophus AS1. Ph.D. thesis, Monash University, Clayton, Australia.
[5] Nunn DN, and Lidstrom ME (1986). J. Bacteriol. 166, 581-590.
[6] Berg DE and Berg CM (1983) Bio/Technology 1, 417-435.
[7] Weaver KE and Tabita FR (1983) J. Bacteriol. 156, 507-515.
[8] Hogrefe C, Romermann D and Friedrich B (1984) J. Bacteriol. 158, 43-48.
[9] Hogrefe C and Friedrich B (1984) Plasmid 12, 161-169.
[10] Whitta S, Sinclair MI and Holloway BW (1985) J. General Microbiology 131, 1547-1549.
[11] Allen L and Hanson RS (1985) J. Bacteriol. 161, 955-962.
[12] Quivey RG Jr and Tabita FR (1984) Gene 31, 91-101.
[13] Schmidhauser TJ, Ditta G and Helinski DR. (1986) In Vectors: A Survey of Molecular Cloning Vectors and Their Uses. Butterworths, London (in press).
[14] Tabita FR, Gibson JL Mandy WJ, Quivey RG Jr. (1986) Bio/Technology 4, 138-141
[15] Fulton GL, Nunn DN and Lidstrom ME (1984). J. Bacteriol. 160, 718-723,
[16] Weaver KE and Tabita FR (1985) J. Bacteriol. 164, 147-154.
[17] Moore AT, Nayudu M and Holloway BW (1983) Gen. Microbiol. 129, 785-799.
[18] Cue D, Beckler GS, Reeve JN and Konisky J (1985) Proc. Natl. Acad. Sci. 82, 4207-4211.
[19] Beckler GS and Reeve JN (1986) Mol. Gen. Genet in press.
[20] Nunn DN and Lidstrom ME (1986) J. Bacteriol. 166, 591-597.
[21] Friedrich B, Kortluke C, Hogrefe C, Eberz G, Silber B and Warrelman J (1986) Biochimie 68, 133-145.
[22] Tatra FR, Gibson JL, Mandy WJ and Quivey RG Jr. (1986) Bio/Technology 4, 138-141.
[23] Simon R, Priefer U and Puhler A (1983) Bio/Technology 1, 784-791.
[24] Holloway BW and Morgan AF (1986) Ann. Rev. Microbiol. 40, 79-105.
[25] Bachmann BJ and Low KB (1980) Microbiol. Rev. 44, 1-56.

GENETICS OF AEROBIC LITHOAUTOTROPHS

B. FRIEDRICH
Freie Universität Berlin, Institut für Pflanzenphysiologie, Zellbiologie und Mikrobiologie, Königin-LuiseStr. 12-16a, 1000 Berlin 33, Federal Republic of Germany.

1. INTRODUCTION

Aerobic lithoautotrophs fix carbon dioxide at the expense of reduced inorganic substrates including ammonia, nitrite, reduced sulfur compounds, iron, carbon monoxide, and hydrogen. This metabolic group of bacteria is physiologically defined and comprises taxonomically diverse species of microorganisms. Nitrifiers, sulfur- and iron-oxidizers are mostly obligately bound to the lithoautotrophic mode of life. They are metabolically extremely specialized and thus experimentally difficult to handle. With the exception of *Hydrogenobacter thermophilus* [1] hydrogenotrophic and carboxydotrophic bacteria are facultative lithoautotrophs which can also use organic substrates. This capability enables them to adopt rapildly to environmental changes in the supply of nutrients. The function of such a versatile metabolism depends on a highly developed regulatory system that keeps the flow of metabolites under a balanced control.

After a period of extensive studies on the physiology and biochemistry of aerobic lithoautotrophs [2] we start to get an insight into the molecular structure of these unconventional organisms. This advancement has been achieved by the application of modern genetic techniques. The present survey will primarily focus on the genetics of energy metabolism in aerobic chemolithoautotrophs. For a detailed discussion of C_1 carbon metabolism the reader is referred to accompanying articles in this symposium.

2. THE ROLE OF PLASMIDS IN LITHOAUTOTROPHIC METABOLISM

The presence of extrachromosomal DNA is widely distributed among lithoautotrophic bacteria. Plasmids have been found in sulfur-, iron-, and ammonia-oxidizing strains, in carboxydotrophic species as well as in hydrogenotrophs (Table 1). The most striking observation was the discovery of extremely large plasmids, termed megaplasmids, in hydrogen-oxidizing strains. The molecular mass of plasmids indigenous to *Thiosphaera pantotropha* [5], *Paracoccus denitrificans* [6], and species of *Alcaligenes* [3, 4, 15] exceeds one tenth of the *Escherichia coli* chromosome. Moreover, it is noteworthy that even closely related species differ considerably in number and sizes of plasmids.

Most of the plasmids are cryptic. So far a single plasmid-encoded lithoautotrophic trait has been conclusively identified. It is hydrogen-dependent autotrophic growth in some *Alcaligenes* spp. [3, 4]. One has to be careful before generalizing this observation. *A. eutrophus* CH34, a typical member of the two hydrogenases-containing *Alcaligenes* strains, harbors two medium-sized plasmids that determine multiple resistances to heavy metal salts (Table 1] but apparently no lithoautotrophic function [12]. The frequent loss of hydrogen autotrophy in *Nocardia opaca* suggested plasmid-borne lithoautotrophic genes [13]. However, there is accumulating evidence for a chromosomal location of hydrogenase and carbon dioxide fixation genes in this host. Thallium resistance was recently assigned to one of the plasmids in *N. opaca* lb [14].

In an early study Pootjes [16] suggested the involvement of plasmid DNA in hydrogen-oxidizing ability of *Pseudomonas facilis*. A previous characterization of Tn5-induced hydrogenase mutants from *P. facilis* strain DSM550 did not support this proposal [17]. Recently, we reexamined a variety of *P. facilis* wild type strains for extrachromosomal DNA and surprisingly found a fairly heterogenous plasmid pattern (Fig. 1). Non-revertible hydrogenase mutants with plasmid deletions were isolated from the two plasmids-harboring strain (Fig. 1, lanes h). These preliminary results indicate that there may exist isolates of *P. facilis* with plasmid-located lithoautotrophic functions and others that bear the respective genes on the chromosome. Apparently the organisms have evolved a variety of genetic structures to maintain the genes required for the expression of lithoautotrophic metabolism.

Table 1. Plasmids in different groups of lithoautotrophic bacteria

Species	Number	Plasmid Size (kb)	Function	Reference
T. pantotropha	2	110;>450	Cryptic	5
T. versutus	2	106;>450	Cryptic	6
T. ferrooxidans	3	7;16;>23	Cryptic	7, 8
T. acidophilus	4	18;27;76;>90	Cryptic	8, 9
T. intermedius	1	36	Cryptic	9
Nicrobacter X14	1	114	Cryptic	18
P. denitrificans	2	754;>450	Cryptic	6
A. eutrophus	1	450	H_2-oxidation	4, 3
			CO_2-fixation	10
			Denitrification	
A. eutrophus CH34	2	163	Ni Hg Co resistance	12
		238	Zn Ca Hg Co resistance	
A. hydrogenophilus	2	357	Cryptic	15
		419	H_2-oxidation	
P. carboxydovorans	1	129	Cryptic	6
N. opaca lb	2	174	Cryptic	13
		140	Thallium resistance	14

3. TRANSFER AND EXPRESSION OF LITHOAUTOTROPHIC FUNCTIONS IN OTHER HOSTS

The ability to grow autotrophically with hydrogen could be transferred by cell contact from *N. opaca* lb to non-autotrophic recipients such as *Rhodococcus erythropolis* [13]. Since plasmid-cured mutants were still Aut^+ and acted as donors of the lithoautotrophic character, it was assumed that the genes coding for hydrogen autotrophy reside on the chromosome as an independent self-transmissible genetic element [14].

The transfer of large pieces of chromosomal DNA including autotrophic genes has been reported for *Xanthobacter autotrophicus* GZ29. Cell contact was necessary to achieve the transfer [19]. However, no indigenous plasmid could be detected in this bacterium [6]. Thus, the nature of the gene transfer is still unclear.

Hydrogen-oxidizing ability (Hox) encoding megaplasmids, designated pHG, are self-transmissible by conjugation at a high frequency of approximately 10^{-3} per donor. This transfer is possible between different strains and species of *Alcaligenes* [4, 15]. The only exception so far known is plasmid pRMB1 indigenous to *A. eutrophus* ATCC 17706 which requires the mobilization by a broad host range plasmid [20].

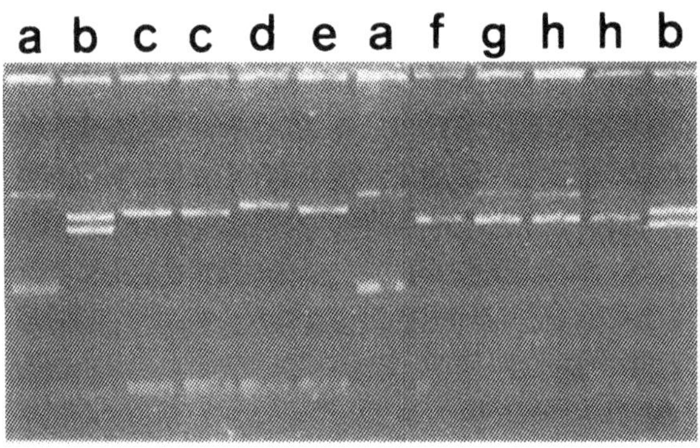

Figure 1. Plasmid pattern of *Pseudomonas facilis* strains. Lane c, strain DSM550; lane d, ATCC15376; lane e, deletion mutant of ATCC15376; lane f, DSM649; lane g, DSM620; lane h, Hox^- deletion mutant of DSM620; plasmids pHG1 (450 kb) and RP4 (54 kb), lane a; pMo128 (163 kb) and pMo130 (238 kb, lane b) were used as size standards.

Preliminary attempts to transfer hydrogen autotrophy from *A. eutrophus* H16 to Gram-negative non-lithoautotrophic bacteria were unsuccessful, indicating a limited host range of plasmid pHG1. The selection for pHG1 and its mobilization were improved by introducing into the megaplasmid the kanamycin (Km) resistance-conferring transposon Tn5 in addition to a Mob function derived from the promisuous plasmid RP4 [21]. This pHG1::Tn5-Mob containing *A. eutrophus* donor gave rise to Km resistant transconjugants [23] of the herbizide degrading non-lithoautotrophic recipient *A. eutrophus* JMP134 [22]. Contrary to the parent strain the transconjugants grew autotrophically with hydrogen, expressed soluble NAD-linked as well as particulate hydrogenase and contained ribulosebisphosphate carboxylase activity [23].

In a previous study we have shown that regulation of hydrogenase synthesis can be altered by plasmid exchange between *A. eutrophus* H16 and *A. hydrogenophilus* [15]. Recently, the latter strain was used as a donor in intergeneric conjugal transfer of hydrogen autotrophy to *Pseudomonas oxalaticus* . Transconjugants of this non hydrogen-oxidizing but carbon dioxide-fixing recipient grew with hydrogen autotrophically and expressed both soluble and particulate hydrogenases [24].

This survey shows that there are several naturally occurring gene transfer mechanisms in lithoautotrophs which may promote the transmission of lithoautotrophic traits even among distantly related bacterial groups.

4. LITHOAUTOTROPHICALLY DEFECTIVE MUTANTS

To dissect the genetics of lithoautotrophy the availability of mutants is essential. Lithoautotrophically defective mutants have been isolated from facultative bacteria. The main advantage is, that with the facultative species, a mutation in lithoautotrophic function is not lethal. Besides conventional techniques one has made use of the powerful transposon mutagenesis. It has been employed for the isolation of mutants impaired in hydrogen- [25, 23, 17, 26, 28] and sulfur-oxidizing [25, 5] ability. Although the sulfur-oxidizing bacteria are well studied from the physiological standpoint, little is known about their biochemical mechanism of thiosulfate cleavage [29]. The phenotypic characterization of mutants so far isolated (Table 2) promisses new insights into sulfur-oxidation. The highly pleiotropic Sox$^-$ phenotype of mutants from *T. pantotropha* suggests the involvement of molybdenum in thiosulfate oxidation.

Due to extensive knowledge on the physiology and biochemistry of hydrogenase [2, 30], the genetics of H_2-oxidation has made major advancements. *Rhizobium japonicum* ,a member of the facultative hydrogen autotrophs, has received a lot of research attention since the hydrogen-oxidizing enzyme, hydrogenase, recycles the hydrogen produced by nitrogenase, supplying additional energy for dinitrogen-fixation [26]. Seven classes of *Rhizobium japonicum* mutants altered in hydrogen-

Table 2. Transposon-induced mutants from sulfur-oxidizing bacteria

Mutagenized species	Transposon used	Relevant phenotype[a]	Reference
T. novellus	Tn501	Reduced ability: to oxidize sulfur compounds and to fix CO_2	27
T. versutus	Tn1721	Auxotrophs, but no mutants with impaired autotrophy	27
Ts. pantotropha	Tn5-Mob	1. Sox$^-$ 2. Sox$^-$ Hox$^-$ Nit$^-$ 3. Sox$^-$ For$^-$ Nir$^-$ Xan$^-$	5

[a] Inability to grow with: thiosulfate and sulfide, Sox$^-$; hydrogen, Hox$^-$; nitrite anaerobically, Nit $^-$; formate, For $^-$; nitrate, Nir $^-$; xanthine, Xan$^-$.

oxidizing ability have been described [30]. Among these Hup⁻ isolates there are candidates for structural gene mutations, defects in the incorporation of hydrogenase into the membrane, hypersensitivity to oxygen-mediated represon of hydrogenase synthesis, and constitutive formation of hydrogenase and ribulosebisphosphate carboxylase. Two groups of mutants were identified, that in addition to lacking hydrogenase activity lacked symbiotic nitrogen-fixing activity presumably as a result of a lesion in heme synthesis, a regulatory gene or the assembling of Fe-S clusters.

A well, if not the best studied hydrogenotroph, is *A. eutrophus*. Hydrogenase mutants derived therefrom have been characterized on the basis of physiological, biochemical, immunological, and genetic investigations. The results are listed in Table 3. Ten phenotypically different classes of hydrogenase mutants are known. The first two classes represent regulatory Hox mutants that have lost the ability to form soluble and particulate hydrogenase due to a mutation on the megaplasmid or the chromosome, respectively [32]. The latter mutation confers to the organism a highly pleiotropic phenotype. We assume that these Hno$^-$ mutants are deficient in a protein which dominantly controls the expression of a variety of genes including those for hydrogenase. According to recent results Hox$^-$ mutants of class three form extremely unstable catalytically hydrogenase [Kortlüke et al., manuscript submitted for publication]. They are possibly lacking a component essential for both hydrogenases such as Fe-S clusters or nickel in its active configuration. Mutants of class four, five, and six have scaped from the Hox regulatory pattern characteristic to the parent. Thus HoxTr mutants from *A. eutrophus* H16 form hydrogenase at an elevated temperature [33], Hox ^{c} mutants from *A. hydrogenophilus* do not require for hydrogenase induction [34] and Hox $^{Ose-}$ isolates from *A. eutrophus* ATCC17707 are insensitive to oxygen repression [35]. Mutants with mutations in the structural genes for the soluble hydrogenase (Hos$^-$) or particulate hydrogenase (Hop$^-$) are represented by classes seven and eight [32]. Strains forming one of the two hydrogenases constitutively (Hos^c and Hop^c) have recently been isolated by means of reversion from Tn5-induced Hox$^-$ mutants of the first class. Restriction mapping and Southern blot analysis indicated that the constitutive phenotype could be due to initiation of *hox* gene transcription at a transposon-borne promoter [Kortlüke et al., manuscript submitted for publication].

With the exception of a single *hox* gene all of the *hox* mutations mapped on the pHG plasmid. This has been shown by examining the phenotype of transconjugants after exchange of plasmids containing various lesions (Table 3) between *A. eutrophus* , plasmid-free derivatives and *A. hydrogenophilus* [32, 15].

5. ORGANIZATION AND ISOLATION OF HYDROGENASE GENES

Although large plasmids have been detected in *R. japonicum*, a correlation of hydrogenase genes with plasmid content has not been possible [36]. Cantrell et al [37] were the first who constructed a genomic DNA library of *R. japonicum* to clone hydrogenase genes (*hup*). Transconjugants that contained *hup* specific DNA were identified by complementation of Hup$^-$ mutants. The cosmid clone pHU1 led to a return of hydrogenase activity when introduced into some of these Hup$^-$ mutants. By analyzing Tn5-insertions into the chromosome, Haughland et al. [26] concluded that hydrogenase-encoding genes reside on a 15-kb DNA sequence of the cosmid insert. They suggested an organisation of *hup* genes in at least two, but probably three transcriptional units.

The situation in *A. eutrophus* is more complex due to the fact that this bacterium contains two distinct hydrogenase proteins. The soluble enzyme is a heterotetramer [39], the particulate hydrogenase a

Table 3. Phenotypes of hydrogenase mutants from *Alcaligenes eutrophus*.

Class	Pheno-type	Characteristics	Genetic location	Reference
1	Hox^-	Lack of hydrogenases	Plasmid	32
2	Hno^-	Impaired in Hox, nitrate assimilation/dissimilation and other things	Chromosome	32
3	Hox^L	Devoid of catalytically active Hox	Plasmid	*
4	HoxTr	Temperature resistant Hox formation	Plasmid	33
5	Hox^c	Constitutive Hos and Hop synthesis	Plasmid	34
6	Hox^{Ose-}	O_2-insensitive Hox formation	Plasmid	35
7	Hos^-	Lack of soluble hydrogenase	Plasmid	32
8	Hop^-	Lack of particulate hydrogenase	Plasmid	32
9	Hos^c	Constitutive Hos synthesis	Plasmid	Unpublished
10	Hop^c	Constitutive Hop synthesis	Plasmid	*

* Kortlüke, C., C. Hogrefe, G. Eberz, A. Pühler, and B. Friedrich, manuscript submitted for publication.

heterodimer [40]. Although the two proteins have Fe-S clusters and a nickel component in common, immunological studies suggest that the peptides are the product of six independent structural genes [23]. Moreover, it has been shown that *hox* gene expression is subject to a sophisticated molecular control including derepression by energy limitation [38], temperature-sensitivity [33] and oxygen repression [35].

Mapping and isolating of *hox* genes from *A. eutrophus* were facilitated by their plasmid location [32]. Based on the preparative procedure for the isolation of pure megaplasmid DNA [41], it was possible to identify on the plasmid sites of insertions and deletions associated with a specific Hox phenotype. All of the plasmid-linked *hox* lesions were found to map within a region of 80-kb on plasmid pHG1 (Fig.2; Kortlüke et al., manuscript submitted for publication). The positions of Tn5-insertions perfectly correlated with the *hox*-containing DNA sequences cloned into the suicide vector pSUP202 [42]. From complementation and marker exchange studies with various Hox mutants we concluded that a *hox* regulatory gene coding for the temperature-sensitivity and the energy-dependent control resides on a 4.8-kb EcoRI/BamHI fragment. This sequence contains an entire transcriptional unit. A *hox* gene that is required for the synthesis of hydrogenase appears to be located at the border of this fragment within an

EcoRI sequence of 8.6-kb (Fig. 2). At least a portion of the soluble hydrogenase (*Hos*) -specific *hox* genes occurs within EcoRI fragments of 11.6-kb and 1.0-kb. An incomplete sequence of a third *hox* gene coding for particulate hydrogenase (*Hop*) has been identified on a 11.0-kb EcoRI insert. It is interesting to note that this particular fragment is the genetic locus of plasmid-encoded *cfx* genes [10]. Not all of the phenotypic classes of *Hos*⁻ mutants could be complemented by the cloned *hox* genes [42]. Thus other hybrid plasmids will have to be isolated to obtain all the genes involved in hydrogen oxidation.

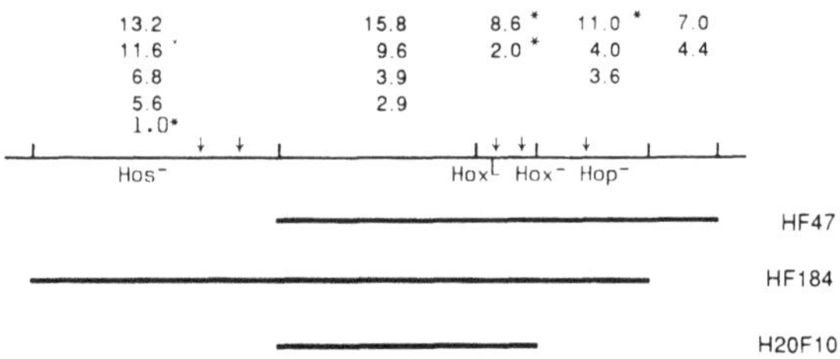

Figure 2. EcoRI restriction map of megaplasmids with deletions and insertions. The size of missing fragments is given in kilobase pairs. Numbers, listed in a row are fragments whose sequential order is still unknown. Fragments with insertions are marked (*). The corresponding mutant phenotype is indicated by arrow. The solid lines mark plasmid deletions.

The chromosomally located *hox* gene whose mutation led to the occurrence of pleiotropic Hno⁻ mutants (Table 3) has recently been cloned on a 20-kb EcoRI fragment from a genomic library of *A. eutrophus* (D. Römermann and B. Friedrich, unpublished result). In parallel cloning experiments, conducted with DNA from *P. facilis*, a cosmid clone has been found that does not only complement Hno⁻ mutants of *P. facilis* but also those of *A. eutrophus* (J. Warrelmann and B. Friedrich, unpublished result). This observation is in accordance with the hypothesis that the *hno* gene determines a general gene expression factor which may also be relevant to other lithoautotrophs.

Recently a gene bank from DNA of *A. hydrogenophilus* was constucted using a broad host range cosmid vector. *hox* genes were identified by transferring the hybrid cosmids into *P. oxalaticus* OX1. One of the Hox⁺ transconjugants expressed particulate hydrogenase whereas soluble hydrogenase activity could not be detected [43].

Attempts to complement Cfx⁻ and Hox⁻ mutants from the hydrogen-methanol autotroph *Xanthobacter* sp. strain H4-14 with a genomic clone bank revealed that *cfx* and *hox* genes are not linked in this organism [28].

REFERENCES

[1] Shiba H, Kawasumi T, Igarashi Y, Kodama T, Minoda Y (1985). Arch. Microbiol. 141: 198-203.
[2] Bowien B., H. G. Schlegel. (1981). Ann. Rev. Microbiol. 53: 405-452.
[3] Andersen, K., R. C. Tait, W. R. King. (1981). Arch. Microbiol. 129: 384-390.

[4] Friedrich, B., C. Hogrefe, H. G. Schlegel. (1981). J. Bacteriol. 147: 198-205.
[5] Chandra, T. S., C. G. Friedrich. (1986). J. Bacteriol. 166: 446-452.
[6] Gerstenberg, C., B. Friedrich, H. G. Schlegel. (1982). Arch. Microbiol. 133: 90-96.
[7] Holmes, D. S., J. H. Lobos, L. H. Bopp, G. G. Welch. (1984). J. Bacteriol. 157: 324-326.
[8] Martin, P. A. W., P. R. Dugan, O. H. Tuovinen. (1981). Can. J. Microbiol. 27: 850-853.
[9] Davidson, M. S., A. O. Summers. (1983). Environm. Microbiol. 46: 565-572.
[10] Klintworth, R., M. Husemann, J. Salnikow, B. Bowien. (1985). J. Bacteriol. 164: 954-956.
[11] Römermann, D., B. Friedrich. (1985). J. Bacteriol. 162: 852-854.
[12] Mergeay, M., D. Nies, H. G. Schlegel, J. Gerits. P. Charles, F. v. Gijsegem. (1985). J. Bacteriol. 162: 328-334.
[13] Reh, M., H. G. Schlegel. (1981). J. Gen. Microbiol. 126: 327-336.
[14] Sensfuss, C., M. Reh, H. G. Schlegel. (1986). J. Gen. Microbiol. 132: 997-1007.
[15] Friedrich, B. C. G. Friedrich, M. Meyer, H. G. Schlegel. (1984). J. Bacteriol. 158: 331-333.
[16] Pootjes, C. F. (1977). Biochem. Biophys. Res. Commun. 76: 1002-1006.
[17] Warrelmann, J., B. Friedrich. (1986). J. Gen. Microbiol. 132: 91-96.
[18] Kraft, J., E. Bock. (1984). Arch. Microbiol. 140: 79-82.
[19] Wilke, D. (1980). J. Gen. Microbiol. 117: 431-436.
[20] Behki, R. M., G. Selvaraj, V. N. Iyer. (1983). Can. J. Microbiol. 29: 767-774.
[21] Simon, R. (1984). Mol. Gen. Genet. 196: 413-420.
[22] Don, R. H., J. M. Pemberton. (1981). J. Bacteriol. 145: 681-686.
[23] Friedrich, B., C. Kortlüke, C. Hogrefe, G. Eberz, B. Silber, J. Warrelmann. (1986). Biochimie 68: 133-145.
[24] Umeda, F., H. Min, M. Urushihara, M. Okazaki, Y. Miura. (1986). Biochem. Biophys. Res. Commun. 137: 108-113.
[25] Srivastava, S., M. Urban, B. Friedrich. (1982). Arch. Microbiol. 131: 203-207.
[26] Haughland, R. A., M. A. Cantrell, J. S. Beaty, F. J. Hanus, S. A. Russel, H. J. Evans. (1984). J. Bacteriol. 159: 1006-1012.
[27] Davidson, M. S., P. Roy, A. O. Summers. (1985). Appl. Environm. Microbiol. 49: 1436-1441.
[28] Lehmicke, L. G., M. E. Lidström. (1985). J. Bacteriol. 162: 1244-1249.
[29] Kelly, D. P. (1985). Microbiol. Sci. 2: 105-109.
[30] Maier, R. J. (1986). CRC, Critical Rev. in Biotechnology 3: 17-38.
[31] Friedrich, C. G., O. Meyer, T. S. Chandra. FEMS, Microbiol. Lett. (in press).
[32] Hogrefe, C., D. Römermann, B. Friedrich. (1984). J. Bacteriol. 158: 43-48.
[33] Friedrich, C. G., B. Friedrich. (1983). J. Bacteriol. 153: 176-181.
[34] Schlegel, H. G., M. Meyer. (1985). Arch. Microbiol. 141: 377-383.
[35] Cangelosi, G. A., M. L. Wheelis. (1984). J. Bacteriol. 159: 138-144.
[36] Cantrell, M. A., Hickok, R. E., H. J. Evans. (1982). Arch. Microbiol. 131: 102-106.
[37] Cantrell, M. A., R. A. Haughland, H. J. Evans. (1983). Proc. Nat. Sci. USA 86: 181-185.
[38] Friedrich, C. G. (1982). J. Bacteriol. 149: 203-210.
[39] Schneider, K., H. G. Schlegel, K. Jochim. (1985). Europ. J. Biochem. 138: 533-541.
[40] Schink, B., H. G. Schlegel. (1979). Biochim. Biophys. Acta 567: 315-324.
[41] Hogrefe, C., B. Friedrich. (1984). Plasmid 12: 161-169.
[42] Eberz, G., C. Hogrefe, C. Kortlüke, A. Kamienski, B. Friedrich. (1986). J. Bacteriol. (in press).
[43] Yagi, K., H. Min, M. Urushira, Y. Manabe, F. Umeda, Y. Miura. (1986). Biochem. Biophys. Res. Commun. 137: 114-119.

MOLECULAR BIOLOGY AND REGULATION OF MICROBIAL CO_2 FIXATION

F. Robert TABITA, Janet L. GIBSON, Yves JOUANNEAU, Deane L. FALCONE, and A. Melinda RAINEY.
Center for Applied Microbiology and Department of Microbiology, The University of Texas at Austin, Austin, Texas, 78712-1095 U.S.A.

1. INTRODUCTION

Purple nonsulfur photosynthetic bacteria (Rhodospirillaceae) are probably the most versatile microorganisms known, since they are capable of five distinct modes of metabolism, including photoautotrophic and chemoautotrophic growth with CO_2 as the sole source of carbon [1]. The first step of CO_2 assimilation is catalyzed by the enzyme ribulose 1, 5-bisphosphate carboxylase/oxygenase (RuBPC/O). Because of the ability of this protein to catalyze the key reactions of both photosynthesis and photorespiration, considerable interest has been focused on the enzymology and molecular regulation of RuBPC/O. *Rhodopseudomonas (Rhodobacter) sphaeroides* ,a typical member of the Rhodospirillaceae, has been shown to contain two structurally and immunologically distinct forms of RuBPC/O, form I and form II [2-4]. Form I is a hexadecamer composed of eight large and eight small subunits; whereas form II is an aggregate of large subunits and resembles the catalytically active dimer isolated from *Rhodospirillum rubrum* in that it lacks small subunits [5]. The mode of regulation and significance of two forms of RuBPC/O in *R. sphaeroides* and *R. capsulata* [6, 7] is not understood at this time. It is the purpose of this paper to review aspects of the molecular biology and control of the form I and form II RuBPC/O in *R. sphaeroides* .

2. REGULATION OF EXPRESSION OF RUBPC/O IN *R. SPHAEROIDES*

Low levels of RuBPC/O activity are found in cells grown with oxidized electron donors, such as malate [2, 5]. However, cells initially grown on malate and subsequently transferred to a 1.5% CO_2- H_2 atmosphere showed a marked increase in the level of RuBPC/O (Fig. 1). This response is similar to that seen with *Rs. rubrum* [6] and, as already pointed out for *Rs. rubrum* , the CO_2 percentage in the H_2-CO_2 gas mixture is critical for obtainung maximal levels of RuBPC/O. Cultures of *R. sphaeroides,* sparged with 2.5% CO_2 in H_2, displayed activities about two-thirds lower than those grown under 1.5% CO_2. As indicated in Fig. 1A, the increase of RuBPC/O specific activity occurred in the midlogarithmic phase of growth and slowed as cells entered the stationary phase.

It was important to establish whether this increase in carboxylase activity was a result of new protein synthesis or activation of preformed enzyme. Thus, quantitative immunological methods were employed

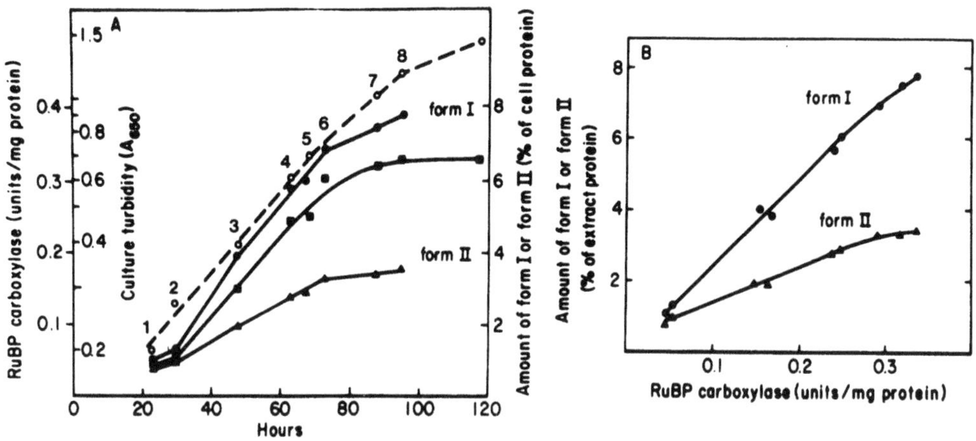

Figure 1. A) Time course of RuBPC/O protein accumulation and activity in extracts of cells growing on 1.5% CO_2 in H_2. B) Relationship between the relative amounts of form I and form II antigen and RuBPC/O specific activity.

to distinguish between these possible alternatives since antisera to the native form I or form II RuBPC/O does not cross-react with the heterologous enzyme [2-4, 8]. Rocket electroimmunoassay (EIA) and competitive enzyme-linked immunosorbent assays (ELISA) indicated that photolithotrophic growth under CO_2 limitation promotes the accumulation of both form I and form II RuBPC/O polypeptides inside the bacteria (Fig. 1). This was also shown by 2-dimensional electrophoresis [8]. There is a close correlation between the increase in specific activity and the accumulation of form I and form II carboxylase proteins (Fig. 1B), which is expressed as a percentage of all the extract protein. From the accumulated percentages of the form I and form II enzymes, it can be calculated that the average specific activity of each enzyme was about 2.7 U/mg of carboxylase protein, in good agreement with the activities reported for the pure enzyme [2].

The EIA and ELISA determinations were also employed to examine the relative levels of form I and form II RuBPC/O during the transition from photoheterotrophic growth in the malate medium to photolithotrophic (H_2/CO_2-dependent) growth in a mineral salts medium. The ratio of form I- form II (on a protein basis) doubled the very early stages of photolithotrophic growth and then reached an equilibrium when form I RuBPC/O became twice as abundant as form II RuBPC/O. This ratio was reversed in cells grown on malate as carbon source (Table 1). These results indicate that RuBPC/O synthesis is not totally repressed in heterotrophically-grown cells, yet form II RuBPC/O is twice as abundant on a protein basis. Upon transfer to the photolithotrophic growth mode, form I RuBPC/O became the more abundant carboxylase species (on a protein basis), indicating that the enzymes are synthesized at different rates; it was calculated that the rate of form I RuBPC/O synthesis is twice that of form II RuBPC/O on a molar basis [8]. These results further suggest that the two enzymes present in *R. sphaeroides* are regulated independently.

Table 1. Comparison of RuBPC/O levels in malate and H_2/CO_2-grown *R. sphaeroides*

Carbon source	Specific Activity (U/mg protein)	Amt. of antigen (% extract protein) form I	form II	form I-form II ratio
Malate	0.032 ± 0.01	0.49 ± 0.05	0.92 ± 0.04	0.53
1.5% CO_2 in H_2	0.37 ± 0.05	8.4 ± 0.8	4.0 ± 0.5	2.10

3. ISOLATION, CLONING, AND EXPRESSION OF FORM I AND FORM II RUBPC GENES

To study the apparent differential regulation of form I and form II RuBPC/O at the molecular level, it will be necessary to employ the specific RuBPC/O genes in subsequent investigations. Accordingly, we embarked upon the isolation of the RuBPC/O structural genes. The form II gene was isolated first since the *Rs. rubrum* gene, when used as a heterologous probe in southern hybridization experiments, strongly hybridized to a 3 kb *Eco*RI fragment from *R. sphaeroides* chromosomal DNA [9]. This fragment was subcloned into pBR322 and also inserted behind the *lac* promoter in the expression vector pUC8. Restriction endonuclease analysis of these subclones revealed a single *SalI* site, a pair of PstI sites, a number of sites for *Sac*II and no sites for the enzymes *Hin*dIII, KpnI, and *Pvu*II. Hybrid plasmid subclones in both orientations were examined for possible expression of RuBPC/O activity in *E. coli*. The plasmids pRQ52 and pRQ53 were examined under induced and non-induced conditions, using the gratuitous inducer IPTG, since the form II gene was behind the *lac* promoter in plasmid pUC8. Extracts prepared from maximally induced cultures showed high levels of enzymatic activity with the fragment ion orientation I (pRQ52), and low, but reproducible levels of enzyme were found in extracts obtained from *E. coli* cells containing pBR322 subclones with the form II gene in orientation I (Table 2). In addition, antisera to purified *R. sphaeroides* form I and form II RuBPC/O showed that the enzymatic activity expressed in *E. coli* was indeed due to the synthesis and assembly of form II polypeptides [9]. Electrophoretically purified combinant form II RuBPC/O was shown to be functionally and chemically identical to the *R. sphaeroides* enzyme [10].

A library of cloned *R. sphaeroides* DNA was screened by colony hybridization for form I RuBPC/O sequences using the form II gene and the *Anacystis nidulans* large and small subunit genes (plasmid pCS75) as heterologous probes [11]. Plasmid DNA from positive clones was isolated, restricted with *Eco*R1 and subjected to Southern hybridization using isolated fragments of pRQ52 and pCS75 as probes. A plasmid, designated pJG336, exhibited the expected hybridization pattern (based on prior

Table 2. Expression of cloned RuBPC/O genes in *E. coli*

Plasmid vector	Hybrid plasmid	Orientation of insert	Specific activity of RuBPC/O (nmol CO_2 fixed/min/mg protein)
pBR322	pRQ2	I	2.60
pBR322	pRQ5	II	0.16
pUC8	pRQ52	I, induced 2.5 h	148.70
pUC8	pRQ52	I, uninduced	11.30
pUC8	pRQ53	II, induced 2.5 h	0.29
pUC8	pRQ53	II, uninduced	0.40
pUC8	pJG29	I, uninduced	6.00
pUC8	pJG29	I, induced 4 h	67.00
pUC8	pJG29	I, induced 12 h	187.00
pUC8	pJG45	II, uninduced	0.00
pUC8	pJG45	II, induced 4 h	0.00

studies with restricted *R. sphaeroides* DNA). Digestion of pJG336 with several restriction enzymes and subsequent hybridization analysis with ^{32}P-labeled RuBPC/O fragments of pCS75 and pRQ52 revealed a single 4 kb *Sma*I hybridizing band, suggesting that the form I genes were contained entirely within this fragment [11]. The *Sma*I fragments, in the region of 4 kb, were eluted from an agarose gel and ligated into *Sma*I digested pUC8. Restriction mapping of two clones, pJG29 and pJG45, showed that both contained the desired 4kb *Sma*I fragment but the inserts were in opposite orientations. A partial restriction map of pJG29 is shown (Fig. 2). Cell extracts of *E. coli* containing either pJG29 (orientation I) or pJG45 (orientation II) were assayed for RuBP-dependent carboxylase activity.

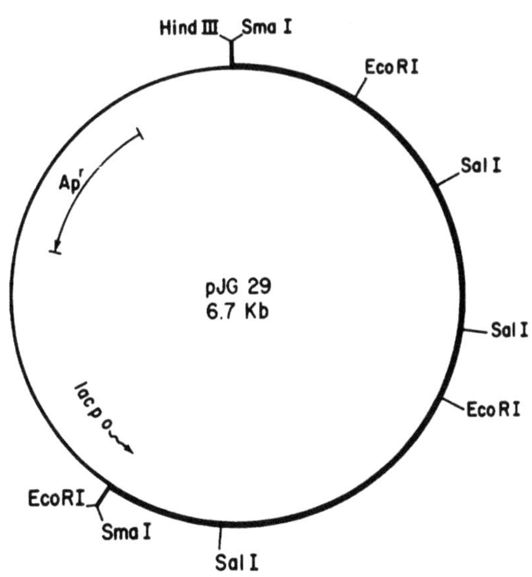

Figure 2. Partial restriction map of plasmid pJG29.

Following a 4 h incubation with IPTG to induce expression from the *lac* promoter of pUC8, substantial activity was detected in extracts of pJG29, which reached even higher levels after a 12 h induction period (Table 2). Expression of the form I RuBPC/O was absolutely dependent on transcription from the *lac* promoter, as virtually no enzyme was produced in cells harboring pJG45, regardless of induction. The subunits of form I RuBPC/O produced in *E. coli* exactly comigrated with *R. sphaeroides* form I large and small subunits. Furthermore, both subunits were immunoreactive with antibodies directed against denatured form I RuBPC/O in Western immunoblot analysis [11].

The cloning of the form I and form II RuBPC/O genes makes possible direct comparison of their nucleotide sequences. The unique opportunity to compare two functionally and structurally distinct RuBPC/O coding sequences from the same organism should provide valuable insight into the evolution of this enzyme and possibly the origins of the small subunit. Of interest also are the noncoding regions upstream of the structural genes which should contain sequences important in regulation and expression of the form I and form II RuBPC/O of *R. sphaeroides* .

4. TRANSPOSON MUTAGENESIS AND REPLACEMENT OF INACTIVATED FORM II RUBPC/O GENE

Recent studies, summarized above, suggest that the form I and form II RuBPC/O genes are regulated independently. To study the physiological significance of the two carboxylases, it would be desirable to obtain mutants which lack one or the other enzyme. Site specific recombination of the wild-type allele for the form II gene with Tn5-inactivated gene was deemed a feasible approach. This has been successfully employed to elucidate the function (or lack thereof) of the cyt-c_2 in photosynthesis of *R. capsulata* [12]. An insertion into the form II-containing 3 kb EcoRI fragment was found to block the synthesis of active enzyme and immunoreactive material in *E. coli* [13]. This Tn5 inactivated form II RuBPC/O gene was then conjugally transferred to *R. sphaeroides* via the broad host-range vector pRK290. Gene replacement through double-reciprocal recombination then allowed for the isolation of form II carboxylase mutants. Four isolates, when subjected to Southern hybridization of *Eco*RI digested DNA probed with the [^{32}P]- labeled form II RuBPC/O gene, showed the presence of the inactivated form II:: Tn5 fragment (8.7 kb) with no chromosomal allele of the form II carboxylase fragment. Extracts prepared from autotrophically-grown mutant strains showed no form II protein when examined by rocket EIA (Fig. 3), confirming the genetic studies.

Interestingly, comparisons of the photolithotrophic growth rates and total enzymatic activities indiicated that the mutants grew somewhat slower than the wild-type and the form II mutant strains exhibited about 2-2.5-fold lower specific activity. Further studies are in progress to assess the physiological significance of mutants lacking form II RuBPC/O; certainly these studies confirm the independent regulation of form I and form II RuBPC/O synthesis and also indicate that the form II enzyme is not required for autotrophic growth.

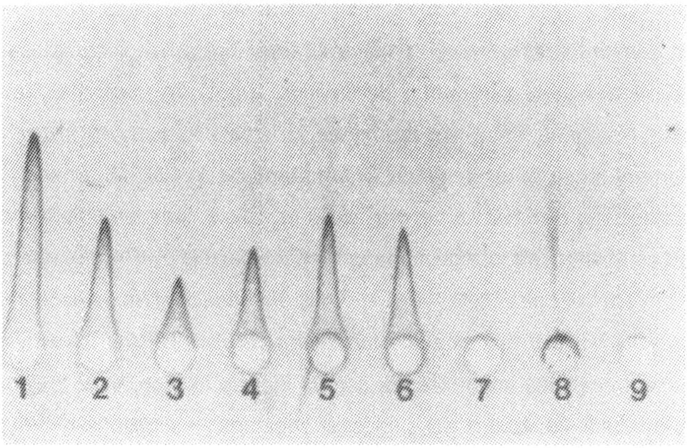

Figure 3. Rocket EIA analysis of extracts of form II-mutant strain 11-6 (lanes 7-9) compared to wild-type strain HR (lanes 4-6); form II RuBPC/O standards (lanes 1-3).

5. ISOLATION OF REGULATORY GENES BY COMPLEMENTATION OF AUT⁻ MUTANTS

Several mutants of *R. sphaeroides* have been isolated which are incapable of growth under photoautotrophic growth conditions [14], yet are capable of growth under photoheterotrophic conditions with malate as a carbon source. These mutants (Aut^-) which have been isolated via chemical and biological mutagenesis procedures, produce basal levels of RuBPC/O under photoheterotrophic conditions. The mutants, however, are incapable of increasing the level of RuBPC/O to the levels normally found in carbon-starved or derepressed wild-type *R. sphaeroides* cells. Mutant strain KW25/11 is typical; studies have shown that this strain is incapable of photolithotrophic and chemolithotrophic growth because of an apparent defect in its ability to derepress the synthesis of form I RuBPC/O [14]. To isolate the wild-type allele of the defective gene, we attempted to complement this mutant with a portion of a *R. sphaeroides* wild type (strain HR) chromosomal library constructed in the broad-host-range cosmid cloning vector pVK102. In four independent matings, several complemented transconjugants were found and plasmid DNA was purified from one transconjugant from each mating. The pVK102 hybrid was then separated from indigenous *R. sphaeroides* plasmids by transformation of *E. coli* RM1058 to tetracycline resistance. This plasmid DNA was identical in three of the four strains, and the plasmid from the fourth complemented strain lacked only a 4 kb *Hin*dIII fragment. This initially isolated complementing plasmid pKWA200 was characterized and a partial restriction map generated [15]. To reduce the size of the complementing fragment, a *Hin*dIII cassette with unique internal *Pst*I and *Sal*I sites was employed. This cassette could be used in either pACYC177or pACYC184 to easily screen inserts. Inserts to be tested for the ability to complement strain KW25/11 were then transferred to pVK102 by transferring the entire *Hin*dIII cassette. Eventually, it was found that the complementing gene or site is located on a 2.7 kb *Sal*I - *Pst*I fragment. Subsequent experiments employed recombinant deficient mutants with the same phenotype as KW25/11 in complementation assays. The 2.7 kb fragment was equally effective in complementing many of these strains, indicative of complementation in *trans*. Several recombination deficient carbondioxide fixation regulatory mutants, however, were not complemented by this fragment. Thus, we attempted to isolate sequences from the original library which would complement these other mutants. One mutant, strain EMS45, phenotypically Aut$^-$, was chosen for further study. The pVK102 library was used to isolate complemented transconjugants and plasmid DNA purified as above. This complementing DNA (in plasmid pWRA100) was shown to possess a distinct restriction pattern from pKWA200 (Fig. 4). Thus, the results of these experiments indicate that there are at least two distinct loci that regulate the synthesis of RuBPC/O. Exactly how many genes regulate the synthesis of enzymes important for carbon dioxide fixation, what they code for, and how the genes and gene products interact, is the subject of ongoing investigations.

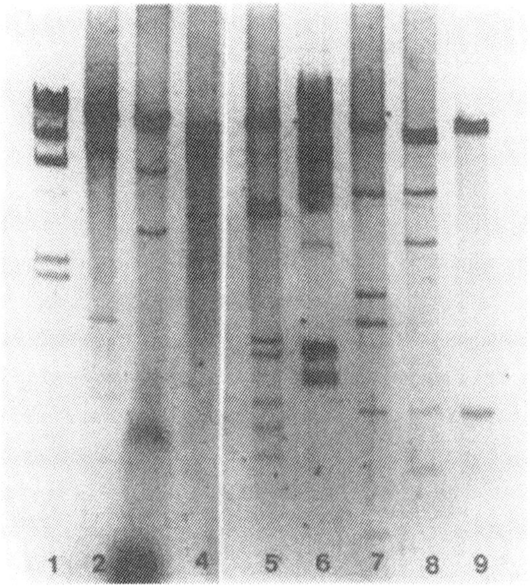

Figure 4. Restriction digests of plasmids pWRA100 (lanes 2, 4, 7), pKWA200 (lanes 3, 5, 8), pVK102 (lane 9); λ*Hin*dIII digests (lanes 1, 6). *Eco*RI digest (lane 2, 3); *Pst*I digest (lane 4, 5); *Bgl*II digest (lanes 7, 8, 9).

ACKNOWLEDGEMENTS

The work reported here was supported by NIH grant GM24497, USDA grant 83 - CRCR - 1 - 1344 and Robert A. Welch grant F - 691.

REFERENCES

[1] Madigan MT, Gest H. (1979) J. Bacteriol. 137, 524-530.
[2] Gibson JL, Tabita FR. (1977) J. Biol. Chem. 252, 943-949.

[3] Gibson JL, Tabita FR. (1977) J. Bacteriol. 131, 1020-1022.
[4] Gibson JL, Tabita FR. (1985)J. Bacteriol. 164, 1188-1193.
[5] Tabita FR, Sarles LE, Quivey RG Jr, Weaver KE, Waddill FE. (1983) in Schlesinger, D. (Ed.). Microbiology -1983, Am. Soc. Microbiol., Washington, DC, pp. 148-154.
[6] Gibson JL, Tabita FR. (1977) J. Bacteriol. 132, 818-823.
[7] Shively JM, Davidson E, Marrs BL. (1984) Arch. Microbiol. 138, 233-236.
[8] Jouanneau Y, Tabita FR. (1986) J. Bacteriol. 165, 620-624.
[9]Quivey RG Jr, Tabita FR. (1984) Gene 31, 91-101.
[10] Tabita FR, Gibson JL, Mandy WJ, Quivey RG Jr. (1986) Bio/Technology 4, 138-141.
[11] Gibson JL, Tabita FR. (1986) Gene (in press).
[12] Dardal F, Shinta C, Applebaum J, Davidson E, Prince R. (1986) Proc. Natl. Acad. Sci. USA 83, 2012-2016.
[13]Tabita FR. (1985) Abst. V. Int. Symp. on Photosynth. Prok., 359.
[14] Weaver KE, Tabita FR. (1983) J. Bacteriol. 156, 507-515.
[15] Weaver KE, Tabita FR. (1985) J. Bacteriol. 164, 147-154.

MOLECULAR BIOLOGY OF METHANOL OXIDATION

M.E. LIDSTRÖM, D.N. NUNN, D.J. ANDERSON, R.L. STEPHENS and M.G. HAYGOOD
Center for Great Lake Studies, University of Wisconsin/Milwaukee, 600 E.Greenfield, Milwaukee, WI 53204, USA

1. INTRODUCTION

In recent years it has become apparent that the development of genetic techniques in the obligate methanotrophs would greatly enhance studies of metabolism in these bacteria. Gene isolation and characterization are needed to help define the functions involved in key biochemical systems, and basic information concerning the organization and expression of genes is essential to understanding the molecular mechanisms that underlie physiological responses. In addition, the development of genetic manipulation capabilities is important for commercial exploitation of methanotrophic metabolism, and the development of gene probes will aid ecological studies.

Although the incentives to develop genetic techniques in the methanotrophs are clear, this task has proven difficult. Spontaneous mutation frequencies are low, and these are not increased substantially by treatment with mutagens [1,2]. The reason for lack of response to mutagens is not known, but it has been suggested that it may reflect the lack of an SOS repair system [3], as has been reported for *Haemophilus* [4]. Mutant isolation is further complicated by the fact that most methanotrophs grow poorly on plates and many show low plating efficiencies. In addition, most grow only on methane on plates, and so it is not possible to isolate mutants in C-1 metabolism. Finally, no reliable high frequency gene transfer systems are available.

In order to circumvent the problems involved in carrying out genetics in obligate methanotrophs, we have utilized an approach involving recombinant DNA techniques and surrogate genetics. That is, we have utilized alternate hosts for cloning and characterizing genes, and have employed broad host range cloning vectors to manipulate and study these genes in methanotrophic hosts. Using this approach, we were able to study nitrogen fixation (*nif*) genes from *Methylosinus* sp. 6. A hybridization probe carrying the *nif* structural genes from *Klebsiella pneumoniae* was used to identify a clone carrying a portion of the analogous genes from *Methylosinus* 6 [5]. This DNA fragment was then used to generate *Methylosinus* 6 mutants defective in nitrogen fixation, by marker exchange of Tn5 insertions within the clone [6]. This approach is useful, in that it allows genetic studies without the necessity of standard mutagenesis and mutant selection techniques. However, it can only be used with nonessential genes and depends upon the availability of hybridization probes.

We are now following a more broadly useful surrogate genetics approach to study genetics of methanotrophs. Since the facultative methanol-oxidizers and the obligate methanotrophs carry out many similar biochemical reactions, it is theoretically possible to use methanol-oxidizers as host systems to study common methylotrophic functions. We have used *Methylobacterium* AM1 (formerly *Pseudomonas* AM1) to develop genetic techniques, obtain baseline genetic data concerning C-1 genes and develop gene probes and mutant complementation systems. This information and these techniques are now allowing us to study C-1 genes from methanotrophs.

2. ISOLATION AND COMPLEMENTATION OF METHANOL OXIDATION (MOX) MUTANTS IN *METHYLOBACTERIUM* AM1

We have focused our initial genetic studies on the methanol oxidation system, since it is a key step for all methane and methanol-utilizers, it is highly conserved biochemically and antigenically amongst the gram-negative methylotrophs, and it is complex, involving multiple functions [7]. The methanol oxidation system in *Methylobacterium* AM1 is shown in Fig. 1.

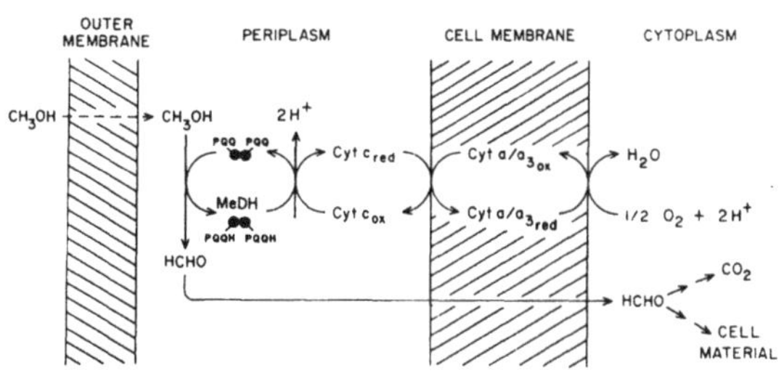

Figure 1. The proposed methanol oxidation system in *Methylobacterium* AM1. PQQ, pyrolloquinoline quinone; MeDH, methanol dehydrogenase; Cyt, cytochrome.

2.1. Mutant isolation

Methanol oxidation mutants were isolated using a direct selection technique, which took advantage of the fact that the methanol dehydrogenase of *Methylobacterium* AM1 can oxidize allyl alcohol to allyl aldehyde, a toxic compound. Cells were either exposed to ultraviolet light or nitrous acid and then plated onto agar plates containing allyl alcohol [8]. The colonies that grew were tested for growth on methanol and methylamine, and 25 mutants unable to grow on methanol but capable of normal growth on methylamine were used for further studies [8].

2.2. Mutant complementation

The mutants were placed into complementation groups using a genomic clone bank constructed in the broad host-range cosmid, pVK100 [8]. Five different clones were identified, and subcloning and *Tn5* mutagenesis further divided the mutants into a total of 10 complementation groups, Mox A1, A2, A3, B, C, D, E, F, G, and H (Table 1). Detailed maps of the clones and subclones are presented in ref. 8.

The genetic data available suggest that groups of C-1 genes are loosely clustered on the *Methylobacterium* AM1 chromosome, as shown in Fig. 2. Mox A1, A2, A3 and B are present on the

Table 1. The ten genes necessary for methanol oxidation in *Methylobacterium* AM1 [8,9].

Clone	Gene	MeDH Activity	MeDH Protein	Cyt c Protein	MeDH PQQ	Proposed Function
HindAB	$moxA_1$	-	+++	+++	none	PQQ attachment
	$moxA_2$	-	+++	+++	trace	PQQ attachment
	mox A_3	-	+++	+++	altered	PQQ attachment
	moxB	-	-	-		Regulation
HindCD	moxC	-	+/-	+/-	trace	Regulation
	moxD	-	-	+++, larger**		Transport/ stability
HindE	moxE	-	-	+/-, larger**		Regulation
HindFG	moxF	-	-	+++		MeDH structural gene
	moxG	+++	+++	-	+++	Cyt c structural gene
HindH	moxH	-	+/-	+/-	trace	Regulation

* Only *mox*F is definitively identified.
** Approximately 2 Kd larger than wild-type cyt c_L with spectra red-shifted 2.5-3 nm.

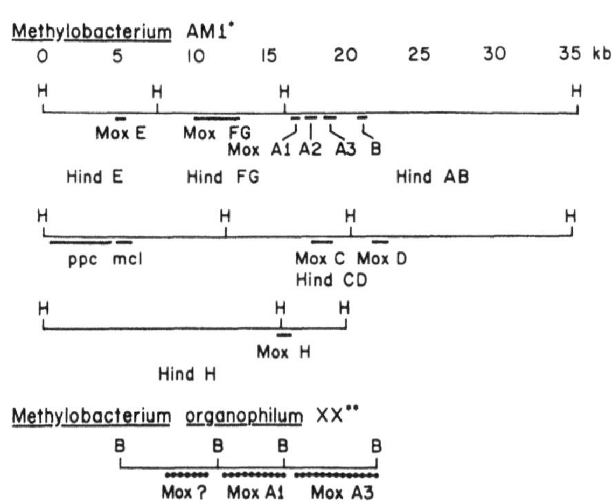

Figure 2. Summary of the known C-1 genes in *Methylobacterium* AM1 and comparison to the known Mox genes in *Methylobacterium organophilum* XX. H, HinDIII; B, BstEII.
* data from ref. 8 and B. Holloway, personal communication
** data from ref. 11.

*Hind*AB fragment, within 6 kb of each other. Mox C and D are present on the HindCD fragment within 8 kb of each other. The HinCD fragment is adjacent to another *Hind*III fragment previously reported tocontain a gene for malyl CoA lyase (mcl) [10]. We have now shown that this fragment also contains a gene for the C-1 PEP carboxylase. In addition, recent data from the laboratory of B. Holloway (personal communication) suggest that the HindFG fragment is located between the HindE and HindAB fragments on the chromosome. Therefore, the available data suggest that seven Mox genes are contained within 17 kb of DNA, in the following order: Mox E, FG, A1, A2, A3, B and that four C-1 genes are contained within 25 kb of DNA in the following order: ppc (PEP carboxylase), mcl, Mox CD. The distance between these two C-1 regions on the chromosome is not known. In addition, a third Mox region exists, comprising the HindH fragment. These data are consistent with results reported for *Methylobacterium organophilum* XX, in which Mox A1 and A3 were found to be linked to a third unknown Mox gene ([11]; Fig. 2).

Although several C-1 genes are grouped together, our data with *Tn5* insertions suggest that within these loosely clustered C-1 genes, only Mox F and G appear to be cotranscribed, and the others may all be contained within separate transcriptional units. Work is in progress to define the transcripts for each of these C-1 genes.

3. CHARACTERIZATION OF THE TEN MOX MUTANT CLASSES IN *METHYLOBACTERIUM* AM1

3.1. Identification of the methanol dehydrogenase structural gene: Mox F

The structural gene for the methanol dehydrogenase 60 Kd subunit was identified by random cloning of small open reading frame fragments into the vector pORF5 [8]. Fragments containing open reading frames of the methanol dehydrogenase gene produced tripartite protein fusions of beta-galactosidase

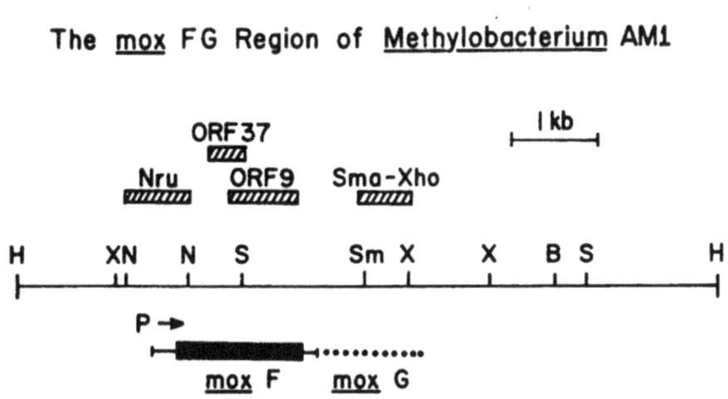

Figure 3. Detailed map of the Mox FG region of *Methylobacterium* AM1, showing the probes used for DNA-DNA and RNA-DNA hybridizations. H, *Hin*DIII; X, *Xho*I; N, *Nru*I; S, *Sal*I; Sm, *Sma*I; B, *Bam*HI; P, promoter.

and methanol dehydrogenase, which were recognized using antisera directed against each protein. Details are presented in ref. [8]. Two open reading frame fragments were identified, ORF9 and ORF37, that mapped to the Mox F gene (Fig. 3). From the orientation of the fragments, the direction of transcription could be determined, as shown in Fig. 3.

3.2. Phenotypic characterization of the Mox mutant classes

Representatives of each Mox class were characterized phenotypically by the following criteria: 1) presence of methanol dehydrogenase protein, by Western blots; 2) presence of methanol dehydrogenase activity in cell-free extracts; 3) absorbance spectrum of purified (inactive) methanol dehydrogenase; 4) reduced versus oxidized difference spectrum of partially purified cytochrome c_L; and 5) size and relative quantity of the cytochrome c_L.

The results are summarized in Table 1. Some mutant classes lacked methanol dehydrogenase protein, some contained inactive protein with an altered PQQ spectrum, one contained activity in vitro but not in vivo, some lacked detectable cytochrome c_L, and some contained the cytochrome, but with an altered spectrum [9]. These phenotypes have been used to propose functions for each of the Mox classes, which include the structural genes for the methanol dehydrogenase (Mox F) and the cytochrome c_L (Mox G), three functions involved in positioning or attachment of the PQQ cofactor to the methanol dehydrogenase apoprotein (Mox A1, A2 and A3), four functions that result in pleiotrophic mutants, which may be regulatory (Mox B, C, E and H) and one function which may involve processing, transport or stability of both the methanol dehydrogenase and the cytochrome c_L (Mox D). These are described in more detail in ref. 9. It should be emphasized that at present, only the Mox F genotype has been rigorously determined.

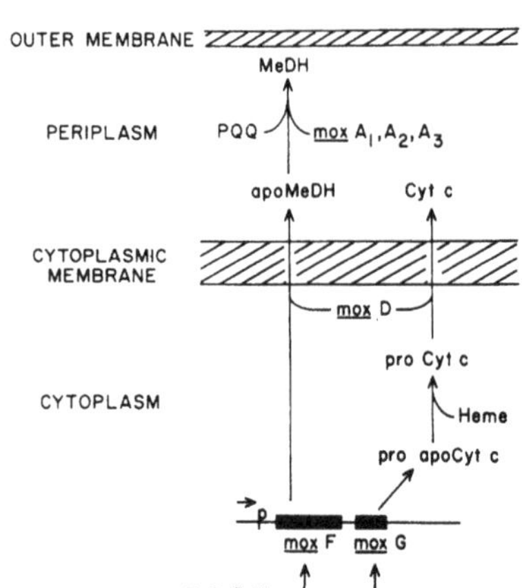

Figure 4. A model for the methanol oxidation system in *Methylobacterium* AM1. Abbreviations as in Fig. 1.

3.3. Working hypothesis for the Mox system in *Methylobacterium* AM1.

Fig. 4 shows our current working model for the Mox system in *Methylobacterium* AM1. We propose that four genes (the Mox B, C, E, and H classes) are involved in regulation of expression or in the stability of both the methanol dehydrogenase gene (Mox F class) and the cytochrome c_L gene (Mox G class), and perhaps other Mox genes. Evidence from Bacterium W3A1 [12] suggests that synthesis of the methanol dehydrogenase is cotranslational, and our evidence from the mutant classes supports that hypothesis for *Methylobacterium* AM1 as well. However, the synthesis and assembly of the soluble cytochrome *c* probably occurs in the cytoplasm. We propose that the Mox D class involves a function that is necessary either for transport, processing or stability of the methanol dehydrogenase and the cytochrome c_L, and that the PQQ cofactor is added to the methanol dehydrogenase apoprotein in the periplasm, involving functions of the Mox A1, A2 and A3 classes. Work is in progress to test this model. Our first approach has involved transcriptional analysis of the Mox regions.

4.TRANSCRIPTIONAL ANALYSIS OF THE MOX FG REGION

The model presented above suggests that as many as four functions may be involved in regulating the Mox system in *Methylobacterium* AM1. If so, the regulatory system may be complex, and may involve a cascade system. In order to define this regulatory network, we have initiated transcriptional studies of the Mox FG region.

4.1. Northern blots

The 330 bp open reading frame fragment of Mox F (ORF37; see Fig. 3) has been used as a hybridization probe to study Mox FG transcripts. Total RNA was isolated by the method of Xhu and Kaplan [13], from *Methylobacterium* AM1 cells grown on either methanol or succinate, separated on agarose gels and blotted to nitrocellulose paper. The RNA blots were hybridized to the ORF37 probe by published methods [13]. The ORF probe identified a transcript of approximately 2700 bp in size, that was approximately ten times more abundant in the blots from methanol-grown cells than those from succinate-grown cells. This transcript is of a sufficient size to encode both the Mox F and G class gene products. These data suggest that the genes of classes Mox F and G are probably cotranscribed and closely spaced on the HindFG fragment, and are subject to transcriptional regulation. In addition, the mRNA identified in methanol blots was a major transcript, visible on the RNA gels as a discrete band, suggesting that the transcript may be unusually stable. Two other probes, the 5' 0.7 kb *Nru*I fragment and the 3' 0.45 kb *Sma*I-*Xho*I fragment will also be used for Northern blots to define the transcripts further.

4.2. Definition of the Mox FG promoter

Understanding the transcriptional regulation of the Mox FG region requires characterization of the promoter region controlling expression of these genes. The Northern blots suggest that the transcriptional start site should lie near the 5' *Xho* I site (Fig. 3). We have constructed a broad host-range promoter probe vehicle (pDA4629) to clone and characterize promoters in *Methylobacterium* AM1 (Fig. 5). This plasmid allows the efficient cloning of small DNA fragments and their introduction into a variety of hosts. Promoter activity is coupled to beta-galactosidase production, so that regulation can be studied in the parent strain without disruption of the parent gene.

We have cloned the 0.7 kb *Nru*I fragment and the 1.55 kb *Xho*I-*Sal*I fragment (Fig. 3) into pDA4629. The *Nru*I fragment shows no promoter activity in *Methylobacterium* AM1, but the *Xho*I-*Sal*I fragment contains activity. Studies are in progress to further define this region and to characterize the Mox FG promoter activity in *Methylobacterium* AM1 grown under a variety of cultural conditions.

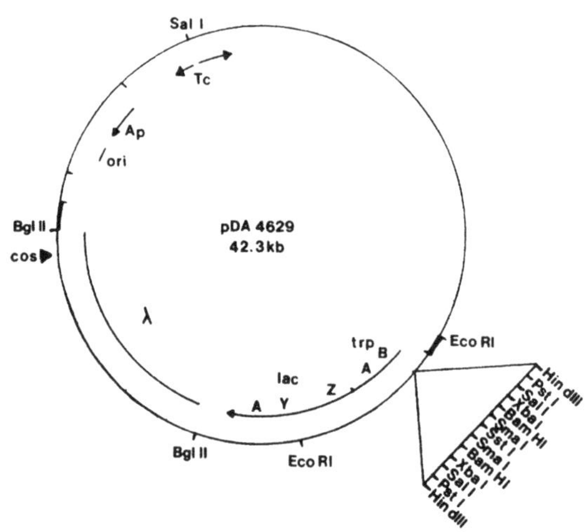

Figure 5. The broad host-range promoter probe cosmid pDA4629 constructed from pVK100, pBR322 and lambda. Ap, ampicillin resistance gene of pBR322; Tc, tetracycline resistance gene of pVK100; ori, origin of replication of pBR322.

5. THE MOX SYSTEM IN METHANOTROPHS

We have shown that the Mox system is complex in *Methylobacterium* AM1. However, it is not known whether similar complexity exists for this system in the obligate methanotrophs. In the cases where the methanol dehydrogenase has been purified from methanotrophs, it seems similar in biochemical terms to the enzyme from *Methylobacterium* AM1, and antigenic analyses suggest it is structurally similar as well [14]. The isolation of Mox mutants and the presence of cloned Mox genes now offer the possibility of determining whether similar genes exist in the methanotrophs. We have initiated these studies with the Mox F gene, since this seems the most likely to be conserved at both gene and protein levels.

5.1. DNA-DNA hybridization analysis with the Mox F gene probe

Total genomic digests of five different obligate methanotrophs were separated on agarose gels, and the gels were dried [15]. The gels were hybridized to the Mox F ORF9 probe at stringencies allowing approximately 25% base pair mismatch, and exposed to X-ray film. Discrete bands were identified in all cases, as shown in Table 2. These data suggest that the Mox F genes in *Methylobacterium* AM1, and the Type I, Type II and Type X methanotrophs were sufficiently conserved to allow cloning and identification using colony blots.

Table 2. Size of restriction fragments from methanotrophs that hybridize to the *Methylobacterium* AM1 *mox*F probe (ORF9).

Strain	Restriction Enzyme	Fragments (kb)
Methylosinus sporium 5	*Eco* RI	4.2 2.25 (major)
Methylomonas albus BG8	*Eco* RI	5.1 (major) 1.8
Methylomonas sp. A1 (marine strain)	*Eco* RI	2.4
Methylosinus trichosporium OB3b	*Pst* I	7.6
Methylococcus capsulatus Bath	*Pst* I	5.5

5.2. Cloning of the putative Mox F gene from *Methylococcus capsulatus* Bath

The ORF9 probe detected a 5.5 kb *Pst* I fragment from *Methylococcus capsulatus* Bath. We have used the cosmid pHC79 to construct a genomic clone bank of a *Pst*I partial digest of *Methylococcus capsulatus* Bath DNA. Colony blots of this bank have been screened with the ORF9 probe, and a clone has been identified that contains a 5.5 kb *Pst*I fragment which hybridizes specifically to ORF9. We are now characterizing this putative Mox F -containing fragment. We have also subcloned this fragment into a broad host-range vector, and this construction will be used to confirm the identity of the gene by complementation of a *Methylobacterium* AM1 Mox F mutant.

6.SUMMARY

The isolation and characterization of the genes involved in methanol oxidation in *Methylobacterium* AM1 have opened the way to detailed study of this key physiological step in a variety of methylotrophs. Not only has this genetic study uncovered previously undetermined Mox functions, it has provided the raw material for defining a complex, tightly coordinated regulatory system, and for isolating and characterizing nine promoters involved in this system. It will be extremely interesting to compare structure-function analyses of these promoters. The advantage of the genetic approach to systems of this type is that definitive data concerning molecular mechanisms can be obtained.

The discovery that at least one of the genes of the Mox system is homologous to discrete DNA fragments in a variety of obligate methanotrophs is a key finding, as it has provided the opportunity to isolate and characterize a central C-1 gene in methanotrophs. Since the broad host-range promoter cloning vehicle we have constructed is functional in the obligate methanotrophs, it will now be possible to clone the Mox F promoter region from these strains and use both beta-galactosidase activity and Northern blots to study transcriptional regulation of the methanol dehydrogenase structural gene in methanotrophs grown under a variety of cultural conditions. It will also be possible to structurally characterize these promoter regions, providing the first information on promoter structure in a methanotroph.

Although surrogate genetics approaches of the type described here have their limitations, they have great value for initiating genetic studies in difficult systems. By developing techniques and characterizing the system in *Methylobacterium* AM1 first, it is now possible for the first time to determine details of C-1 gene order, structure and transcriptional regulation in obligate methanotrophs.

REFERENCES

[1] Williams E, Shimmin F, Bainbridge BW. (1977). FEMS Microbiol. Lett. 2, 293-296.
[2] Harwood JH, Williams E, Bainbridge BW. (1972) J. Appl. Bacteriol. 35, 99-104.
[3] Williams E, Shimmin MA. (1978) FEMS Microbiol. Lett. 4, 137.
[4] Kimball RF, Boling ME, Perdue SW. (1977) Mutat. Res. 44, 183-196.
[5] Toukdarian AE, Lidström ME. (1984) J. Bacteriol. 157, 925-930.
[6] Toukdarian AE, Lidström ME. (1984) J. Bacteriol. 157, 979-983.
[7] Duine JA, Frank J, Jongejan JA, Dijkstra M. (1984) Microbial Growth on C-1 Compounds. Crawford RL, Hanson RS, eds. ASM Press, Washington, D.C.
[8] Nunn DN, Lidström ME. (1986) J. Bacteriol. 166, 581-590.
[9] Nunn DN, Lidström ME. (1986) J. Bacteriol. 166, 591-597.
[10] Fulton GL, Nunn DN, Lidström ME. (1984) J. Bacteriol. 160, 718-723.
[11] Allen L, Hanson RS. (1985) J. Bacteriol. 161, 955-962.
[12] Davidson VL, Neher JW, Cecchini G. (1985) J. Biol. Chem. 260, 9642-9647.
[13] Xhu YS, Kaplan S. (1985) J. Bacteriol. 162, 925-932.
[14] Anthony C. (1982) Biochemistry of Methylotrophs. Academic Press, London.
[15] Meinkoth J, Wahl G. (1984) Anal. Biochem. 138, 267-284.

DIVERGENCE OF METHANOGENS, CONSERVATION OF THE *His* I GENE SEQUENCE IN ALL THREE BIOLOGICAL KINGDOMS AND THE STATUS OF *METHANOBACTERIUM THERMOAUTOTROPHICUM*

John N. REEVE, Gregory S. BECKLER, James W. BROWN, David S. CRAM, Elizabeth S. HAAS, Paul T. HAMILTON, Christina J. MORRIS, Bruce A. SHERF, Clifford F. WEIL

Department of Microbiology, Ohio State University, Columbus, OH 43210, USA

1.INTRODUCTION

Methanogens are archaebacteria [1]. It was therefore somewhat surprising when it was shown that methanogen-derived genes could function in eubacterial species [2]. This has offered the opportunity of using gene cloning in *Escherichia coli* to investigate the basic structure of methanogen genes [3, 4] and to use *E. coli* to synthesize methanogen enzymes involved in methanogenesis [5] and nitrogen fixation [6].

The methanogen genes easiest to isolate are those which complement auxotrophic mutations in eubacterial species [7-10]. Using this technique, it is possible to isolate genes from different methanogens which complement the same mutation in *E. coli* and which therefore should be functionally homologous genes. This procedure has also been used to isolate genes from eucaryotic species, particularly from lower eucaryotes such as *Saccharomyces cerevisiae* [11, 12] in which intron occurrence is relatively rare. The *his*4 gene of *S. cerevisiae* [13] and *his*3 [14] gene of *Neurospora crassa* are expressed in *E. coli* and complement mutations in the *E. coli his* I/E locus. As the DNA sequences of these eucaryotic *his* genes [13, 14] and of the eubacterial *his*I/E genes of *E. coli* and *Staphylococcus typhimurium* [15] were known, it was of interest to isolate and sequence a methanogen gene which could also complement mutations in the *E. coli his*I/E locus. The structural relationships of genes, which encode enzymes with the same function in all three biological kingdoms, could then be determined.

2.CONSERVATION OF *HIS*I GENE SEQUENCES

Cloning and sequencing the *his*I gene from *Methanococcus vannielii* and its comparison to the DNA sequences of the *his*I/E genes of *E. coli* and *S. typhimurium* [5] and *his*4 gene of *S. cerevisiae* [13] have been reported [16]. In Figure 1 and Table 1 this comparison is extended to include the *his*3 gene of *N. crassa* [14]. In *E. coli* and *S. typhimurium*, the *his*I gene is part of a bifunctional gene fused to a sequence identified, by mutations, as the *his*E locus [17].

It appears that gene fusion must have occurred in these enterobacterial species to produce the *his*IE locus.

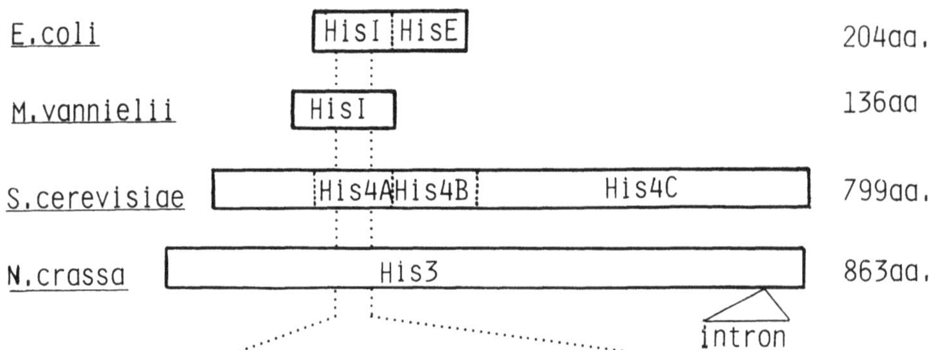


```
M.& E.  -FS------W-KGE-S-N---V-----DCD-D-LL-------G--CH-G--SCF-
M.& S.  Y-S-SRN-IW-KGE-S-N-QK-L----DCD-DAL-FIVEQ-----CH---MSCF-
M.& N.  Y-S--R---W-KG--S----------DCD-DAL-F-V-Q--G--CH-----CF-
S.& N.  Y-SR-R---W-KG-TSG----L--IS-DCD-DALKF-V-Q----FCHL----CFG
S.& E.  --SR-----W-KGETSGN------I--DCD-D-L-----------CH--T-SCFG
E.& N.  --SR----LW-KG-TSG-----V-I--DCDND-L--------G--CH---S-CFG
   All  --S------W-KG--S----------DCD-D-L-----------CH-----CF-
```

Figure 1. Comparison of *his*I sequences. The boxes represent the length of the open-reading frames and are positioned to indicate the regions of homology. The *N. crassa his3* gene contains a 59bp-intron which interrupts amino-acid 800, a proline residue. The regions between amino-acid residues, 65 through 120, of the *M. vannielii* sequence are shown in pairwise comparisons. M = *Methanococcus vanielii*, E = *Escherichia coli*, S= *Saccharomyces cerevisiae* and N= *Neurospora crassa*. All = amino acids present in all sequences, including the *Salmonella typhimurium* polypeptide (not shown [15]). The standard single amino-acid code is given at the locations where the same amino-acid occurs in the two *his*I polypeptides being compared.

Table 1. Conservation of the *His*I gene and protein

Sequences compared[a]	% Identical [b]	
	Amino acids	Nucleotides
M. vannielii/E. coli	32	40
M. vannielii/S. cerevisae	36	47
M. vannielii/N. crassa	27	40
S. cerevisiae/N. crassa	53	52
S. cerevisiae/E. coli	34	45
E. coli/N. crassa	33	45
E. coli/S. typhimurium	91	81

[a] Sequence data are taken from references [13-17].
[b] Comparisons relate only to the *his*I regions of the *his*I/E, *his*3 and *his* 4 genes.

An additional fusion event must have taken place during the evolution of the two fungal species as the *his4* gene and *S. cerevisiae* and *his3* gene of *N.crassa* encode enzymes with at least three separte catalytic activities [14, 18]. In *S. cerevisiae* the *his4* gene has been subdivided into *his4*A, *his4*B and *his4*C regions which are functionally equivalent to the *E. coli his* I, *his*E and *his* D genes [13, 15, 18]. In contrast to the eubacterial and eucaryotic genes, the archaebacterial *his*I gene from *M. vannielii* does not appear to encode a multifunctional enzyme [16]. There is, however, conservation of amino acids in the polypeptides encoded by the genes from all three biological kingdoms. In the region of the polypeptides shown, as sequences, in Figure 1 there are 14 amino-acid residues conserved in all the enzymes which include three cysteine residues located in two conserved dipeptides (cys-his and cys-phe) and in a conserved tripeptide (asp-cys-asp). When the full length of the *his*I sequences are compared it is clear, however, that there has been very extensive divergence of these genes. The percentage of conserved amino acids, in pairwise comparisons, (Table 1) range from 26 to 36% for cross-kingdom comparisons, to 53% conservation of amino acids in the two eucaryotic enzymes, to 91% identity of amino acids in the two eneterobacterial enzymes.

3. CONSERVATION OF METHANOGEN GENES

As shown in Table 1, the two enterobacterial *his*I genes are very similar [15]. When comparisons of DNA sequences of functionally homologous genes are made, within the enteric group of eubacteria, DNA homologies ranging from 75-95% are found. For example, the five *trp* operon genes of *E. coli* and *S. typhimurium* have homologies ranging from 75% to 84% [19]. The extent of homology in gene sequences decreases significantly when comparisons are made of more divergent eubacterial species; the DNA sequences of the *trp*D genes of *E. coli* and *B. subtilis* are only 44% homologous. Table 2 shows the extent of homology in functionally equivalent genes from different methanogenic species. The *his*A genes from two methanococci, *M. vannielii* and *M. voltae* [8], are more closely related to each other than are the two *pur*E genes from *Methanobrevibacter smithii* and *Methanobacterium thermoautotrophicum* ΔH [10]. The evolutionary distance between *M. smithii* and *M. thermoautotrophicum* ΔH would appear to be of the same magnitude as that between *E. coli* and *B. subtilis* or that between *S. cerevisiae* and *N. crassa*. Although the methanococci are more closely related to each other they are apparently not so closely related as are the members of the enterobacterial group of eubacteria.

4. STATUS OF *METHANOBACTERIUM THERMOAUTOTROPHICUM*

A large percentage of research focussed on the biochemistry and composition of methanogens is undertaken with organisms named *Methanobacterium thermoautotrophicum* [20-23]. This name does not, however, signify one species and although this unfortunate situation has already been

Table 2. Conservation of methanogen genes.

Species compared	Gene[a]	% Identical[b]	
		Amino acids	Nucleotides
M. vannielii/M. voltae	*his*A	67	66
M. smithii/M. thermo-autrophicum ΔH	*pur*E	45	53

[a] Sequence data are taken from references [8-10].

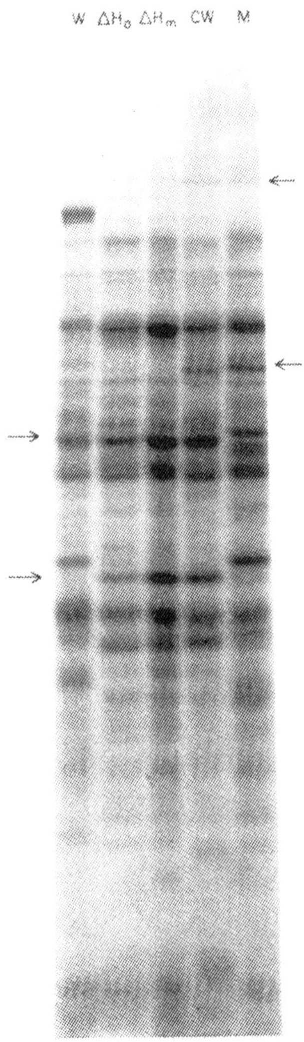

Figure 2. SDS-PAGE separation of polypeptides from different strains of *M. thermoautotrophicum.* Polypeptides in lysates of strains Winter(W) [24], ΔH (ΔH$_O$ from The Ohio State University and ΔH$_M$ from the Massachessetts Institute of Technology), CW (isolated from the Columbus, Ohio sewage system by Clifford Weil) and Marburg (M) were separated by electrophoresis through a 15% polyacrylamide gel and stained with Coomassie blue. Examples of polypeptides with the same mobilities in strains CW and M or in CW and ΔH are indicated by arrows to the right or left of the figure respectively.

documented [22], the obvious potential for confusion and problems remains. As cloning and sequencing of methanogen genes becomes more widespread it is essential to clearly identify which *M. thermoautotrophicum* is being used as the source of DNA. Figures 2 and 3 are presented to underline this point. The patterns obtained by SDS-PAGE separation of polypetides from lysates of four different *M. thermoaurotrophicum* isolates are shown in Figure 2. Clearly there are both similarities and differences. Strain CW, which we isolated from the Columbus sewage treatment facility and identified as *M. thermoautotrophicum* on the basis that it conforms exactly with the descriptions of *M. thermoautotrophicum* [20, 21], was included to demonstrate that this name identifies a group of related but not identical organisms.

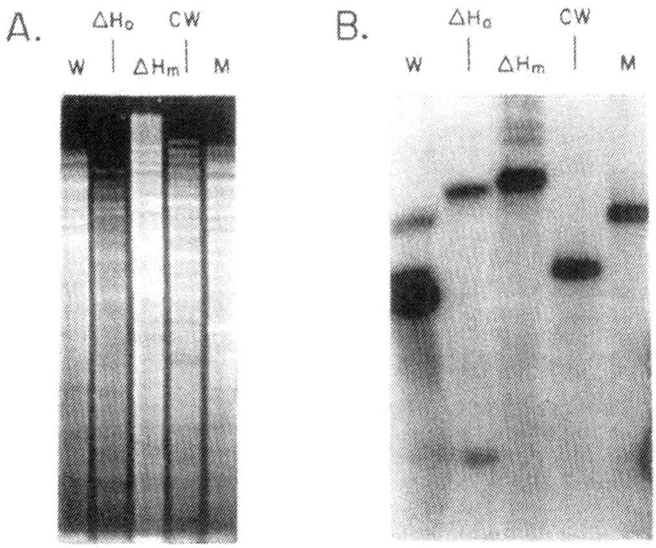

Figure 3. Restriction digests and Southern blots of *M. thermoautotrophicum* DNAs.
A. Fragments of DNA produced by *Bam*H1 digestion of genomic DNAs from strains Winter (W), ΔH (ΔH_O and ΔH_M), CW and Marburg (M) were separated by agarose gel electrophoresis and stained with ethidium bromide.
B. A *Bam*H1 fragment of DNA, cloned from strain Winter was used to probe the digests shown in Figure 3A showing that this sequence is present in the genomes of all the *M. thermoautotrophicum* strains but located on differently-sized restriction fragment. The strains ΔH_M and ΔH_O are identical, the slight difference in electrophoretic mobilities apparent in this figure is an artifact.

Strain CW contains polypeptides which co-migrate only with polypeptides from strain Marburg (arrows to the right of the Figure) or only with polypeptides from strain ΔH (arrows to the left of the Figure) or are common to all strains of *M. thermoautotrophicum.* Similarly when DNAs from different *M. thermoautotrophicum* strains are digested with the same restriction enzyme, each digest results in a different pattern of restriction fragments (Fig. 3A). These organisms are however sufficiently related that when a DNA fragment, cloned from *M. thermoautotrophicum* Winter [24], was used in a Southern blot procedure to probe the genomic digests shown in Figure 3A, the sequence was found to be present in the genomes of all the *M. thermoautotrophicum* strains (Figure 3B). All cloned DNA sequences, so far used as hybridization probes, have been found in the genomes of all the different *M. thermoautotrophic* strains however always located on differently-sized restriction fragments indicating that during the divergence of these organisms changes have occurred in the DNA sequences of most restriction sites. Sequencing of DNA fragments from strains W and ΔH , identified as homologous DNA by hybridization, has shown that these DNAs are, in fact, 87% homologous. This is within the range of sequence homologies found for DNAs from different enterobacterial species and is similar to the estimate previously reached for the relatedness of *M. thermoautotrophicum* strains [22]. It would now seem appropriate to consider the proposal for different names [25] or to insist that the epithets (ΔH, W, Marburg, etc.) always be included in both the written and spoken presentations.

ACKNOWLEDGEMENTS

Research at The Ohio State University, investigating the molecular biology of methanogens, is supported by the U.S. Department of Energy, Environmental Protection Agency and Office of Naval Research under contracts AC02-81ER10945, CR810340 and N00014-86-K-0211 respectively.

REFERENCES

[1] Woese CR and Wolfe RS. (1985) The Bacteria 8: 561-564.
[2] Reeve JN, Trun JJ, and Hamilton P. (1982) In "Genetic Engineering of Microorganisms for Chemicals" Plenum Publishing, N.Y. pp. 233-244.
[3] Reeve JN, Hamilton PT, Beckler, GS, Morris CJ and Clarke CH. (1986). System. Appl. Microbiol. 7 : 5-12.
[4] Hamilton PT, and Reeve JN. (1986). In "Microbiology 1986," ASM Press, Washington, D.C. 20006. pp. 338-348.
[5] Allmansberger R, Bollschweiler C, Konheiser U, Muller B, Muth E, Pasti G, and Klein A. (1986). System. Appl. Microbiol. 7 : 13-17.
[6] Souillard N, and Sibold L. (1986). Mol Gen Genet 203 : 21-28.
[7] Wood AG, Riedborg AH, Cue DR, Whitman WB, and Konisky J. (1983). J. Bacteriol 156 : 19-29
[8] Cue D, Beckler GS, Reeve JN and Konisky J. (1985) Proc. Natl. Acad. Sci. U.S.A. 82 : 4207-4211.
[9] Hamilton PT and Reeve JN (1985). Mol. Gen. Genet. 200 : 47-59.
[10] Hamilton PT and Reeve JN. (1985) J. Mol Evol. 22 : 351-360.
[11] Struhl K, Cameron JR and Davis RE (1976). Proc. Natl. Acad Sci. U.S.A. 73 : 1471-1476.
[12] Ratzkin B and Carbon J. (1977). Proc. Natl. Acad, Sci. U.S.A. 74 : 487-492.
[13] Donahue TD, Farabaugh PJ and Fink GR. (1982). Gene 18 : 47-59.
[14] Legerton TL and Yanofsky C. (1985) Gene 39 : 129-140.
[15] Bruni CB, Carlomagno MS, Formisano S, Paolella G and Fink GR. (1986) Mol. Gen. Genet. 203 : 389-396.
[16] Beckler GS and Reeve JN. (1986) Mol. Gen. Genet. 204 : 133-140.
[17] Chiariotti, L Alifono P, Carlomagno MS and Bruni CB. (1986) Mol. Gen. Genet 203 : 382-388.
[18] Jones CW and Fink GR. (1982) In "The Molecular Biology of the Yeast Saccharomyces", Cold Springs Harbor Laboratory, Cold Spring Harbor, N.Y. 11724, pp. 181-299.
[19] Horowitz H, vanArsdell J, and Platt JT. (1983) J. Mol Biol. 169 : 775-797.
[20] Zeikus JG and Wolfe RS. (1972) J. Bacteriol 109 : 707-713.
[21] Balch WE, Fox GE, Magrum LJ Woese CR and Wolfe RS. (1979) Microbiol. Rev. 43 , 260-296.
[22] Brandis A ,Thauer RK and Stetter, KO. (1981) Zbl. Bakt. Hyg. I. Abt. Orig. C2 : 311-317.
[23] Daniels L, Sarling R and Sprott GD. (1984) Biochim Biophys. Acta 768 : 113-163.
[24] Winters J. (1983) System Appl. Microbiol 4 : 558-563.
[25] Prevot AR. (1980) Bull. de l'Institut Pasteur 78 : 217-265.

APPLIED ASPECTS OF ONE-CARBON METABOLISM

THE POTENTIAL OF ONE-CARBON COMPOUNDS AS FERMENTATION FEEDSTOCKS

J.D. LINTON AND H.G.D. NIEKUS
Shell Research Limited, Sittingbourne Research Centre, Sittingbourne, Kent ME9 8AG
and Gist-Brocades NV, Delft, The Netherlands

1. INTRODUCTION

Micro-organisms have been reported to produce a wide range of chemicals from one-carbon compounds [1-5]. These include amino-acids, organic acids like ctric, butyric and acetic as well as exopolysaccharides, single cell protein, co-factors, vitamins and enzymes. However, few attempts have been made to examine critically processes based on one-carbon feedstocks in comparison with those already operating commercially on carbohydrates. In this introductory paper to the session on "Applied Aspects of One-Carbon Metabolism" an attempt will be made to compare one-carbon feedstocks with those traditionally used in the fermentation industry on the basis of price, availability, stoichiometry of conversion and process requirements. The price-volume and price-productivity relationship for various chemicals will be examined in order to arrive at realistic estimates of the productivities and costs that will have to be attained for bioconversion from methanol to be competitive with existing processes. Both aerobic and anaerobic conversion of methanol will be discussed and the relative merits of each assessed. This general analysis will help to define the types of chemicals that offer the best prospects for bio-conversions of one-carbon compounds. The economics of production of specific chemicals will not be discussed in detail as this area will be covered by other speakers in this session.

2. CHOICE OF ONE-CARBON FEEDSTOCK FOR FERMENTATION PROCESSES

Methane, methanol and synthesis gas could all be considered as potential feedstocks for the fermentation industry. Synthesis gas is a mixture of H_2 and CO and it can be produced from coal, natural gas or biomass. The technology for the conversion of synthesis gas into a wide range of chemicals currently produced from oil is well developed. The next speaker Dr. Drent will discuss the potential and timing of synthetic processes from synthesis gas. At present, with the exception of South Africa, synthesis gas is not produced on a large scale for subsequent conversion to fuels and chemicals. It is produced on a relatively small scale (compared with that which would be required for fuel production) from natural gas and coal for the captive synthesis of methanol, acetic acid, acetic anhydride and other derivatives.

As a fermentation feedstock methanol has a number of important advantages over methane and synthesis gas. It is a commodity chemical of high purity, it is relatively cheap, easy to transport and store. Methanol is miscible with water and the hazard of explosions is considerably less than that with the gaseous feedstocks. In addition, methanol is more oxidized than methane and consequently both the oxygen demand and the amount of heat generated during aerobic fermentations will be less; both these parameters may have an important impact on process economics. For the reasons outlined above, methanol is generally regarded as the most suitable feedstock for both aerobic and anaerobic fermentations [1-7].

The current world production of methanol is approximately 12 x 10^6 t/y and the traditional outlets for methanol are only expected to show a modest rate of growth (3-5%/year) over the next 5 years [7,8]. However, there has been considerable speculation regarding the use of methanol as a fuel, either blended in gasoline as an octane booster or for use neat. The fuel related uses have not developed as anticipated and at present there is a large overcapacity for methanol production. Although a large number of plans have been shut down many new ones have come on stream and more are expected [8] so the overcapacity is likely to continue for a number of years (Figure 1). This will tend to keep the methanol price low.

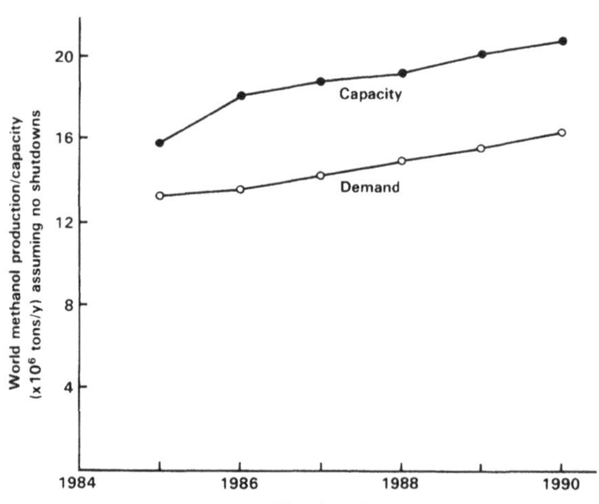

Fig 1
The imbalance between world capacity and demand for methanol assuming no further shutdowns. (based on Coxon, 1985)

Moreover, methanol prices are expected to be linked to energy prices [8] and these are also expected to remain low. It should be emphasised that a future development of fuel uses for methanol could reverse the situation completely since a 1% addition of methanol to gasoline would utilise most of the surplus capacity available [9].

3.COMPARISON BETWEEN METHANOL AND TRADITIONAL FERMENTATION FEEDSTOCK PRICES

When considering methanol as a potential fermentation feedstock it is necessary to compare it with those traditionally used in the industry. The world sugar price over the past 15 years has fluctuated dramatically whereas the price of methanol has remained relatively steady (Figure 2). At first sight this looks very encouraging, because on a weight basis the price of methanol is considerably lower than that of sugar. However, most of the world sugar production (approximately 100 x 10^6 t/y) is traded directly between producer and consumer or between countries and only 30% is traded on the free market. The price of the latter is disproportionately affected by shortages and surpluses and is not a true reflection of the cost of sugar to the fermentation industry [10]. Similarly, the EEC gross intervention price for sugar is considerably higher than world prices. However, the fermentation industry in the EEC qualifies for various rebates and the actual cost of sugar will depend on the specific chemical being produced and whether or not it is exported outside the EEC.

Molasses is also widely used in the fermentation industry. On the basis of sugar content the price of molasses has been similar to that of methanol (Figure 2). However, molasses cannot be stored for long periods, its composition is variable and it contains particulate material that has to be removed from the product. The latter will add to the cost of down-stream processing.

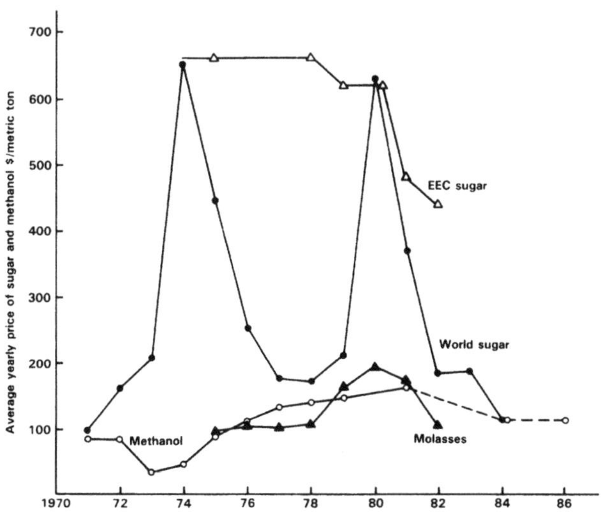

Fig 2

The price of sugar and methanol from 1970 to 1984. The international sugar agreement daily price ● was obtained from the sugar year book 1984; the EEC gross intervention price, Δ (Brown, 1983); the US list price for methanol, o (Abshire *et al.*, 1980; and the ECN News. Molasses (cane) ▲ selling price, New Orleans (Brown, 1983), the molasses price has been adjusted assuming a 50% sugar content.

A wide range of factors may influence the price of these feedstocks and the relative price differential will vary from time to time and from country to country. The difficulties experienced in forecasting feedstock prices coupled to changes in their relative price means that it is not justifiable to plan a fermentation process based solely on methanol, because of its present price advantage. Instead, a flexible feedstock policy that allows the use of the most favourable feedstock available at a given time and location, would be advisable. The vulnerability of being tied to one feedstock is illustrated by the BP process for SCP from n-paraffins [11] and this lesson has not been lost on the fermentation industry as is clear from the policy adopted by Philips Petroleum for SCP production from a range of feedstocks including methanol, sugar and whey [6].

4.DEFINITION OF TARGET AREAS FOR METHANOL CONVERSIONS

What types of chemicals offer the best prospects for methanol conversions? The most important attribute of methanol as a feedstock is its price. Consequently, methanol may be used to best effect in processes where the feedstock costs are a significant proportion of production costs. In general, the higher the cost of a given chemical the smaller the impact feedstocks costs are likely to have on process economics.

It is well known [12] that the price of a wide range of chemicals is inversely related to the volume in which they are produced. This relationship also applies to the fermentation industry [13] and it shows that virtually all fermentation products cost more than $2/kg because of the high capital and production costs associated with fermentation processes Figure 3.

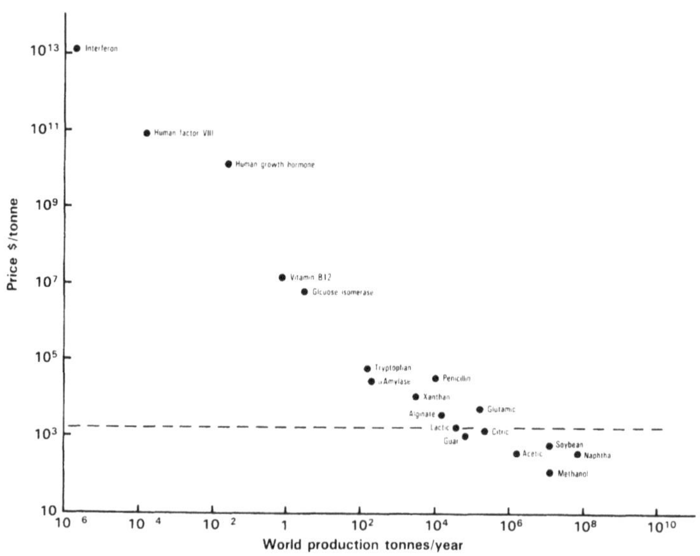

Fig 3
The relationship between the price of various chemicals and the volume in which they are produced. (Data taken from Dunnill, 1982; Tong and Cannell, 1983).

Processes favouring the use of methanol are therefore likely to be those for the production of bulk chemicals that command a low price. Competition in this area is strong from chemical synthesis for the production of simple molecules like ethanol and acetic acid. For complex molecules like natural oils and protein there is a strong competition from agriculturally-derived commodities. In the production of chemicals that have a moderate value (e.g. amino acids) the price difference between sugar and methanol has to be large in order to justify the use of methanol as a feedstock. Moreover, other factors such as product concentration, productivity and down-stream processing then assume as much significance as feedstock costs. For high-value products, feedstock costs are a minor part of overall process costs and the use of methanol for the production of these chemicals will depend on whether it offers other process advantages. For example, methylotrophs could offer advantages in the production of heterologous proteins using the powerful promoter of the methanol oxidase system in yeast.

5.BIOCONVERSIONS OF METHANOL

The microbial conversion of methanol can be carried out aerobically or anaerobically. The economic constraints of these processes differ and these will be examined separately.

5.1 Aerobic fermentation

On the basis of known biochemical pathways, stoichiometric equations can be written for the aerobic production of acetic acid, citric acid, glutamic acid, exopolysaccharide and a lipid from methanol and glucose (Table 1). The yields of these products, when expressed on a carbon-carbon basis are similar from glucose and methanol. However, the oxygen requirement for methanol conversion can be considerably higher than that from glucose. Thus, for the production of acetate from methanol the oxygen demand is higher than that for the production from glucose by a factor of 2.5, whereas this factor increases to 10 for exopolysaccharide production. The higher oxygen demand is accompanied by a corresponding increase in the heat output of these fermentations (Table 1). Consequently, in processes where maximum productivity is already limited by mass transfer and cooling, for example in the production of exopolysaccharide from glucose, these problems will be considerably more severe with methanol as the feedstock. For each product it will be necessary to determine whether the relative price advantage of methanol is sufficient to offset the disadvantages inherent in this reduced feedstock as outlined above. Examples where methanol offers an advantage for the aerobic production of specific chemicals will be covered by Professor Tani and Drs. Thill and Stroman.

5.2 Anaerobic fermentation

There are two groups of micro-organisms capable of utilizing methanol anaerobically. The first group are the methanogens that convert it into methane and the second group, which are of more interest from the point of view of chemicals production, are the acidogens. The latter group have only been discovered comparatively recently [15] and the physiological and energetic requirements for cell

Table 1 - Stoichiometry of metabolite production from glucose and methanol showing the relative oxygen demand, free energies and yields of various chemicals.

Product	Carbon source	Overall Stoichiometry (Assuming that excess reducing equivalents are oxidised to water)	- $G^{0'}$ (kJ)	Yield Y substrate ($g\ g^{-1}$)	YO_2 ($g\ g^{-1}$)
Acetate	Methanol	$3CH_3OH + 2.5O_2 = CH_3COOH + CO_2 + 4H_2O$	1254	0.625	0.75
	Glucose	$0.5C_6H_{12}O_6 + O_2 = CH_3COOH + CO_2 + H_2O$	531	0.67	1.88
Citrate	Methanol	$6CH_3OH + 4.5O_2 = C_6H_8O_7 + 8H_2O$	2280	1.0	1.33
	Glucose	$C_6H_{12}O_6 + 1.5O_2 = C_6H_8O_7 + 2H_2O$	829	1.07	4.0
Glutamate	Methanol	$6CH_3OH + NH_3 + 4.5O_2 = C_5H_9O_4N + CO_2 + 9H_2O$	-	0.76	1.02
	Glucose	$C_6H_{12}O_6 + NH_3 + 1.5O_2 = C_5H_9O_4N + CO_2 + 3H_2O$	-	0.82	3.06
Hexose-exopoly-saccharide	Methanol	$6.14CH_3OH + 3.21O_2 = [C_6H_{10}O_5] + 0.14CO_2 + 7.28H_2O$	1550	0.82	1.58
	Methane	$12CH_4 + 18O_2 = [C_6H_{10}O_5] + 6CO_2 + 19H_2O$	7760	0.84	0.28
	Glucose	$1.05C_6H_{12}O_6 + 0.3O_2 = [C_6H_{10}O_5] + 0.3CO_2 + 1.3H_2O$	146	0.86	16.3
Lipid (palmitic acid)	Methanol	$24CH_3OH + 13O_2 = C_{16}H_{32}O_2 + 8CO_2 + 32H_2O$	7029	0.33	0.62
	Glucose	$4.16C_6H_{12}O_6 + 1.97O_2 = C_{16}H_{32}O_2 + 8.97CO_2 + 8.97H_2O$	1682	0.34	4.06

synthesis of these organisms are still poorly understood. The only metabolites produced by the various acidogens are acetic acid, butyric acid or both and these products are excreted as a consequence of the energy metabolisms in these organisms. The amount of energy harnessed by these organisms during the conversion of methanol to acetic or butyric acid is small (Table 2). Thus the yield of cells per mol of methanol consumed is low, but the product yields are high. Moreover, because CO_2 is incorporated together with methanol into the product, the yield of acetic acid from methanol exceeds 1 g/g (Table 2) and most of the energy of the substrate is conserved in the acidic product. This results in a large reduction in heat production compared with the aerobic production of acetic acid (Table 1). Most of the above mentioned advantages of using methanol as a feedstock would also apply to the anaerobic production of other well-known fermentation products like ethanol, acetone or butanol. However, no reports of methylotrophic anaerobes capable of producing these products from C_1 compounds have been published.

Table 2 - Yields on a weight or energy basis and free energy of the reactions ($G^{0'}$) from methanol to acetic and butyric acid.

	Yield		
	Weight basis (g/g)	Energy basis	- $G^{0'}$ (kJ)
$1.33CH_3OH + 0.67CO_2 \longrightarrow CH_3COOH + 0.67H_2O$	1.41	0.91	71
$3.33CH_3OH + 0.67CO_2 \longrightarrow CH_3CH_2CH_2COOH + 2.67H_2O$	0.83	0.90	177

5.3 Prospects for the microbial production of acetic acid

Acetic acid is a bulk chemical and therefore its production from methanol meets the criteria described previously concerning economic targets for bioconversions from methanol. As methanol can be converted into acetic acid in high yield by anaerobic methylotrophs it is instructive to examine the prospects for the biotechnological route to this chemical. In order to be competitive with the synthetic route the fermentation process has to be greatly improved in a number of ways. From the relationship between product price and fermentation productivity (Figure 4) it can be concluded that the fermentation process will have to attain a productivity in the region of 25-80 g/l.h. These productivities are exceptionally high for fermentation processes, but they are not impossible. In a recent patent [14] volumetric productivities as high as 14.3 g acetic acid/l.h were reported for immobilized *Clostridium thermoaceticum* cells using glucose as the substrate. Daicel Chemical Industry, the major manufacturer of acetic acid in Japan, has started research on acetic acid production from $H_2 + CO_2$ using anaerobic micro-organisms. However, the best productivity reported was 1.9 g/l.h [18]. This productivity could potentially be improved by using a cell re-cycle system to maintain a high level of biomass in the reactor. The Japanese interest in this area may also be gathered from a large number of patent applications issued on behalf of the Japanese Agency of Industrial Science and Technology dealing with the production of a variety of fermentation products by autotrophic and methylotrophic (facultative) anaerobes. The main factor limiting productivity is the sensitivity of these organisms to their own acidic metabolic products. A theory recently proposed concerning acetic acid toxicity [19] suggests that there may be physiological limits to the extent of the improvement likely to be achieved by mutation or adaptation. The other important area in need of marked improvement is in the recovery of acetic acid

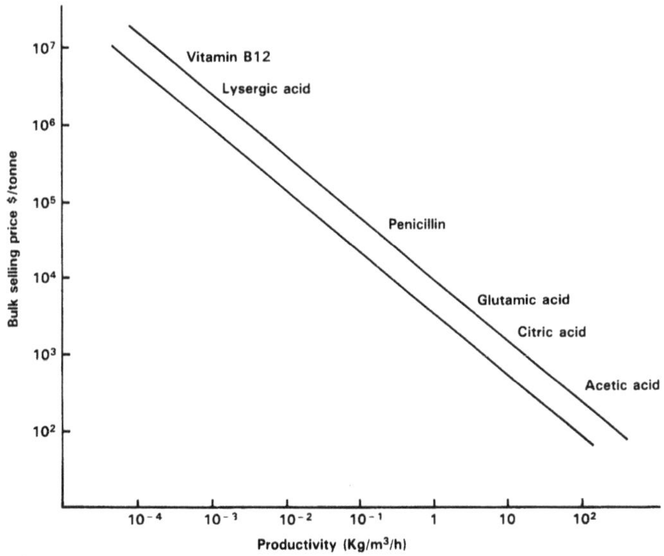

Fig 4
The relationship between the price of various chemicals and the productivity at which they are operated. (based on Andrew, 1983).

from dilute aqueous fermentation broths, (preferably as the acid rather than as a salt). Recovery is directly linked to productivity and a cheap and efficient means of continuously removing acetic acid could be of considerable help in a developing high productivity fermentation. A number of new methods of recovering chemicals from aqueous fermentation broths are being developed, for example super-critical CO_2 and membrane extraction methods. If these new methods are successful then the fermentation process to acetic acid could become competitive with the synthetic route.

6.CONCLUSIONS

Methanol is a relatively cheap commodity chemical and is likely to remain so for the foreseeable future. It can be converted into a wide range of chemicals by micro-organisms. The most important economic incentive for using methanol is its cost relative to traditional fermentation feedstocks, but the extent of the price advantage varies with time and location. Because of its relatively low price methanol may offer advantages over carbohydrate feedstocks for the production of bulk chemicals where feedstock costs represent an important part of the overall costs. However, there is strong competition from both the chemical industry and agriculture in this area. Consequently specific markets have to be identified. The rational for using methanol for the production of high value chemicals is more likely to be based on the special metabolic attributes of methylotrophs than on the relative price of C_1 feedstocks.

REFERENCES

[1] Zeikus, JG (1983) In Organic Chemicals from Biomass (Wise DL, Ed) pp. 359-383. The Benjamin/Cummings Publishing Company Inc, London.

[2] Morinaga Y, Hirose Y (1984) In Methylotrophs: microbiology, biochemistry, genetics. (Hou CT, Ed) pp. 107-118. CRS Press Inc.

[3] Kennedy MJ (1984) Report No. IPD/ISO/1009 Industrial Processing Division, Department of Science and Industrial Research, New Zealand.

[4] Tani Y (1985) Biotechnology and Genetic Egineering Reviews 3, 111-135.

[5] Dijkhuizen L, Hansen TA, Harder W (1985) Tr. Biotechnol. 3, 262-267.

[6] Norell, JR, Dreisker DW, Wegner GH, King AC (1984) Provesteen. Products and processes/substrates methanol, sugars and whey permeate. International Conference and Exhibition, Washington, DC.

[7] Marshall EF (1982) Chemical Technology 6, 358-362.

[8] Coxon R (1985) Methanol world balance. In 1985 World Methanol Conference, Amsterdam. pp. I-I - I-7. Dewitt and Company Inc.

[9] Gaylord W (1985) Current trends in the U.S. methanol market. In 1985 World Methanol Conference, pp. V-1 - V-9. Dewitt and Company Inc.

[10] Brown O (1983) Political and economic aspects of fermentation raw materials markets. Chemistry and Industry 7.2.83. pp. 95-97.

[11] Guidoboni GE, Wislon DR (1983) Economic comparison between carbohydrate and petrochemical feedstocks. Institution of Chemical engineers Symposium. Series No. 79, pp. 416-428.

[12] King PP (1982) Journal of Chemical Technology and Biotechnology 32, 2-8.

[13] Dunnhill P (1983) The Future of Biotechnology. In Biochemical Society Symposium No. 48 Biotechnology 9-23.

[14] Tong GE, Cannell RP (1983) The economics of organic chemicals from biomass. In Organic chemicals from Biomass (Wise DL, Ed) pp. 407-451. The Benjamin/Cummings Publishing Company Inc., London.
[15] Zeikus JG, Lynd LH, Thompson TE, Krzycki GA, Weimer PJ, Hegge PW (1980) Curr. Microbiol. 3, 381-386.
[16] Andrew SPS (1983) Old and new feedstocks - chemistry and commerce. Inst. Chemical Engineers Symposium. Series No. 79 pp. 400-408.
[17] Reed WR (1985) US patent 4, 506, 012.
[18] Anon (1985) Biosynthesis of acetic acid with anaerobes. Chemical Econ. Engineer. Review 17 pp. 13.
[19] Menzel U, Gottschalk G (1985) Arch. Microbiol. 143 pp. 4751.
[20] Abshire AD, Al-Sayyari S, Ratie AK, Kamatari O, Wilson E (1980) CEH Marketing Research Report. Methanol Chemical Economics Handbook - SRI International.

CHEMICAL SYNTHESIS FROM C_1 COMPOUNDS

E. DRENT

Koninklijke/Shell-Laboratorium, Amsterdam (Shell Research B.V.), Postbus 3003, 1003 AA Amsterdam

Carbon monoxide (CO) is a most versatile C_1 building block for the synthesis of a wide variety of chemicals. As syngas (CO/H_2), it can be produced cheaply from low-cost natural gas and, in the longer term, by large-scale coal gasification. In syngas-based chemical production, emphasis is placed on the oxygen contents of the final products and thus CO is used as a source of both carbon and oxygen; this is in contrast with fuel production from syngas, where emphasis is on CH_x contents in the final product and CO is used as the source of carbon.

Chemicals derived from syngas feedstock can be separated into two classes:

i - those from syngas feedstock comprising CO/H_2 alone; and

ii - those from syngas feedstock comprising CO/H_2 plus a third, commonly petrochemically derived, molecule .

With respect to chemicals belonging to class (i), (higher than methanol) oxygenate chemicals cannot be produced selectively direct from syngas alone. However, methanol, which itself can be produced direct from syngas at a very high selectivity, plays a crucial role as a base chemical in selective, indirect, totally-syngas-based routes to industrial chemicals. Specific examples of methanol/syngas-based routes are discussed.

Chemicals belonging to class (ii) are numerous and a survey is given of recent advances in C_1 homogeneous catalysis, which have made possible the efficient and economic synthesis of a diverse range of industrially important chemicals.

1. INTRODUCTION

Over the past decade, much attention has been focussed on C_1 chemistry. The initial interest was triggered by the sharp increases in prices of petroleum products in the 1970's. Only a few years ago, chemical and petroleum industry analysts were predicting that crude oil prices would reach \$ 100 per barrel by 1990. The gasification of coal and the conversion of CH_4, or even biomass, to make carbon monoxide (CO) and hydrogen (H_2) - a mixture called syngas - were considered as a promising avenue for the future production of chemicals.

In the 1970's, the potential economic advantages of C_1 chemistry over petrochemistry were so great that the question of which chemicals should be produced from syngas seemed irrelevant. The projected large-scale production of fuel from coal via syngas would have a significant impact not only on the pricing of chemicals from syngas but also on the logistics of their manufacture.

However, the recent drop in crude-oil prices has induced a marked change in the chemical and oil industry's perspective; the economic gap between C_1 chemistry and petrochemistry has narrowed. Therefore, one has to be more critical in selecting the type of chemicals to be produced from syngas. The utility of syngas will depend to a great extent on the value of the products and the efficiencies of reaction. Ideally, all of the reactants should end up in the product.

Table 1 shows the theoretical efficiency of syngas utilization for the production of some important chemicals. It brings out clearly the desirability of retaining high oxygen to carbon ratios in the products in order for the maximum reactant utilization to be achieved. Furthermore, it illustrates why the decline in price of petroleum-based feedstocks, such as ethylene, has caused their production from syngas to become very unattractive: oxygen, introduced via gasification of coal or CH_4 steam reforming to generate syngas, is completely lost again as H_2O in the process of their manufacture from syngas.

Table 1. Theoretical efficiency of syngas utilization in the production of economically important chemicals.

Product	CO/H_2 usage ratio	Syngas weight retention, %
Methanol	1 : 2	100
Acetic acid	1 : 1	100
Acetic anhydride	1 : 1	85
Ethanol	1 : 2	72
Ethyl acetate	2 : 3	72
Vinyl acetate	4 : 5	71
Propionic acid	3 : 4	80
Methacrylic acid	4 : 5	71
Ethylene	1 : 2	44

With respect to the production of fuels from syngas, weight retention in the products is not <u>the</u> critical issue: energy contents, transportability, etc. are the crucial factors here. Thus, for example, in places like Canada, Mexico, New Zealand and the Middle East, extremely low-cost natural gas is already being, and will continue to be, used to generate transportable liquid fuels.

Naturally, such developments have repercussions farther afield. In this case, methanol derived from natural gas via syngas exported from these countries is having a major impact on methanol pricing in Europe and the U.S.A. In fact, at current prices methanol represents the most cost-effective octane enhancer for gasoline [1]. In turn, trends in the fuels business will have an impact on the production of chemicals. The ready and cheap availability of methanol also makes it an attractive C_1 feedstock for the production of, in particular, all-syngas-based chemicals. Some examples of all-syngas-based routes to important industrial chemicals, with only methanol and syngas as feedstocks, will be given subsequently.

Recent research into syngas utilization has also highlighted a most attractive feature of syngas, in particular of carbon monoxide: its reactivity towards a great variety of petrochemically derived feedstock molecules, enabling oxygen to be introduced into these molecules. A classic example of this type of syngas utilization is the hydroformylation of olefins to yield alcohols or aldehydes ("oxo products"):

$$R^{=} + CO + H_2 \longrightarrow R{-}C{\begin{smallmatrix} /\!\!/ O \\ \diagdown H \end{smallmatrix}} \xrightarrow{H_2} R{-}CH_2OH \qquad (1)$$

The industrial impact of this reaction is best illustrated by the fact that the worldwide production of oxo chemicals is about 5 million t a^{-1}.

It is worth noting here that the recent advantages in C_1 chemistry have largely been attributable to homogeneous catalysis by organometallic catalyst precursors dissolved in the liquid phase. Subtle effects brought about by the use of several complexing ligands enable transition metal ions to promote specific reactions, so that highly selective reactions have become possible.

2. ALL-SYNGAS-BASED CHEMICALS

Synthesis gas (CO/H_2) offers many routes to industrial chemicals. Basically, these routes can be separated into two broad categories: direct-path and indirect-path.

Apart from methane, the only chemical which can be produced at very high selectivity direct from syngas is methanol:

$$CO + 2H_2 \rightarrow CH_3OH \qquad (2)$$

The direct production from syngas of any chemical containing more carbon atoms than oxygen atoms poses great technical difficulties with respect to catalyst requirements. The catalyst should allow complete reduction of CO molecules to yield methylene ($-CH_2$) or methyl (CH_3-) units while, at the same time, some carbon-oxygen bonds must be retained. It is not surprising, therefore, that the many attempts to produce, for instance, ethanol or acetic acid direct from synthesis gas have been fraught by relatively low selectivity.

However, we shall show that methanol, as a syngas product, provides a unique single-carbon building block for the indirect synthesis of higher oxygenates, which cannot be produced in high selectivity direct from synthesis gas alone.

2.1. Acetic acid

The prime example of methanol/syngas-based chemicals is acetic acid, produced via:

$$CH_3OH + CO \rightarrow CH_3COOH \qquad (3)$$

The overall weight retention of syngas in this molecule is 100%. The reaction is catalysed homogeneously by rhodium/methyl iodide, and is performed under 30 bar CO pressure at 175°C [2]. Monsanto have been using the reaction in a commercial process for fifteen years, and it has been accepted as the most cost-effective route to acetic acid manufacture. Worldwide about 1.5 million t a^{-1} of acetic acid is produced via this process.

2.2. Acetic anhydride

A significant extension of the rhodium/iodide catalysis is adopted in a new Tennessee-Eastman process for converting methyl acetate to acetic anhydride [3].

One plant, which has a capacity of 500.000 t a^{-1}, began operation in 1983, and is part of a whole complex based on chemicals from coal. To date, acetic anhydride is the only higher-than-C_1 chemical derived from coal feedstock. Two important factors have determined the economic success of Eastman's chemicals-from-coal project: the location of the plant close to coal mines; and the integration of acetic anhydride manufacture with cellulose acetate production.

The underlying chemistry is summarized as follows:

$$\left.\begin{array}{ll} \text{i. } C \xrightarrow{H_2O/O_2} CO/H_2 \\ \text{ii. } CO + 2H_2 \rightarrow CH_3OH \\ \text{iii. } CH_3OH + CH_3COOH \rightarrow CH_3COOCH_3 + H_2O \\ \text{iv. } CH_3COOCH_3 + CO \rightarrow (CH_3CO)_2O \\ \text{v. } (CH_3CO)_2O + \text{cellulose} \rightarrow \text{cellulose acetate} + CH_3COOH \end{array}\right\} \quad (4)$$

Acetic acid is recycled to reaction 4-iii. Overall, coal-based syngas is thus converted with cellulose to cellulose acetate and H_2O:

$$2CO + 2H_2 + \text{cellulose} \rightarrow \text{cellulose acetate} + H_2O \quad (5)$$

The overall weight retention of syngas in the final product is 85%; 15% is lost as H_2O in reaction 4-iii.

2.3. Ethanol

Indirect methanol/syngas-based routes to ethanol proceed either via homologation:

$$CH_3OH + CO + 2H_2 \rightarrow CH_3CH_2OH + H_2O \quad (6)$$

or via acetic acid hydrogenation:

$$\left.\begin{array}{l} \text{i.} CH_3OH + CO \rightarrow CH_3COOH \\ \text{ii.} CH_3COOH + 2H_2 \rightarrow CH_3CH_2OH + H_2O \end{array}\right\} \quad (7)$$

Reaction (6) is catalysed by homogeneous cobalt/iodide-based catalysts under rather severe reaction conditions [4]: P = 200 bar, T $\approx$ 200°C. With recycle of aldehydes and esters, molar yields close to 90% should be achievable. Bimetallic Ru/Co/I catalysts allow higher reaction rates, but reaction conditions still remain severe. Reaction (6) has so far found no commercial application, not least for reasons of selectivity in relation to activity and the severity of reaction conditions.

The alternative approach to ethanol via acetic acid (reactions 7) allows a very high overall yield, since reactions 7-i and 7-ii can both proceed with high selectivity. BASF has developed the hydrogenation technology for reaction 7-ii, using a heterogeneous catalyst. Disadvantages of this approach are: the separate use of CO and H_2, necessitating CO/H_2 separation technology, and the increased number of reaction steps.

The process economics for both reactions 6 and 7 are such that at present continued development does nog seem warranted. Only where very low-cost natural gas is available on a very large scale might favourable economics arise. Even then, though, competition with ethanol derived via fermentation, together with the tax benefits associated with the latter, remains a significant obstacle to "chemical" ethanol manufacture.

2.4. Ethyl acetate/vinyl acetate

As shown above, alcohol homologation requires severe reaction conditions in order to be able to proceed. Considerable attention has been focussed on the homologation of other substrates. Promising results are being obtained with esters. Shell [5] has developed catalyst systems which enable esters to be converted to their next higher homologue under mild reaction conditions at high selectivity. For example, methyl acetate can be homologated to ethyl acetate via the following stoichiometry:

$$2CH_3COOCH_3 + 2CO + 2H_2 \rightarrow CH_3COOC_2H_5 + 2CH_3COOH \qquad (8)$$

This reaction is catalysed by bimetallic Rh/Ru/I catalyst systems at temperatures as low as 150°C and at a pressure as low as 20 bar. The use of certain catalyst promoters, in particular phosphine oxides, allows both high reaction rates and high selectivities (>90%).

The coproduction of acetic acid in homologation reaction 8 has the important consequence of allowing integration with simple esterification:

$$2CH_3OH + 2CH_3COOH \rightarrow 2CH_3COOCH_3 + 2H_2O \qquad (9)$$

to yield overall:

$$2CH_3OH + 2CO + 2H_2 \rightarrow CH_3COOC_2H_5 + 2H_2O \qquad (10)$$

As a result, a two-step process to a C_4 chemical is possible via reaction 10 using only C_1 feedstocks. In this respect it is worth noting that a C_1 route to ethyl acetate has a considerable advantage over conventional $C_2^{=}/CH_3OH$-based ethyl acetate manufacture, which requires three process steps:

$$\left.\begin{array}{ll} \text{i.} & CH_2^{=} + H_2O \rightarrow C_2H_5OH \\ \text{ii.} & CH_3OH + CO \rightarrow CH_3COOH \\ \text{iii.} & C_2H_5OH + CH_3COOH \rightarrow CH_3COOC_2H_5 + H_2O \end{array}\right\} \qquad (11)$$

Mechanistic studies of reaction 8 have revealed that it is composed of three elementary steps:

$$\left.\begin{array}{ll} \text{i.} & 2CH_3COOCH_3 + 2CO \rightarrow 2(CH_3CO)_2O \\ \text{ii.} & 2(CH_3CO)_2O + H_2 \rightarrow CH_3\underset{\underset{H}{|}}{C}\text{-}(OOCCH_3)_2 + CH_3COO \\ \text{iii.} & CH_3\text{-}\underset{\underset{H}{|}}{C}\text{-}(OOCCH_3)_2 + H_2 \rightarrow CH_3COOC_2H_5 + CH_3COOH \end{array}\right\} \qquad (12)$$

Each of reactions 12 can be demonstrated separately and besides CH_3COOCH_3, both $(CH_3CO)_2O$ and $CH_3\underset{\displaystyle H}{\underset{|}{C}}\text{-}(OOCCH_3)_2$ can also be used as feedstock for ethyl acetate manufacture.

Alternatively, reaction sequence 12 can be halted at any desired product, for example:

$$2CH_3COOCH_3 + 2CO + H_2 \longrightarrow CH_3\underset{\displaystyle H}{\underset{|}{C}}\text{-}(OOCCH_3)_2 + CH_3COOH \qquad (13)$$

provided that a suitable catalyst composition is chosen, for example Rh/CH_3I, promoted by pyridine oxide or phosphine and pyridines [5].

Reaction 13 allows integration with the pyrolysis reaction:

$$CH_3\text{-}\underset{\displaystyle H}{\underset{|}{C}}(OOCCH_3)_2 \xrightarrow{T} CH_2{=}\overset{\displaystyle H}{\overset{|}{C}}\text{-}OOCCH_3 + CH_3COOH \qquad (14)$$

and esterification reaction [9], to yield overall:

$$2CH_3OH + 2CO + H_2 \longrightarrow CH_2{=}\overset{\displaystyle H}{\overset{|}{C}}\text{-}OOCCH_3 + 2H_2O \qquad (15)$$

i.e. a three-step route to vinyl acetate baed on only C_1 feedstocks.

2.5. Propionic acid/methylcrylic acid

As a spin-off of our work in ester homologation to higher-homologue esters we discovered that the homologation of esters can also proceed to produce higher-homologue carboxylic acids, for instance:

$$CH_3COOCH_3 + 2CO + 2H_2 \rightarrow CH_3CH_2COOH + CH_3COOH \qquad (16)$$

This reaction is catalysed by Rh/CH_3I, promoted by protonic acids and phosphine oxides [5]. High yields of propionic acid (~80%), accompanied by relatively small quantities of next-higher-homoloque carboxylic acids, are obtained.

Reaction 16 can be integrated with simple esterification reaction 9, yielding overall:

$$CH_3OH + 2CO + 2H_2 \rightarrow CH_3CH_2COOH + H_2O \qquad (17)$$

i.e. a methanol/syngas-based two-step process to propionic acid.

Propionic acid itself finds application as a grain-preservative, but is also suitable for use as a precursor to methacrylic acid, a large-volume industrial chemical, on condensation with formaldehyde via conventional chemistry:

$$CH_3CH_2COOH + CH_2{=}O \longrightarrow CH_2{=}\overset{\displaystyle CH_3}{\overset{|}{C}}\text{—}C\begin{matrix}{/\!\!\!\,}^{O}\\ \backslash_{OH}\end{matrix} + H_2O \qquad [18]$$

Since formaldehyde is produced commercially on a large scale from methanol:

$$CH_3OH \rightarrow CH_2{=}O + H_2O \qquad [19]$$

methacrylic acid, via reactions 17 and 18, represents a further example of a totally syngas-derived molecule. The whole process of its manufacture from methanol/syngas as feedstock requires three process steps (formaldehyde production not included).

2.6. Ethylene glycol

In theory, ethylene glycol would be the ideal molecule to synthesize direct from syngas, since it contains an equal number of C and O atoms:

$$2\,CO + 3\,H_2 \rightarrow HO\text{-}CH_2\text{-}CH_2\text{-}OH \qquad (20)$$

Much research has been devoted to this reaction at Union Carbide [6]. The reaction has been demonstrated using homogeneous catalysts such as cobalt, rhodium and ruthenium carbonyls. The greatest problem is the very low reaction rate, even at temperatures around or above 200°C. At these temperatures the thermodynamics of reaction 20 are unfavourable, requiring super-high pressures (1000-2000 bar). Even then, selectivity to ethylene glycol hardly reaches 60%. For these reasons we consider the prospects for a commercial direct ethylene glycol process as gloomy.

Not surprisingly, several proposals for indirect syngas-based ethylene glycol processes have been made.

One of the most promising routes proceeds via hydroformylation of formaldehyde. The synthesis of glycol aldehyde by hydroformylation of formaldehyde has been patented by a number of companies, e.g. Monsanto and Shell:

$$CH_2{=}O + CO + H_2 \longrightarrow HO\text{-}CH_2\text{-}C\genfrac{}{}{0pt}{}{\nearrow O}{\searrow H} \qquad (21)$$

This reaction is best catalysed by Rh-based catalysts promoted by certain phosphines (PR_3) and a proton source: yields in excess of 95% can be obtained under mild reaction conditions (70°C, 60 bar). Combined with glycol aldehyde hydrogenation (e.g. with heterogeneous Ni catalysts) and with reaction 19 it allows overall reaction 22:

$$CH_3OH + CO + H_2 \rightarrow HO\text{-}CH_2\text{-}CH_2\text{-}OH \qquad (22)$$

3. CARBON-MONOXIDE-BASED CHEMICALS

An attractive feature of syngas and, in particular, of one of its components, carbon monoxide, is its reactivity towards a great variety of petrochemically derived feedstock molecules.

We have already mentioned a large-scale commercial application of this type of chemistry: hydroformylation of olefins to yield aldehydes or alcohols (reaction 1). In general, this type of chemistry gives access to a great variety of high-added-value chemicals, which via alternative chemistry are accessible either not at all or only with great difficulty. We summarize some of the recent advances in carbonylation chemistry.

3.1. Carbonylation of olefins

HomogenEous, transition-metal-based catalysts have been developed for the carbonylation of olefins [7]:

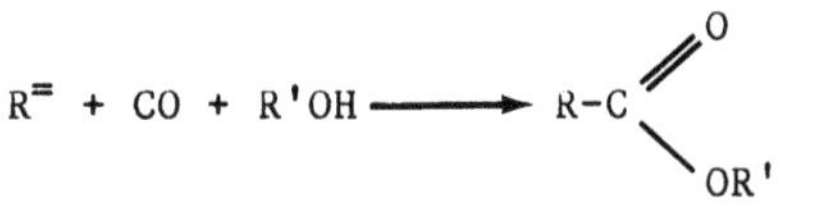

(23)

in which R = allyl, aryl, H or acetyl. A wide range of products -esters, carboxylic acids and anhydrides - can be made in this way, products which cannot easily be manufactured by alternative means such as, for example, oxidation chemistry.

In addition to simple alkenes, functionalized olefins can also be used as a feed, thus giving access to a variety of bifunctional molecules, for example diesters.

Catalyst systems have also been developed for the dicarbonylation of conjugated olefins such as butadiene [8]:

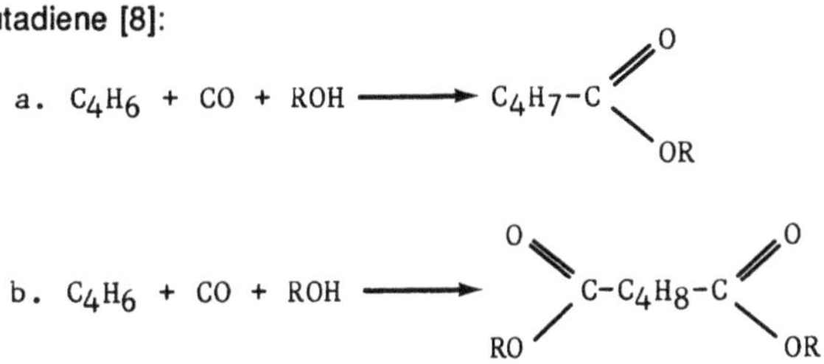

(24)

Diesters are important precursors for synthesis of polyamides such as nylon 6.6.

Certain catalyst systems allow alternating $CO/C_2^{=}$ copolymerization [9]

$$nC2^{=} + nCO \rightarrow -(C\text{-}\overset{\overset{O}{\|}}{C}\text{-}C)_n- \qquad (25)$$

thus yielding high-molecular-weight polyketone polymers, which are inaccessible via other chemistry.

A wide range of unsaturated esters, carboxylic acids and carboxylic anhydrides can be produced via:

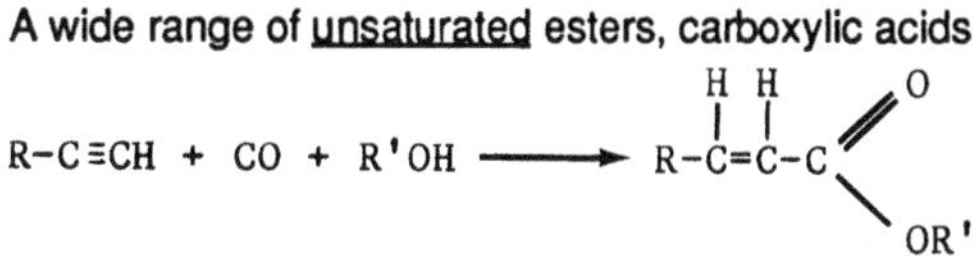

(26)

This reaction forms the basis for commercial acrylic acid production from acetylene [10].

N-acyl-α-amino acids can be synthesized by reductive carbonylation of olefins in the presence of amides [11]:

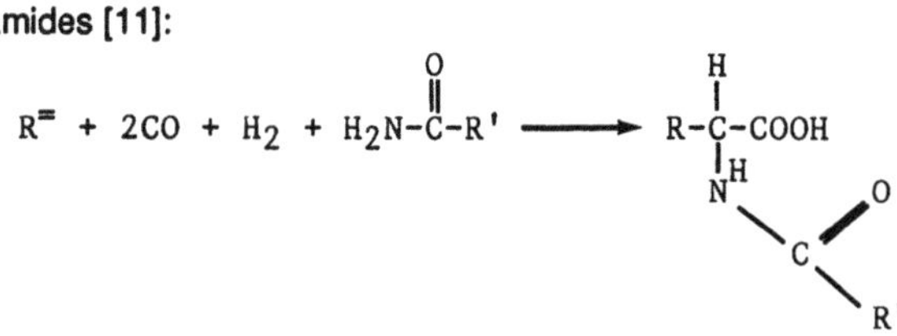

(27)

This reaction is catalysed by $Co_2(CO)_8$ and proceeds via in-situ-generated aldehydes.

3.2. Reductive carbonylation of nitro compounds

Active catalysts have recently been developed for the reductive carbonylation of aromatic nitro compounds to yield carbamates [12], for example:

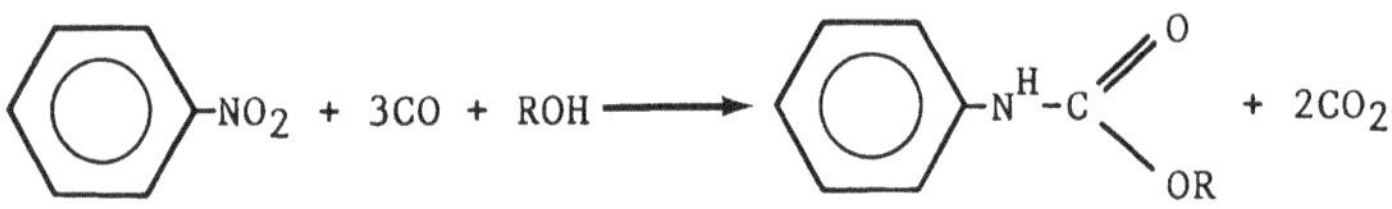

(28)

The reaction is catalysed efficiently by Pd compounds, promoted by bi-nitrogen-containing chelating ligands. Carbamates find application themselves, for example, as insecticides and herbicides or as intermediates to other agrochemicals. One application of potential importance is their conversion to isocyanates by pyrolysis, e.g.

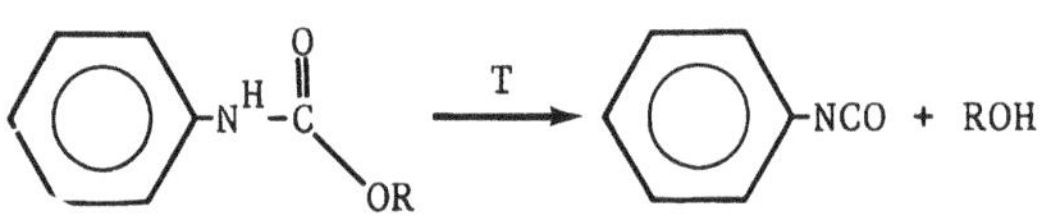

(29)

Integration with reaction 28 thus allows a two-step synthesis to isocyanates from nitro compounds. Traditionally, isocyanates are produced by reaction of amines, derived from nitro compounds, with phosgene, an exceedingly toxic chemical, requiring special handling techniques.

Diisocyanates find large-scale industrial application for the production of polyurethanes.

In addition to the examples above, carbon monoxide is expected to make considerable further inroads in chemical synthesis, not only of large-volume chemicals, but also increasingly of small-volume fine and specialty chemicals, especially those not easily accessible via other chemistry.

4. CONCLUDING REMARK

We hope to have demonstrated that synthesis gas constitutes an important and versatile C_1 building block in chemical synthesis. C_1 chemistry now being developed will play an increasingly important role in reshaping the chemical industry in the decades ahead.

REFERENCES

[1] Chem. and Eng. News (1984), July 16, p.9.
[2] Roth JF, Craddack JH, Herschman A, Paulik FE. Chem. Technol. (1971), 600.
[3] Coover HW, Hart RC. Chem. Eng. Progr. (1982), 72.
[4] Wender I. Catal. Rev. Sci. Eng. 14 (1976) 112.
[5] Drent E. Preprints, ACS Symposium 31 (1986) 97.
[6] Dombek BD. J. Chem. Ed. 63 (1986) 210.
[7] Falbe J (Ed.) 1980. New synthesis with carbon monoxide. Springer-Verlag, Berlin, p.243.
[8] US patent 4 256 909 (to BASF).

[9] Sen A. Chem. Tech. (1986) Jan., p. 48.
[10] Weissermel K, Arpe HJ. 1978. Industrial Organic Chemistry. Verlag Chemie, Weinheim.
[11] UK patent application no. 2 000 132 (by Rhone Poulenc).
[12] European patent application no. 169 650 (by Montedison).

MICROBIOLOGICAL SYNTHESIS FROM C_1-COMPOUNDS: APPLICATION OF SOME METHYLOTROPHIC FUNCTIONS TO SYNTHESIS OF USEFUL CHEMICALS

Y. TANI, T. YONEHARA, Y. SAKAI AND B.D. YOON
Research Center for Cell and Tissue Culture, Faculty of Agriculture, Kyoto University, Kyoto, Japan.

1. INTRODUCTION

Selection of the raw material for the culture substrate is an important factor in biotechnology which uses microorganisms as tools, as well as the pinpointing of the target product and the screening and breeding of microorganisms. C_1- compounds have attracted much attention as the culture substrate for these microorganisms. Of C_1-compounds, the most promising as a raw material for the fermentation industry is considered (at present) to be methanol because of its nature. Thus, methylotrophs offer promise for the production of fine and commodity chemicals, as well as other useful materials. So far, investigations have been carried out to establish the new process of microbiological synthesis by methylotrophs [1]. We reported previously the production of some amino acids and coenzymes, in which the unique metabolism of methylotrophs was utilized [2].

In this paper, recent development of processes for the production of ATP, formaldehyde and cytochrome *c* in our laboratory will be introduced and discussed. In those, the utilization of unique functions of methylotrophic yeasts and bacterium are advantageously demonstrated.

2. PRODUCTION OF ATP FROM ADENOSINE BY METHANOL YEAST

In preceding studies (2,3), we established an ATP production system, from AMP, with Zymolyase- and sorbitol-treated cells of a methanol yeast, *Candida boidinii (kloeckera* sp*)* No. 2201. The system consisted of oxidation of methanol to reduce NAD^+, oxidative phosphorylation of ADP and phosphorylation of AMP by adenylate kinase. It seemed to be still insufficient for industrial production of ATP, because AMP is chemically unstable and commercially expensive as a starting material, Then, we investigated adenine nucleotide formation from adenosine or adenine by sorbitol-treated cells of the yeast.

2.1 Production of ATP

The reaction conditions for ATP production from adenosine were optimized as to initial concentrations of substrates and coenzymes, pH of potassium phosphate buffer, temperature, osmotic pressure and other additives. Efficient ATP production from adenosine was achieved by the addition of pyrophosphate togeher with Mg^{2+}. Under the optimal conditions, 198 mM (100 g/l) of ATP was produced in the conversion rate of 77% from adenosine. ATP production from adenine was not so high under the conditions tested.

Analysis by HPLC showed the accumulation of IMP in the reaction mixture, possibly due to catalysis by AMP deaminase. Among osmoregulatory compounds that plasmolyzed the yeast cells, only sorbitol caused high ATP production activity on addition to the reaction mixture. The sorbitol effect was attributed to the supplying of NADH by the oxidation in parallel with the methanol oxidation. The ability from adenine and adenosine was found in many methanol yeasts belonging to the genera, *Hansenula, Pichia, Candida* and *Torulopsis. C boidinii* had especially high ability to accumulate ATP from adenosine.

Biological production of ATP has been known with substrate level phosphorylation of yeast and with salvage synthesis in *Brevibacterium ammoniagenes*. Our system using the oxidative phosphorylation overwhelmed preceding conventional processes (Table 1).

2.2 Energy efficiency

The sequential conversion system from methanol and adenosine to ATP might involve NADH production in the cytosol and electron transport to mitochondrion by the glycerol phosphate shuttle

Table 1. ATP production by biological system

System	Raw material	ATP(g/l)	Efficiency*	Molar yield (%)
Culture of *B. ammoniagenes*	Adenine, glucose	1.57	-	20
Cell of *B. ammoniagenes*	Adenine, glucose	10.9	0.10	83
Cell of yeasts	Adenine,	102.5	0.34	50
Cell of *E. coli*	AMP, glucose	14.0	0.14	92
Chromatophore of *R. rubrum*	ADP, light	0.4	2.0	80
Chromatophore of *R. capsulata*	ADP, light	-	200.**	-
Cell of blue-green alga	ADP, light	0.17	56.**	17
Acetate kinase	ADP, acetyl-P	10.0	-	100
Adenylate kinase	ADP	0.24	1.3	60
	AMP, methanol	30.0	1.5	60
Cell of *C.boidinii*	Adenosine, methanol	100.0	4.0	77
	Adenine,	4.1	0.20	40

*Gram of ATP produced by 1 g of DCW, protein or chlorophyll

mechanism to reduce FAD bound to the inner membranes. Consequently the P/O ratio in this system is not 3 but 2 [2]. The material balance of ATP production from adenosine can be summarized as the equation below:

$$\text{adenosine} + 3\ P_i + 3/4\text{methanol} + 9/8O_2 \quad \rightarrow \quad \text{ATP} + 3/4CO_2 + 9/2H_2O$$

Based on the free energy changes of respective reactions involved in the system, the theoretical value of the total energy efficiency from methanol to ATP was calculated as 14.3%. In a typical reaction mixture, the practical energy efficiency against the methanol consumption was expressed as 4.25%. Therefore, the ratio of the practical against the theoretical was 30%.

3. PRODUCTION OF FORMALDEHYDE BY METHANOL YEAST

Methanol is now mostly used to produce formaldehyde by a chemical conversion process at 500 to 600°C. The amount is more than 4 million tons and corresponds to 43% of industrially produced methanol. The introduction of a microbiological process for the conversion should enable energy to be saved and side-product formation to be minimized. Alcohol oxidase of methanol yeasts uses molecular oxygen as the electron acceptor and, therefore, does not require regeneration or an exogenous supply of cofactor in the methanol oxidation. Production of formaldehyde from methanol was investigated with immobilized alcohol oxidase and cells of *Hansenula polymorpha* [4,5]. However, improvement of the yeast itself has not been reported. To develop a new process of formaldehyde production from methanol, we studied the improvement of formaldehyde productivity of a methanol yeast, *Candida boidinii* S2.

3.1 Isolation of a mutant having high alcohol oxidase activity

C. boidinii S2, which showed a high productivity of formaldehyde among methanol yeasts tested, was mutagenized to increase the alcohol oxidase activity [6]. The procedure for mutant screening consisted of; 1) induction of alcohol oxidase activity on a methanol plate, 2) catabolite inactivation of alcohol oxidase on a glucose plate, and 3) visualization of alcohol oxidase activity in a colony. One of the mutants, strain AOU-1, showed 1.7 times higher formaldehyde productivity and a higher growth rate on methanol than the wild type strain. Specific activities of methanol dissimilating enzymes were compared between the wild type strain and the mutant (Table 2).

The mutant showed 2.4 times higher alcohol oxidase activity than the wild type strain under the conditions used. Specific activities of other methanol-dissimilating enzymes were, however, essentially the same.

To determine whether the change in alcohol oxidase level of mutant strain AOU-1 is qualitative or quantitative, we purified the alcohol oxidases from the wild type and mutant strains. Both the enzyme subunits showed the same molecular weight of 72.000 on SDS-polyacrylamide electrophoric analysis, showing the same mobility. Both enzymes also showed the same retention time on HPCL-gel filtration. The specific activity and apparant K_m value for methanol of both enzymes were 4.2 U/mg protein and 0.846 mM^{-1}, respectively. Thus, no qualitative changes of the enzyme was detected between the wild type and mutant strains. The high activity of mutant strain AOU-1 could be attributed to a quantitative change and a change in the rate of enzyme synthesis.

Table 2. Activities of methanol-dissimilating enzymes

	Wild type strain	Mutant AOU-1
Total protein (mg)	76.4	79.0
Enzyme activities (U/mg protein)		
Alcohol oxidase	0.384	0.921
Catalase	1355	1471
Formaldehyde dehydrogenase	0.183	0.211
Formate dehydrogenase	0.0757	0.0819

3.2 Optimization of reaction conditions

The reaction conditions for formaldehyde production with cells of mutant strain AOU-1 were optimized. When the reaction was carried out at low temperature below 10°C and low pH of 6.0 under oxygen, formaldehyde was accumulated efficiently in the reaction mixture. Alcohol oxidase is susceptible to H_2O_2, a reaction product, which masks the catalytic sulfhydryl group of the enzyme. Another product of the reaction, formaldehyde, might inactivate the enzyme by forming a Schiff's base with the amino groups of the enzyme. These inhibitory non-catalytic reactions could be fairly suppressed by lowering the reaction temperature. The lower pH for the formaldehyde production than the optimum pH of purified alcohol oxidase can be explained as the consequence of the retarded activitity of formaldehyde and formate dehydrogenases at this pH.

Under optimal conditions and with cell-feeding, 750 mM formaldehyde was obtained after 36-h reaction.

3.3 Formaldehyde production by methanol-limited chemostat cultured cells

The formaldehyde productivity of mutant strain AOU-1 was much improved by methanol-limited chemostat culture. Cells from the chemostat culture at a dilution rate of 0.075 h^{-1} showed the highest productivity. The cells had the highest activity of alcohol oxidase and almost minimum activity of formaldehyde dehydrogenase. Formaldehyde of 950 mM was produced in 9-h reaction with the cells. The chemostat culture improved not only formaldehyde productivity but also cell productivity in comparison with batch culture. A maximum cell productivity of 0.2 $g.l^{-1}.h^{-1}$ and a cell yield of 47% were obtained.

4. CYTOCHROME *C* PRODUCTION BY A KCN-RESISTANT MUTANT OF AN OBLIGATE METHYLOTROPH, *METHYLOMONAS* SP.

Cytochrome *c* is produced as a medicine for heart diseases caused by oxygen deficiency from extracts of horse and cattle hearts and yeast. Methylotrophic bacteria oxidize methanol with methanol dehydrogenase as the first step enzyme, followed by formaldehyde and formate dehydrogenases, from which the electrons flow to cytrochrome *c* [7]. Therefore, methanol could be a more suitable carbon source for the production of cytochrome *c* as to the regulation of the amount of cytochrome *c* than other multicarbon sources: the growth substrate is directly connected with cytochrome *c* in methylotrophs. So far, no attempt to increase the cellular content of cytochrome *c* by means of mutagenesis has been reported. We studied the enhancement of cytochrome *c* productivity of mutant of an obligate methylotroph, *Methylomonas* sp.

4.1 Production of cytochrome *c*

About 150 strains of methylotrophic bacteria were isolated by the usual enrichment culture method. Among them, *Methylomonas* sp. YK 1 was selected as cytochrome *c* producer because of the rapid growth in a methanol medium and a considerably high cellular content of cytochrome *c* [8]. The strain was mutagenized so as to be resistant to metabolic inhibitors related to the function of cytochrome *c*. Table 3 shows mutants having the highest cytochrome *c* content for each inhibitor. Among them, a KCN-resistant mutant, strain YK 8, had the highest cytochrome *c* content. A 1.2 mg KCN-resistant mutant, strain YK 56, was derived from strain YK 8. The content of the mutant was increased about 3 times compared to that of the wild type strain.

Table 3. Cyrochrome *c* content of inhibitor-resistant mutants

Inhibitor	Resistant strain	Cytochrome *c* produced (mg/l of culture medium)
-	YK 1 (wild)	2.97
KCN	YK 8	6.83
Antimycin A	YK 126	5.96
Rotenone	YR 372	5.20
2,4-Dinitrophenol	YD 873	4.75
Pentachlorophenol	YP 1116	4.62
Progesterone	YPR 1385	4.95

The conditions for enhancement of the cytochrome *c* productivity were investigated using mutant strain YK 56. The cellular content of cytochrome *c* increased with the increase in methanol concentration to 2.0%. The highest amount of cytochrome *c* in the culture medium, 9.6 mg/l, was obtained at the methanol concentration of 2.0% after 24- to 30-h cultivation. The amount of cytochrome *c* in the culture medium and the cellular content were high with the medium supplemented with peptone, L-malate and

$FeSO_4.7H_2O$. Based on results of optimization experiments, we constructed a suitable medium for the production of cytochrome *c* with mutant strain YK 56. The medium is composed of 2 g of methanol, 0.35 g of K_2HPO_4, 0.15 g of KH_2PO_4, 0.3 g of NH_4NO_3, 0.1 g of NaCl, 0.02 g of $MgSO_4.7H_2O$, 0.0025 g of $FeSO_4.7H_2O$, 0.2 g of L-malate and 0.3 g of peptone in 100 ml of tap water, pH 7.0. The growth and the amount of cytochrome *c* in the culture medium increased in proportion to the increasing cell concentration and reached the maximum value of 48 mg/l after 54-h cultivation. When methanol or methanol and peptone were fed after 42-h cultivation, the cytochrome *c* productivity increased on further cultivation and reached the maximum at 36-h after the methanol and peptone were fed. The amount of cytochrome *c* reached about 60 mg/l of culture medium. Further improvement of the productivity was achieved by a jar fermentor culture with feeding methanol and the amount of 160 mg/l was obtained after 32-h cultivation.

4.2 Cytochrome c of KCN-resistant mutant

Cells of the wild type strain and mutant strain YK56 were treated by sonication and centrifuged at 24,000x*g* . The resultant supernatant was recentrifuged at 100,000 x*g* and a deep-red supernatant was obtained as soluble cytochrome *c* preparation. After fractionation by ammonium sulfate (35-90% saturation), the cytochrome *c* preparation was applied onto DEAE-cellulose column chromatography. The cytochromes were designated as cytochromes *c*-I, *c*-II, and *c*-III in the order of elution.

Soluble cytochrome *c* of the wild type strain was separated into two fractions; one type, cytochrome *c*-I, was eluted as a non-adsorbed fraction with 10 mM Tris-HCl buffer, pH 8.0, on DEAE-cellulose column, and another type, cytochrome *c*-III, was eluted at around 0.3 M NaCl with a linear gradient of 0-0.4 M NaCl in the same buffer. The proportion of cytochromes *c*-I and *c*-III was 2.1 and 97.9%.

On the other hand, mutant strain YK 56 contained three distinct soluble cytochromes *c*; cytochrome *c*-II, which was eluted at around 0.2 M NaCl in the elution system described above, was found as a new fraction in addition to cytochromes *c*-I and *c*-III of the wild type strain. The proportion of these three soluble cytochromes *c* were 2.4, 71.5 and 26.1%. Amounts of cytochromes *c*-I and *c*-III were similar in both strains, and the mutant produced cytochrome *c*-II as the main component.

Cytochromes *c*-II and *c*-III of mutant strain YK 45 were purified by a procedure involving ammonium sulfate fractionation and DEAE-, CM-cellulose, and Sephadex G-150 column chromatography. Cytochrome *c*-II was obtained as crystals with ammonium sulfate. Both absorption peaks of the α-band of the two cytochromes *c* were at 551.5 nm at room temperature and the ß-band of cytochrome *c*-II had a shoulder at 530 nm. Molecular weights of cytochromes *c*-II and *c*-III were 16,000 and 20,000, and their isoelectric points were 4.1 and 3.5, respectively. Thus, properties of cytochrome *c*-II were clearly different from those of cytochrome *c*-III. This indicates that cytochrome *c*-II of the mutant appeared newly in the cell.

Membrane-bound cytochrome *c* was composed of two, which were similar properties but different from cytochromes *c*-II and *c*-III of soluble fraction.

REFERENCES

[1] Tani Y. (1985) In Biotechnology and Genetic Engineering Reviews, volume 3, pp. 111-135. Intercept Ltd.

[2] Tani Y, Yamada H. (1984) In Microbial Growth on C_1 Compounds, pp. 293-296. American Society for Microbiology.

[3] Tani Y, Yonehara T, Mitani Y, Yamada H. (1984) J. Biotechnol., 1, p. 119-127.

[4] Baratti J, Couderc R, Cooney CL, Wang DIC. (1978) Biotechnol. Bioeng., 20, p. 333-348.

[5] Courderc R, Baratti J. (1980) Biotechnol. Bioeng., 22, p. 1155-1173.

[6] Tani Y, Sakai Y, Yamada H. (1985) Agric. Biol. Chem., 49, p. 2600-2706.

[7] Anthony C. (1975) Biochem. J., 146, p. 289-298.

[8] Tani Y, Yoon BD, Yamada H. (1985) Agric. Biol. Chem., 49, p. 2385-2391.

THE METHYLOTROPHIC YEAST PICHIA PASTORIS AS A HOST FOR HETEROLOGOUS PROTEIN PRODUCTION

G. THILL, G. DAVIS, C. STILLMANN, J.F. TSCHOPP, W.S. GRAIG, G. VELICELEBI, J. GREFF, M. AKONG, D. STROMAN*, R. TORREGROSSA* AND R.S. SIEGEL
SIBIA, La Jolla, CA. and Phillips Petroleum*, Bartlesville, OK.

1. INTRODUCTION

Yeasts have held great promise as hosts for production of economically important proteins and peptides using recombinant DNA technology. Among the heralded features of the bakers yeast, *Saccharomyces cerevisiae*, are its safety (lack of toxicity, pyrogenicity and allergenicity), ability to be grown on an industrial scale, and the vast wealth of information regarding its genetics and molecular biology which has led to the development of a number of expression vector systems. Invariably these vectors are based on the endogenous multicipy plasmid, 2 micron circle, and use a variety of promoter systems, either constitutive or regulated, to drive the expression of the desired gene [1] [2].

However, expression systems in yeasts have suffered from several deficiencies which have been circumvented by the use of alternative systems in both procaryotes and cultured mammalian cells. Unfortunately, the alternatives also have defects to be overcome. Cytoplasmic expression in *Saccharomyces* in general has produced neither the high per cell yields [3] nor the high cell densities which are characteristic of *E. coli* fermentations. There are, however, certain problems associated with expression of foreign proteins in *E. coli.* The major one is the formation of "inclusion bodies" of foreign proteins, basically an insoluble mass of protein which must be renatured for biological activity [4]. Clearly, a yeast production system capable of high volumetric productivity of soluble biologically active proteins would be commercially desirable.

Another weakness demonstrated by *Saccharomyces*, and attributed to all yeast in general, is an inability to perform "correct" glycosylation. In *Saccharomyces*, N-linked glycosylation can consist of a polymer ranging from twenty to one hundred fifty mannose residues in length [5][6]. Such glycosylation may be detrimental to the efficient secretion of proteins into the growth medium, as well as to the biological activity of the secreted heterologous protein. The biotechnology industry has responded to this by using cultured mammalian cells when post-translational modification has been a problem in *Saccharomyces*. Cultured animal cells are capable of producing reasonable levels of biologically active proteins. They do suffer from several drawbacks, nonetheless. Among these are expensive media components, stringent sterility requirements to prevent bacterial contamination, and expensive hardware requirements to maintain growth and production phases [7]. A yeast system capable of discrete glycosylation and efficient secretion of proteins in high cell density batch culture or continuous culture will undoubtedly play a large role in the production of recombinant products.

The ability to move swiftly from research scale to pilot scale protein production is a prerequisite to the developement of a successful system. There are some limitations to current *Saccharomyces* production technology vis-a-vis scale up. The first concerns the use of multicopy plasmids. An idealized system for production of large quantities of a foreign protein must contain elements of regulated expresion. Repression of expression during the development of cell mass in the fermentor minimizes selective pressure against high producing cells. It is clear that a promoter dependent on trans-acting factors for proper repression and induction of expression may not be properly regulated on a high copy number plasmid [8]. Indeed, the higher the vector copy number the less likely is the regulation to be tightly controlled.

Another property of multicopy plasmids which is potentially deleterious to scale-up is mitotic instability [9]. Although a certain proportion of the daughter cells in a culture do not receive the plasmid necessary for survival in the fermentor they do not die immediately since they retain residual amounts of the protein encoded by the selective marker gene. Eventually however, protein turnover and cell division will render the cell dead. Two factors determine whether nonproducing cells represent a severe problem to fermentation. The first is the amount of selective marker protein in the plasmid-deficient cell. This quantity is most likely proportional to plasmid copy number an determines, all other parameters being equal, how long the cell will survive in the culture. The higher the plasmid copy number the greater is the chance that non-producers can represent a significant portion of the cell population in a growing culture [10]. The second factor is the selective disadvantage of the producing cell. The slower the producing cell grows, the larger will be the proportion of non-producers.

One final note regarding optimization of fermentation conditions. The ability of *Saccharomyces* to metabolize both sugars and ethanol leads to oscillations in oxidative metabolism between aerobic and anaerobic phases. These oscillations can lead to instability in fermentation leading to cell death, but more importantly, their idiopathic nature makes it difficult to optimize the production of a given protein in the fermentor [11].

We have developed a second generation yeast expression system based on the methylotrophic yeast *Pichia pastoris.* This organism was developed by the Phillips Petroleum Company for single cell protein production. It is fermentor optimized, capable of high cell density growth (150 g/l dry weight) in continuous culture for long periods of time [12]. It can utilize methanol, as opposed to glucose, as a carbon and energy source. Using the highly expressed [13], tightly regulated alcohol oxidase and dihyroacetone synthase promoters in a regulated expression vector, we can control the synthesis of foreign proteins simply by altering the carbon source of the growth media. Additionally, these expression vectors are single copy integrants and have excellent mitotic stability. The high per cell yields coupled with the high cell density fermentations lead to very high volumetric yields.

This paper describes the molecular biology and fermentation technology of the *Pichia pastoris* yeast expression system for the production of hepatitis B surface antigen.

2. HBsAG PRODUCTION IN PICHIA

Hepatitis B infection represents a significant worldwide health problem. Large numbers of inhabitants in Asia and Africa are at risk. In addition to hepatitis viremia, persistent hepatitis infection is associated with liver cirrhosis as well as liver cancer [14]. As a result, research has centered around effective vaccines against the hepatitis virus. Current prophalaxis is centered around a 22 nm virsus capsid-like protein purified from human serum and used as a vaccine [15]. However, health related and economic factors have made it a necessity for a safer, more economical vaccine based on recombinant DNA technology.

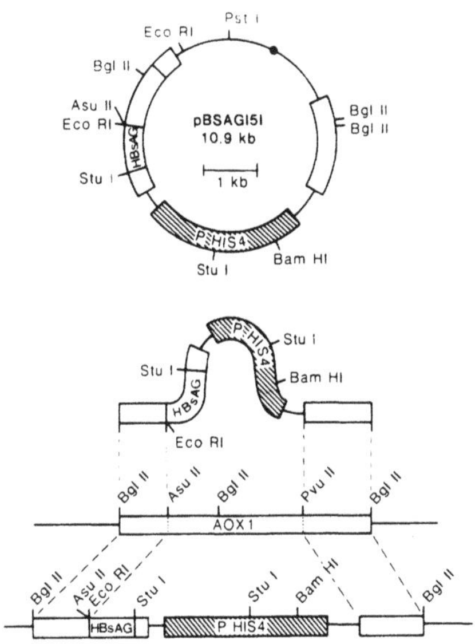

Figure 1.
A *Pichia pastoris* vector for hepatitis B surface antigen expression.

TOP PANEL: The expression vector, detailed in the text, which can integrate into the genome in an "addition" type integration.

LOWER PANEL: A description of recombination events which lead to replacement of the AOX1 structural gene with hepatitis B surface antigen coding sequences.

The surface antigen gene encodes a hydrophobic protein of approximately 24,000 daltons. Approximately 100 monomers assemble into a antigenic particle of around 2.2 megadaltons, 22 nm in diameter [16]. This study employs two assay methods to measure hepatitis B production; the first is a biological assay wherein antigenicity of *Pichia* extracts is compared with a blood derived human standard; the second is an immunoblot in which total monomer is measured using antisera to denatured monomer with purified monomer as standard.

Figure 1 shows a typical expression vector. The alcohol oxidase promoter is truncated immediately upstream of the initiation codon, and synthetic EcoRI linkers added. The length of the promoter fragment is approximately 1.0 Kb. The hepatitis surface antigen gene fragment is composed only of coding sequences. Signals for proper termination of transcription follow the gene and are derived from the alcohol oxidase gene. Also included in this vector are sequences encoding the wild type His4 gene which complement the histidinol dehydrogenase deficiency in *Pichia* host strain GS115, as well as a fragment of *Pichia* DNA further downstream of the AO gene. The vector can be introduced into *Pichia* in two ways. The intact plasmid can integrate into the alcohol oxidase locus or the His4 locus in an "addition type" integration, or the larger BglII fragment can replace the wild type AOX1 gene with hepatitis surface antigen coding sequences and the HIS4 gene. The former case generates a wild type methanol-utilizing strain whereas the latter generates an AOX1 mutant which grows more slowly on methanol. The doubling time on methanol is approximately ten-fold longer since it depends on the activity of a second alcohol oxidase gene, AOX2, which contributes less than 10% of the total methanol oxidase activity in wild type strains. A comparison of these vectors, both integrated at the AOX1 locus, reveals that the AOX1 wild type strain produces 10% of the antigenic hepatitis surface antigen produced by the AOX1 defective strain (Table I). This slower growing strain is much more productive than wild type strain is. Consequently the more productive slower growing strain was used for further investigation.

Table I

A comparison of hepatitis B surface antigen immunoreactivity
in AOX1 wild type and mutant strains

	Wild type	AOX1 mutant
% 22 nm particle	0.3%	2.5%

Cells were grown in batch culture in defined media containing methanol.
Extracts of soluble protein were prepared and measured for immunoreactivity by the AUSRIA kit.
Protein concentrations were measured by the Bradford reagent using Bovine serum albumin as a standard.

3. REGULATED EXPRESSION OF SURFACE ANTIGEN PARTICLE

In order to minimize potential deleterious effects of foreign protein production, we have devised a two stage production system which is easily regulated by the carbon source feed. An inoculum is grown in defined media containing 6% glycerol as sole carbon source. This will allow the batch culture to grow with wild type kinetics to approximately 30% of maximal cell density. As seen in Figure 2, no production of hepatitis surface antigen monomer is detected during this phase. When the glycerol is exhausted, a methanol feed is added, and a time-dependent linear increase of both monomer and AUSRIA detectable particle is seen. These batch fermentations run between 150 and 200 hours and yield AUSRIA levels of 2-3% soluble protein. We have achieved cell densities of approximately 70 g (dry weight) per liter. Volumetric yields are routinely between 400 and 600 mg per liter. These represent significant increases over expression results reported by labs using *Saccharomyces* expression vectors which do not exceed 1 mg/l [1-3]. Most importantly, the product is easily soluble in aqeous buffer and does not require detergent solubilization.

4. SCALE UP OF HEPATITIS B SURFACE ANTIGEN PRODUCTION

In order to demonstrate that *Pichia pastoris* can be used in a bona fide production system, it is necessary to show that it can produce proteins in large quantities. Figure 3 indicates that with a variety of fermentor volumes ranging from one to two hundred forty liters, and at various cell densities, there is no

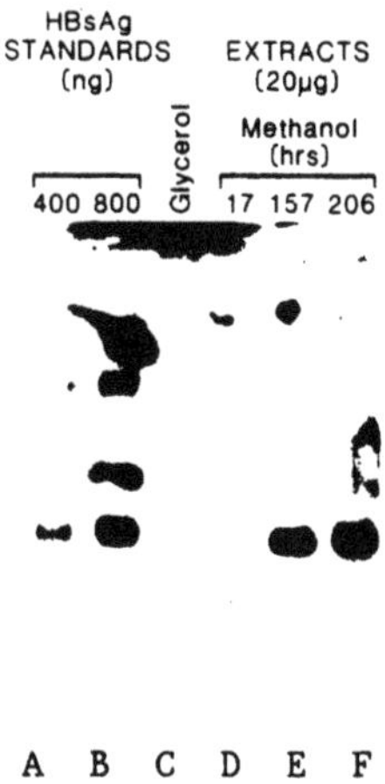

Figure 2

Induction of hepatitis B surface antigen production under different growth conditions

Extracts of soluble protein from disrupted *Pichia* cells grown under varying conditions were separated in 12% SDS polyacrylamide gel electrophoresis. An immunoblot using antisera to denatured surface antigen monomer was used to quantitate expression levels
Lane A 400 ng standard
Lane B 800 ng standard
Lane C extract from glycerol grown cells
Lane D-F extract from methanol grown cells at the indicated times.

loss of hepatitis B surface antigen productivity. In fact it required only a few months to scale up production from the one liter vessel to the two hundred forty liter run. Additionally, these fermentations are extremely reproducible and easy to control.

We believe that three factors contribute to the extreme stability of this organism. The biggest component is the organism itself which has been genetically selected for optimal growth, fermentor stability and maximal feed efficiency. Another important factor is the use of integrated vectors. These vectors are a component of the host's genetic information and cannot be easily lost without destruction of the host cell itself. This greatly reduces the likelihood of mitotic instability of the vector contributing to the generation of host cells which do not produce the desired protein. Thirdly, the conditional nature of the expression system makes it difficult for nonproducing cells to become a significant proportion of the growing cells. Although it is likely that for some products high level expression will lead to decreased growth rates in the production strain relative to nonproducing cells, the regulatable system developed for *pichia pastoris* does not allow for expression of the foreign gene until the final few generations of cell growth. This eliminates the potential for competition by nonproducing cells since, in the absence of foreing gene expression, growth rates should be uniform. The shift to methanol accomplishes two things. The first is expression of the product and second, decreased growth rate. This slower growth rate leads to a constant accumulation of product without the cells rapidly entering stationary phase. This also limits the number of cell divisions which occur during the expression phase since two doublings are theoretically all that are possible.

As a demonstration of the stability of this system twenty-five single isolates were obtained from nonselective plates at the end of the largest scale run (240 l). Each isolate was regrown and tested for

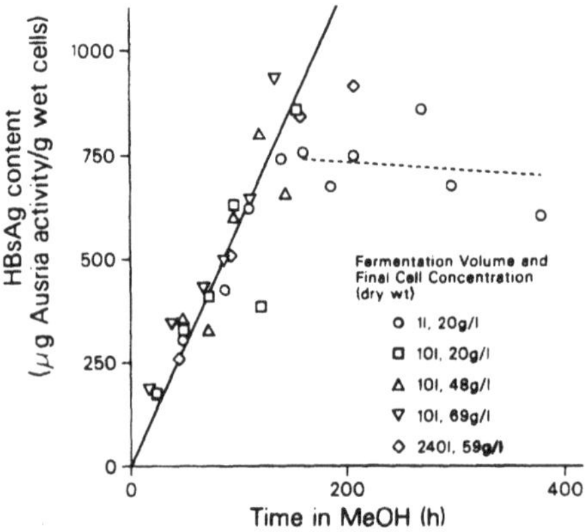

Figure 3
Time course of hepatitis B surface antigen production.

Cells were grown in defined media at the densities and volumes indicated. Soluble extracts were prepared and assayed for 22 nm particle.

hepatitis B surface antigen production. Table II indicates that all isolates were not only producers but they were all high level producers. The success of this system derives from a proven industrial organism to which vector development and fermentation technology has been added to complement the production capabilities of the organism.

Table II
Analysis of 25 isolates from large fermentation run.

Colony Number	HBsAg content (μg/g cells)	HBsAg concentration of extract (% soluble protein)
1	680	1.6
2	770	2.3
3	770	2.6
4	810	2.5
5	720	2.2
6	890	2.7
7	810	2.4
8	640	2.0
9	840	2.4
10	770	2.6
11	860	2.8
12	680	2.3
13	810	2.4
14	560	1.6
15	730	2.0
16	800	2.3
17	760	2.5
18	800	2.6
19	670	2.1
20	620	2.0
21	740	2.3
22	940	2.6
23	820	2.2
24	690	2.1
25	680	1.8
Mean	750 ± 90	2.3 ± 0.3

Colonies were isolated from non selective plates and grown for seven days under inducing conditions. Surface antigen was measured by the AUSRIA method.

5. CHARACTERIZATION OF THE HEPATITIS B SURFACE ANTIGEN

In addition to measuring its reactivity with antisera raised against purified 22 nm particles, we have characterized the hepatitis B surface antigen produced in *Pichia pastoris* by other techniques. Figure 4 shows the profile of the antigen on 20-50% sucrose equilibrium gradients. The majority of the particle sediments to a density similar to the native blood borne particle (ρ = 1.16 g/ml). This density is characteristic of a high molecular weight protein-lipid complex containing approximately 30% lipid. Interestingly, when fractions from this gradient were assayed for monomer, the 24 Kd protein was found only in the particle containing fractions (data not shown), indicating that all detectable monomer is particle associated. This high degree of particle formation contrasts with the situation found in *Saccharomyces* where monomer production is abundant but very little particle is formed [3].

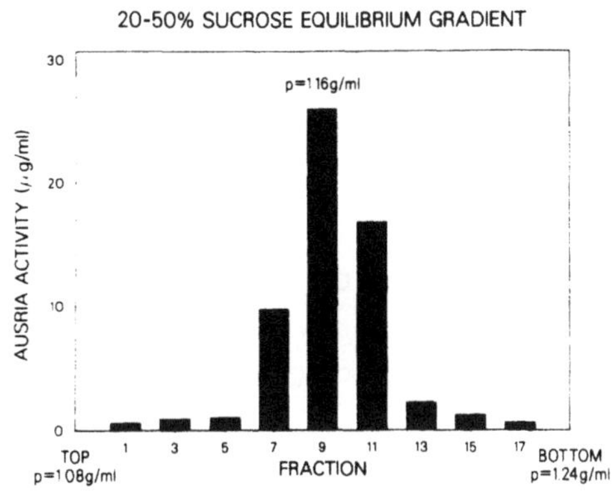

Figure 4:
Hepatitis B surface antigen characterization on sucrose equilibrium gradient.

A soluble extract was subjected to centrifugation on a 20-50% sucrose gradient. Surface antigen was detected using the AUSRIA assay.

6. SUMMARY

We have developed a second generation yeast expression system based on the methylotropic yeast *Pichia pastoris*. The particular organism used in these studies was a growth optimized strain developed by Phillips Petroleum Company. Using tightly regulated, highly expressed promoters derived from methanol regulated genes we have developed integrated vectors which are the basis for a conditional expression system based on regulation of expression by carbon source feed. The majority of growth takes place on the repressing carbon source, glycerol. Expression occurs when the glycerol is exhausted and the cells begin to metabolize methanol as carbon source. Site-specific integration into the AOX1 locus deletes the primary alcohol oxidase gene leading to decreased growth kinetics on methanol. During this slow growth phase (150-200 h), product accumulates in a linear fashion. The high per cell yields and high cell density lead to very high volumetric yields. This process yields approximately 0.5 g of immunoreactive material per liter and can be easily scaled up. This type of productivity has not been seen in yeast systems to date. Additionally, the system works for other products such as *E. coli* ß-galactosidase which accumulates to between 10-20% of the soluble protein in *Pichia pastoris*. This represents a realistic yeast production system which can be used to produce a variety of proteins for industrial or pharmaceutical use.

ACKNOWLEDGEMENTS

The authors would like to thank Benjie Davis for critical comments on the manuscript and Pat Thomas for help in the manuscript preparation.

REFERENCES

[1] DeWilde M, Cabezon T, Harford N, Rutgers T, Simoen E, and Wijnerdaele F van. (1983) "Eighteenth Congress Joint IABDS/WHO Symposium on Standardization and Control of Biologicals Produced by Recombinant DNA Technology, Geneva Switzerland, Vol 59 Pp. 99-107.
[2] Bitter GA and Egan KM. (1984) Gene 32, 263-274.

[3] Hitzeman RA, Chen CY, Hagie FE, Patzer EJ, Lio C, Estell DA, Miller JV, Yaffe A, Kleid DG, Levinson AD and Opperman H. (1983) Nucleic Acids Research 11, 2745-2763.
[4] Halling SM and Smith S. (1985) Biotechnology 13, 715-726.
[5] Ballou CE. (1982) "The Molecular Biology of the Yeast *Saccharomyces*" (Strathern J, Jones E, Broach J). Cold Spring Harbor Laboratory, Vol. II Pp. 335-360.
[6] Williams RS, Trumbly RJ, MacGoll R, Trimble RB and Maley F. (1985) The Journal of Biological Chemistry 260, 13334-13341.
[7] Brunt J van. (1986) Biotechnology 4, 505-510.
[8] TaitKamradt AG, Turner KJ, Kramer RA, Elliott QD, Bostian SJ, Thill GP, Rogers DT and Bostian KA. (1986) Molecular and Cellular Biology 6, 1855-1865.
[9] Murray AW and Szostak JW. (1983) Cell 34, 961-970.
[10] Srienc F, Campbell JL and Bailey JE. (1986) Biotechnology and Bioengineering 28, 996-1006.
[11] Parulekar SJ, Semones GB, Rolf MJ, Lievense JC and Lim HC. (1986) Biotechnology & Bioengineering 28, 700-710.
[12] Brunt J van. (1986) Biotechnology 4, 395-401.
[13] Ellis S, Brust P, Waters A. Koutz PJ, Harpold M and Gingeras TR. (1985) Molecular and Cellular Biology 5, 1111-1121.
[14] Francis DP. (1983) Infectious Diseases Review 5, 322-335.
[15] Peterson DL, Roberts IM and Vyas GN. (1977) Proceedings of the National Academy of Sciences USA 74, 1530-1534.
[16] Tiollais P, Charnay P, Vyas GN. (1981) Science 213, 406-411.

SUBJECT INDEX

N

AUTHORS INDEX

www.ingramcontent.com/pod-product-compliance
Ingram Content Group UK Ltd.
Pitfield, Milton Keynes, MK11 3LW, UK
UKHW020633260926
475829UK00003BA/73
* 9 7 8 9 0 2 4 7 3 4 5 9 7 *